SOLID STATE PHYSICS LITERATURE GUIDES
Volume 5

BIBLIOGRAPHY OF MAGNETIC MATERIALS AND TABULATION OF MAGNETIC TRANSITION TEMPERATURES

Solid State Physics Literature Guides

Prepared under the auspices of the Research Materials Information Center,
Oak Ridge National Laboratory

General Editor: T. F. Connolly
Solid State Division
Oak Ridge National Laboratory*
Oak Ridge, Tennessee

Volume 1: Ferroelectric Materials and Ferroelectricity—1970

Volume 2: Semiconductors—Preparation, Crystal Growth, and Selected Properties—1972

Volume 3: Groups IV, V, and VI Transition Metals and Compounds—Preparation and Properties—1972

Volume 4: Electrical Properties of Solids—Surface Preparation and Methods of Measurement—1972

Volume 5: Bibliography of Magnetic Materials and Tabulation of Magnetic Transition Temperatures—1972

---

*Oak Ridge National Laboratory is operated by Union Carbide Corporation for the U.S. Atomic Energy Commission.

SOLID STATE PHYSICS LITERATURE GUIDES
Volume 5

BIBLIOGRAPHY OF MAGNETIC MATERIALS AND TABULATION OF MAGNETIC TRANSITION TEMPERATURES

Compiled by
T. F. Connolly
and
Emily D. Copenhaver

Research Materials Information Center
Solid State Division
Oak Ridge National Laboratory
Oak Ridge, Tennessee

IFI/PLENUM • NEW YORK-WASHINGTON-LONDON • 1972

Library of Congress Catalog Card Number 74-133269
ISBN 0-306-68325-3

© 1972 IFI/Plenum Data Corporation
A Subsidiary of Plenum Publishing Corporation
227 West 17th Street, New York, N.Y. 10011

United Kingdom edition published by Plenum Press, London
A Division of Plenum Publishing Company, Ltd.
Davis House (4th Floor), 8 Scrubs Lane, Harlesden, London, NW10 6SE, England

All rights reserved

No part of this publication may be reproduced in any form
without written permission from the publisher

Printed in the United States of America

Preface

This referenced compilation of magnetic transition temperatures represents (with the Addendum) papers actually received by the RMIC through May 1972 and consists of two lists (alphabetical by compounds), one for Curie and one for Néel temperatures. Where different values appeared in the literature for a single compound, all are listed with separate references given for each. There is no attempt at critical evaluation, which, except for a few well-studied and well-characterized materials, would hardly be worth the effort. All that one can say for most of the compounds is that for a given material with a certain (or all too often uncertain) history of preparation and treatment, stoichiometry, homogeneity, and chemical or structural purity a magnetic transition was indicated at the temperature(s) listed. Only when the reasons for different values are explicitly stated in the literature do they appear as brief comments in the body of the lists.

In order to include the most recent data, and to eliminate the delay involved in recomposition of the lists, an addendum is provided. While this requires the perusal of two lists rather than one, it does ensure that the compilation represents the entire RMIC collection at the moment of going to press.

The 2478 references are restricted to those papers specifying a Curie or Néel temperature and do not reflect the complete magnetics literature even for the materials listed. To offset this limitation a separate listing of books, reviews, and other compilations on magnetic materials is provided, and because of the interest shown by recent inquiries to RMIC, another bibliography, entitled "Monochromator and Analyzing Crystals," which includes references concerning the preparation and properties of crystals employed in both neutron and x-ray diffraction.

At this writing, there are over 80,000 searchable references in the RMIC collection on solid-state inorganic materials science, and the coverage of the field is good back to 1960, although many earlier references are included. Still, there will be omissions and errors in compilations drawn from the collection, and any pointed out to us will be corrected in future editions. (Such corrections should be sent to the Center and not to the publisher.)

The timeliness of these compilations, as well as our ability to answer daily inquiries, depends very largely on the continued receipt by the Center of all papers, reprints, reports, and preprints within our scope. These should be mailed to

T. F. Connolly
Research Materials Information Center
Oak Ridge National Laboratory
P. O. Box X
Oak Ridge, Tennessee 37830

Availability of Documents

U. S. Government contractor reports, usually identified by an alphanumeric report number, can be purchased from

    National Technical Information Service
    U. S. Department of Commerce
    Springfield, Virginia 22151

and often on request from the issuing installation.

USAEC Reports are also available from

    International Atomic Energy Agency
    Kaerntnerring A 1010
    Vienna, Austria

    National Lending Library
    Boston Spa, England

Monographs and reports of the National Bureau of Standards are for sale by

    Superintendent of Documents
    U. S. Government Printing Office
    Washington, D. C. 20402

Theses, listed as Dissertation Abstracts + number, are available in North or South America from

    University Microfilms
    Dissertation Copies
    P. O. Box 1764
    Ann Arbor, Michigan 48106

and elsewhere from

    University Microfilms, Ltd.
    St. John's Road
    Tylers Green
    Penn, Buckinghamshire
    England

Contents

Tabulation of Curie Temperatures 1
Tabulation of Néel Temperatures 31
Addendum 61
    Curie Temperatures 61
    Néel Temperatures 65
References 69
Recent Reviews, Conferences, and Compilations on Magnetic Materials 163
Monochromator and Analyzing Crystals 173

CURIE TEMPERATURE

Material	Temperature Curie (°K)	Reference
Ag-5 at. % Eu	6.5 ± 0.5	481
$Ag_{0.5}In_{0.5}Cr_2Se_4$	50	1790, 1890
$Ag_{0.5}La_{0.5}Fe_{12}O_{19}$	708	365
AlNiCo (22Co-2Ti)	1143	745
Au-5 at. % Eu	6.5 ± 0.5	481
AuFe	300	23
Au−Fe		
at.% Fe:		
13	≈40	2090
17	≈148	2090
20	≈210	23
22	≈220	2090
Au−20.1 at% Mn	377	2113
Au_4Mn	371.1	1059
	371	1611
	363	414
	~360	1408
Au_5Mn_2	100	1325
Au_2MnAl	252−258	699
	220	48
$Au_2Mn_{2-x}Al_x$		1026
x = 0.6	314	
x = 1.0	147	
$Au_4(Mn_{1-x}Cr_x)$		
x = 0.2	325	1719
x = 0.2	≈340	2153
x = 0.4	≈250	2153
x = 0.6	≈180	2153
x = 0.6	225	1719
x = 0.8	≈45	2153
x = 0.8	150	1719
Au_2MnIn	140	884
Au_4V	43	569
	~50	1294
	~53	1396
	56	1697
	~57	1408
$Au_{80}V_{19}Fe_1$	43	1697
$Au_{80}V_{17}Fe_3$	57	1697
$Au_{80}V_{15}Fe_5$	66	1697
$Au_{80}V_{13}Fe_7$	67	1697
$Au_{80}V_{11}Fe_9$	26	1697
$Au_4(V_{0.5}Mn_{0.5})Au_4Cr$	~200	1408
$BaAl_xFe_{12-x}O_{19}$		1804
x = 0	718	
x = 2	618	
x = 4	514	
x = 6	414	
x = 8	314	
$Ba_2Co_2Fe_{28}O_{46}$	740 ± 4	1712
$BaFe_2Co_{1.3}Fe_{4.7}O_{23.7}F_{1.3}$	718	1080, 1112
Ba_2FeMoO_6	334	231
	≈355 (ferri.)	1823
$BaFeO_{2.95}$	~190	1746
$BaFeO_3$	180	831
$BaFe_{12}O_{19}$	738	745
	740	1391
$BaFe_{12}O_{19}F_2$	710	1108
$BaFe_{0.5}Re_{0.5}O_3$	316	1104
$BaFe_{12-x}Sc_xO_{19}$		1058
x = 1.2	≥ 500	
x = 1.8	~600	
$Ba_3FeU_2O_9$	123	1684
$BaMn_{0.5}Re_{0.5}O_3$	136	1104
$BaO_{0.75}SrO_{0.25}\cdot 6Fe_2O_3$	713	1822
+ 1.11 mole B_2O_3	713	1822
+ 1.51 mole Al_2O_3	563	1822
+ 0.65 mole Ga_2O_3	618	1822
$Ba_{0.50}Sr_{1.50}FeMoO_6$	413	231
$Ba_{0.60}Sr_{1.40}FeMoO_6$	410	231
$Ba_{0.84}Sr_{1.16}FeMoO_6$	399	231
$Ba_{1.42}Sr_{0.58}FeMoO_6$	372	231
$Ba_{0.4}Sr_{1.6}Zn_2Fe_{12}O_{12}$	400	1558
$Ba_{0.6}Sr_{1.4}Zn_2Fe_{12}O_{22}$	323	1733
$Ba_{1.6}Sr_{0.4}Zn_2Fe_{12}O_{22}$	358	1733
$Ba_{1.0}Sr_{1.0}Zn_2Fe_{12}O_{22}$	333	1733
$Ba_5Ti_3Mg_2Fe_{12}O_{31}$	391 ± 3	2150
$Ba_5Ti_3Zn_2Fe_{12}O_{31}$	310 ± 5	2150
$BaVO_3$	<77 (possible ordering)	2093
$Ba_2Zn_2Al_{2.5}Fe_{9.5}O_{22}$	370	1292
$BaZn_xFe_{7-x}O_{11}$	700x + 80	496
$Ba_2Zn_2Fe_{12}O_{22}$	~400	1292
	373	1733
$Ba_2Zn_2Fe_{28}O_{46}$	705 ± 3	1712
$Ba_4Zn_2Fe_{36}O_{60}$	673 ± 2	1002
$BaZn_{0.5}Ir_{0.5}Fe_{11}O_{19}$	563	1126
$Bi_{1-x}Ca_xMnO_3$		
0.2 ≤ x < 0.35	90	708
x = 0.9	114	1690
x = 0.85	126	1690

Material	Temperature Curie (°K)	Reference	Material	Temperature Curie (°K)	Reference
$Bi_{0.3}Ca_{2.68}Pb_{0.02}Fe_{3.65}$ $V_{1.35}O_{12}$: CoO		1922	$Cd_xMn_{3-x}O_4$		767
CoO wt%:			x = 1.0	<4	
0	501 ± 3		x = 0.8	5	
0.03	502 ± 3		x = 0.6	15	
0.06	500 ± 3		x = 0.4	24.5	
Bi-MnBi	240	810	CeAg	9	1274
(0.50 wt. % Mn)			$CeAl_2$	8	695
$BiMnO_3$	103	14	$CeAl_4$	9	1184
	110	601	$Ce_{1.2}Cd$	~12	937
$Ca_3Fe_2(Fe_{1.5}V_{1.5})O_{12}$	493	571	$CeCl_3$	0.345	1498
$Ca_2(FeMo)O_6$	≈400 (ferri.)	1823	$CeCo_3$	78	417
			$CeCo_5$	687	444
				737	442
$CaFe_4O_7$:3.0 mol % Gd_2O_3	393	453		647	630
$CaFe_{0.5}Re_{0.5}O_3$	538	1104		464	504
$Ca_3Fe_3VGeO_{12}$	319	2160	Ce_2Co_7	151	630
$Ca_3Fe_{3.3}V_{1.3}Ge_{0.4}O_{12}$	424	2160	Ce_2Co_{17}	1083	417
$Ca_{0.86}La_{0.14}Fe_{12}O_{19}$	718	1100		1068	537, 759
$CaO \cdot 2Fe_2O_3$	403	838	$CeFe_2$	226	630
$Ca_3Sb_{1.5}Fe_{3.5}O_{12}$	55	1092, 1094, 1095		221	257
				230	2037
				235	286
$CdCr_2S_4$	86	434		878	542
	85	594	$CeFe_5$	228	349
	84.5	507, 510	$CeFe_7$	93	386
	84	1392, 1828		(probably Ce_2Fe_{17}; see ref. 2037)	
	84	2959			
	(Raman spectra)		Ce_2Fe_{17}	91	630
	80	1637	$CeFeO_3$	719	1491
	97	563	$Ce_{1-x}Gd_xRu_2$	*<5	628
$CdCr_2Se_4$	129 ± 2	933	Ce_5Ge_3	<4.2	906
	129.5	510	$(Ce_{1-x}La_x)_3Al_{11}$		2040
	≈129	1873	x = 0	6.2	
	121	1267	x = 0.1	5.4	
	127.67 ± 0.005	1745	x = 0.3	3.7	
	128	1392	x = 0.95	1.3	
	130	434, 594	(Ce-mischmetal)Co_5	~500	504
	130	2059	$Ce_2Mg_3(NO_3)_{12} \cdot 24H_2O$	0.0019 ± 0.0001	1403
	(Raman spectra)			(see also ref. 1539)	
	133	951	Ce_2Ni_7	48	954
	140	976	Ce_3S_4	<10?	1819
	142	563	Ce_3Se_4	10	1619
$CdFe_2$	782	330	$Ce_{1-x}Tb_xRu_2$		1863
			x = 0.2	4.7	
			x = 0.4	12.3	
			$Ce_{0.33}Th_{0.67}Al_3$	5	1781
			$Ce_{0.67}Th_{0.33}Al_3$	6	1781

CURIE TEMPERATURE

Material	Temperature Curie (°K)	Reference	Material	Temperature Curie (°K)	Reference
Co	1382	754	$CoFe_2O_4$	673–769	683
	1387.20	596	$(Co_{1-x}Fe_x)_2P$		1612
	1388 ± 2	300	x = 0.2	∼140	
	1390	444	x = 0.3	∼286	
	1390	1910	x = 0.4	∼407	
	(thermal expansion)		x = 0.6	∼453	
			$(Co_{0.38}Fe_{0.62})_2P$	459	1369
	1395 ± 5	753	$CoFe_{0.5}Rh_{1.5}O_4$	160	281
	1400	1030	$CoFe_{2-x}Rh_xO_4$		281
	1383.4-1407	1296	x = 0	790	
	(depending upon extrapolation)		x = 1.0	355	
			x = 1.5	160	
$Co_2Al_3B_6$	406	916	$Co_{0.75}Fe_{0.25}S_2$	≈140.7	2083
$Co_{20}Al_3B_6$	409	553	$Co_{1+x}Fe_{2-2x}Ti_xO_4$		281
$CoAlFeO_4$	420	382	x = 0.1	755	
$CoAs_xS_{2-x}$		2130	x = 0.5	450	
x = 0	≈122		x = 0.8	200	
x = 0.10	≈90		$Co_{1.5}Fe_{1.0}Ti_{0.5}O_4$	450	281
x = 0.25	≈12		$Co_{1-x}Mn_x$		1725
CoB	477	302	x = 0.35	≅140	
(but not ferromagnetic according to ref. 318)			Co_2MnAl	697	1648
			Co_2MnGa	694	1648
			Co_2MnGe	905	1648
Co_2B	429	318	$CoMnO_3$	391	137
	433	393		394	1836
Co_3B	747	318	$Co_{3-x}Mn_xO_4$		1806
$Co_2BaFe_{14.2}In_{1.8}O_{27}$	571	663	x = 1.2	≈205	
$Co_2BaFe_{16}O_{27}$	753	663	x = 1.0	≈183	
$Co_2Ba_{3-x}Sr_xFe_{24}O_{41}$		1984	x = 0.8	≈188	
x = 0.4	683		$Co_{1.8}Mn_{1.2}O_4$	191	525
x = 2.4	638		$CoMn_2O_4$	95-105	1359
$CoCr_2O_4$	95	1136, 1155	$Co_xNi_{1-x}Fe_2O_4$		2149
			x = 0.075	871	
	95.9 ± 1	1502	x = 0.7	831	
	96	1462	x = 0.97	786	
	97	962, 1109, 2001	Co–20 at% P	≈573	2027
				118–124	2096
	98	53	x = 0.9	≈92	1930
	≈98	2028	$CoSe_xS_{2-x}$		2130
	100	1303	x = 0	≈122	
$CoCr_2S_4$	235	1054	x = 0.10	≈90	
	238	53	x = 0.25	≈40	
	240 ± 5	939	$(Co_{1-x}Mn_x)_2P$		
	227	1642	x = 0.2	∼200	1369
$Co_{3.5}CuFe_{0.5}Ce$	>993	1709	x = 0.2	∼218	1612
$Co_{1-x}Cu_xRh_2S_4$		1465	x = 0.3	∼386	1612
x = 0.5	∼100		x = 0.5	583	1369
$CoF_25HF·6H_2O$	∼246	1223	x = 0.5	∼583	1612
$Co_{16}Fe_4Al_3B_6$	623	1500	x = 0.6	∼445	1612
$CoFeCoO_4$	450	316			

Material	Temperature Curie (°K)	Reference
Co_2MnSi	985	1648
Co_2MnSn	829	1648
	811	1756
$CoO_{1-x}Zn O_xFe_2O_3$		1556
$x = 0$	~800	
$x = 0.2$	~695	
$x = 0.4$	~550	
$x = 0.6$	~380	
CoPt	813	436
$CoRh_{0.35}Cr_{1.65}S_4$	200	1054
$CoRh_{1.5}Cr_{0.5}S_4$	~50	1054
$CoS_{1.97}$	122	824
CoS_2	110	53
	116	600
	124	966, 1084
	130	963
$Co(S_xSe_{1-x})_2$		
$x = 0.95$	~94	1301, 1302
$x = 0.9$	≈92	1930
$x = 0.90$	~42	1301, 1302
$CoSe_xS_{2-x}$		
$x = 0$	≈122	
$x = 0.10$	≈90	
$x = 0.25$	≈40	
$CoSe_{0.25}S_{1.75}$	~17	1373
$CoSiF_6 \cdot 6H_2O$		272
heating	259 ± 1	
cooling	246 ± 2	
$Co_{21}Sn_2B_6$	425	553
Co-Ti		
at. % Ti:		
21.4	38	1476
28.0	42	1727
29.0	44	1727
32.8	17	1727
$Co_{3+x}Ti_{1-x}$		1827
$x = 21.4$	38	
$x = 23.0$	<4.2	
Co_2VO_4	158	993, 1113
	160	333
CoV_2O_4	145	993, 1109
$Co_3V_2O_8$	10	2157
CoVSb	58	2165, 2166
$Co_{0.7}Zn_{0.3}Fe_2O_4$	~591	1434
$CoZrF_6 \cdot 6H_2O$	≈246 ± 2	2098

Material	Temperature Curie (°K)	Reference
Cr_3As_2	213	263
	223	478
	243	1337
	257	478
$CrBe_{12}$	~50	1210
	50?	1047
$CrBr_3$	36	428, 691, 1005
	32.844	1371
	32.5	2138
	32.56	1734
	37	1871
$Cr_{0.39}Co_{0.61}S_2$	275	2096
$Cr_{0.25}Co_{0.75}S_2$	335	2096
$Cr_{0.18}Co_{0.82}S_2$	300	2096
$Cr_{0.05}Co_{0.95}S_2$	170	2096
$Cr_{0.09}Co_{0.91}S_2$	210	2096
Cr-Fe		372
20% Fe	81 ± 5	
18.2% Fe	50 ± 5	
$Cr_{1-x}Fe_xO_2$		1814
$x = 0$	389	
$x = 0.1$	391	
$x = 0.2$	389	
$x = 2$	423	
$x = 9.5$	427	
$x = 18.1$	428	
$Cr_{0.75}Fe_{0.25}Sb$	80 (transition "possible")	1467
$CrGe_2$	98	936
	100	859
$Cr_{11}Ge_{19}$	86	1328
CrI_3	68	1871
$Cr_{0.4}Mn_{0.6}As$	160	1394
$CrMn_2B_4$	440	956
$CrMn_2O_4$	65	1359
$Cr_{1/3}NbS_2$	~160	1695
$Cr_{2/3}NbS_2$	65	1695
$Cr_{1/3}NbSe_2$	105	1695
$Cr_{0.25}NbSe_2$	79	2151
CrO_2	378	446
	386.5	521
	389	1393
	391	588
	393	1443
	393 (ac conductivity)	1651
	398	778
	≃399	800

CURIE TEMPERATURE

Material	Temperature Curie (°K)	Reference	Material	Temperature Curie (°K)	Reference
CrO_2(+0-20 at. % Te)	389	335	$CsNiF_3$	111	1086
$Cr_9Pd_{71}Si_{20}$	60	1629		130	1948
Cr-70% Pt	687	860	$Cu_{0.5}Co_{0.5}Cr_2S_4$	~235	1054
CrS	303	53	$CuCrMnO_4$	45	209
$CrS_{1.17}$	310	687	$CuCr_2O_4$	133	53
$CrS_{1.194}$	305	930		135	1453
Cr_2S_3			$CuCrRhSe_4$	255	456
rhomb.	120	1032		270	979
trig.	110	1032	$CuCrRh_{0.6}Sn_{0.4}Se_4$	140	1899
	833	1347	$CuCr_2S_4$	364	1935
Cr_5S_6	303	181		375	951, 1203
	~305	1656			
	613	1347		378	1067
$Cr_{1/3}TaS_2$	~170	1695		420	1105, 1106
$Cr_{1/3}TaSe_2$	~120	1695			1166
CrTe	239	681	$CuCrS_{4-x}Cl_x$		
	279	181	x = 0.2	~370	
	308	680	x = 1.	~218	
	329	13	$CuCr_2Se_4$	414.5	1935
	334	436		429	951
	335	827		432	1203
	337	1860		433	1140
	340	1600		460	456
film	343	1766	$CuCr_2Se_3Br$	274	984, 985
	345	1308		354	1203
Cr_2Te_3	172	1580	$CuCr_2Se_{4-x}Br_x$		1166
	182 ± 2	2080	x = 0.8	~300	
	303	1330	x = 0.2	~400	
Cr_3Te_4	329	13	$CuCr_2Se_3Cl$	383	1203
	325 ± 2	2080	$CuCr_2Se_xS_{4-x}$		1067
	300	1580	x = 1	~353	
$Cr_3 \sqcup Te_4$	325	1776	x = 2	~360	
Cr_5Te_6	327	1580	x = 3	~393	
	327 ± 2	2080	x = 4	~433	
Cr_7Te_8		1610	$CuCr_2Te_4$	329	1203
disordered	361			344.6	1935
ordered	350			365	1105, 1106
$CrTe_{1-x}Sb_x$		1566			
x = 0.25	~290		$CuCr_2Te_3Br$	282	1203
x = 0.50	~220		$CuCr_2Te_3I$	281	1203
$Cr_{1-x}V_xO_2$		367		294	984
x = 0.00	389.5		$CuFe_{2(1-x)}Ga_{2x}O_4$		806
x = 0.02	385.5		x = 0.10		
x = 0.05	381.5		slowly cooled	681	
x = 0.08	344		quenched	679	
$CsFeF_3$	≈58	1824	x = 0.50		
	60	1235, 1240	slowly cooled	409	
	62 ± 2	1365	quenched	313	

Material	Temperature Curie (°K)	Reference	Material	Temperature Curie (°K)	Reference
$CuFe_2O_4$	~723	908	Cu-Ni	346	212
slowly cooled	743	615, 806	24 at. % Cu	346	212
quenched	718	615, 806	44 at.% Ni	4.2	1894
	768	1446	$Cu_{0.55}Ni_{0.45}$		2120
$CuFe_2O_{4-x}$	723–763	429	quenched	≈5	
$CuFe_5O_8$			annealed	10	
(40% Fe_3O_4)	~618	908	$Cu_{0.5}Ni_{0.5}$		2120
	630	1446	quenched	38.8	
$Cu_{0.5}Fe_{2.5}O_4$	623–673	1911	annealed	62.5	
	623	1802	$Cu_{0.45}Ni_{0.55}$		2120
(low O prep.)	823	1911	quenched	134	
$Cu_{1.65}Fe_{1.35}O_{3.35}F_{0.65}$	603	1119	annealed	133	
$Cu_{2.0}Fe_{1.0}O_{3.0}F_{1.0}$	658	1112	$CuNi_{0.5}Mn_{1.5}O_4$	150	209, 258
$CuK_2Cl_4 \cdot 2H_2O$	0.88	739, 1761	$CuRb_2Cl_4 \cdot 2H_2O$	1.02	739
$Cu_{1-x}Mg_xFe_2O_4$		950	$CuRh_xCr_{2-x}Se_4$		1790
(x = 0.2)	~898–1023		x = 0	460	
(x = 0.8)	~878–983		x = 0.5	350	
(depending on quench temp)			x = 1	220	
$CuMg_{0.5}Mn_{1.5}O_4$	57	209, 258, 1893	x = 1.6	50	
			x = 1.8	15	
Cu–22.5 at.% Mn	206	2013	x = 1.9	≈5	
CuMnAl	433	468	$CuRhMnO_4$	35	209
12.5 at. % Cu			$CuRh_2O_4$	830	914
42.5 at. % Mn			$CuSiF_6 \cdot 6H_2O$	0.07	293
45% Al			$CuV_xCr_{2-x}S_4$		1747
Cu_2MnAl	~583	483	x = 0.25	267	
	600	1648	x = 0.75	46	
	630	1613	x = 1.75	8	
$Cu_{0.8}Mn_{0.2}Fe_2O_4$	635	2137	$Cu_{0.4}Zn_{0.6}Fe_2O_4$	288	2137
oxygen–treated	707	2137	oxygen–treated	359	2137
$Cu_{0.35}Mn_{0.65}Fe_2O_4$	592	2137	$Cu_{0.5}Zn_{0.3}Fe_{2.2}O_4$		2077
oxygen–treated	608	2137	heated in air; quenched	480	
Cu_2MnIn	500	1613	heated in air; slowly cooled	558	
	520	1648			
$CuMn_2O_4$	25	1113	$Cu_{0.5}Zn_{0.5}Fe_2O_4$		2077
$Cu_{1.5}Mn_{1.5}O_4$	80	209	heated in air; quenched	346	
Cu_2MnSn	528	1613	heated in air; slowly cooled	420	
$Cu_{0.55}Mn_{0.25}Zn_{0.2}Fe_2O_4$	541	2137			
oxygen–treated	583	2137	$Cu_{0.8}Zn_{0.2}Fe_2O_4$	525	2137
$Cu_{0.2}Mn_{0.2}Zn_{0.6}Fe_2O_4$	318	2137	oxygen–treated	634	2137
oxygen–treated	339	2137	$Cu_{0.9}Zn_{0.3}Fe_{1.8}O_4$		2077
$Cu(NH_4)_2Br_4 \cdot 2H_2O$	1.73	739	heated in air; quenched	463	
	1.74	1918	heated in air; slowly cooled	573	
	1.74 ± 0.01	1568			
normal	1.789 ± 0.001	1671	Dy	77	1847
deuterated	1.804 ± 0.001	1671		(at 16 kOe)	
	1.831 ± 0.001	2012		84.7	531
$Cu(NH_4)_2Cl_4 \cdot 2H_2O$	0.70	739, 2030	(continued)		

CURIE TEMPERATURE

Material	Temperature Curie (°K)	Reference	Material	Temperature Curie (°K)	Reference
Dy	85	1, 214, 1960	DyN	21	512
				22	522
	85, 87	198		26	355
	87	24, 294, 1941		17.6	1614
			DyNi	48	802
	87 ± 1	1011	$DyNi_2$	32	802
	108	926		30	736
$DyAl_2$	53	695	$DyNi_3$	69	538
	62	412, 1338	$DyNi_5$	15	802
	70	1194	Dy_2Ni_7	81	954
	48 (elec. resist.)	1972	Dy_2Ni_{17}	168	712
				604	1181
Dy_3Al_2	76	581, 1214	$DyNi_{2-x}Al_x$		1800
	~105	1471	x = 0.16	42	
$Dy_3Al_{0.5}Fe_{4.5}O_{12}$	490	1129, 1130	x = 1.80	83	
			$Dy(OH)_3$	3.50	1000
DyC	165	1038		15	41
$Dy_{1.3}Cd$	105	937	$DyOs_2$	~25	350
$DyCo_2$	159	736	$Dy_{5.07}Pd_{1.93}$	4.2	1024
	≈150	257	$DyPt_2$	≈14	257
$DyCo_3$	450	417	$DyRh_2$	≈26	257
$DyCo_5$	1125	444	Dy_5Si_4	140	949
	966	442	Dy–Th (85:15)	119	1914
Dy_2Co_7	647	630			
Dy_2Co_{17}	1152	417	Dy–5% Y	91	926
	1189	537, 759	$Dy_xY_{1-x}Fe_2$		998
Dy_4Co_3	55	1151	(x = 0.8)	~618	
Dy_2CrSbO_7	16	511	(x = 0.5)	~590	
$Dy_{0.6}Er_{0.4}Al_2$	49	1194	DyZn	144	1295
$DyFe_2$	638	330, 630	Er	19	536, 444
	635	1821, 1944		18.2	1789
	663	542		19.6, 20	24, 296
$DyFe_3$	600	630		20	198, 585, 711
	605	1821			
Dy_2Fe_{17}	363	386	$ErAl_2$	13	1023
	362	630		14	27
	380	1821		14.5	957
Dy_6Fe_{23}	524	630		21	695
	545	1821		24	1194, 1800
$Dy_3Fe_5O_{12}$	551.4	239		12	1338
	552	1314	$Er_3AlFe_4O_{12}$	420	1129, 1130
	563	430			
	563 ± 2	741	$ErAu_4$	≈5	2146
$Dy_3Ga_{0.25}Fe_{4.75}O_{12}$	520	1129, 1130	ErC	50	1038
			$Er_{1.2}Cd$	14	937
Dy–10% Ho	80	1014	$ErCo_2$	39, 36	306
$DyIr_2$	23	41		36	736
$DyMn_5$	430	875		≈35	257
Dy_6Mn_{23}	**443**	204		35	1424

Material	Temperature Curie (°K)	Reference	Material	Temperature Curie (°K)	Reference
$ErCo_3$	401	417, 442	$ErNi_5$	13	802
$ErCo_5$	986	442	Er_2Ni_7	67	954
	1050	444	Er_2Ni_{17}	166	712
$ErCo_6$	1050	1261		602	1181
Er_2Co_7	644	630	$ErNi_{0.90}Al_{1.10}$	16	1800
	670	1261	$ErNi_{0.24}Al_{1.76}$	25–27	1800
Er_2Co_{17}	1186	417	$ErNi_{0.16}Al_{1.84}$	31	1800
	1193	537, 759	$ErNi_{0.10}Al_{1.90}$	33	1800
	1160	1261	$ErNi_{0.30}Al_{1.70}$	23.5–28.5	1800
Er_3Co	7	1797	$ErNi_{0.03}Al_{1.97}$	22	1800
Er_4Co_3	25	1151	$ErNi_{2-x}Al_x$		1800
$ErCo_2$-$ErAl_2$		742	x = 0.17	30	
10 mol % $ErAl_2$	86		x = 1.90	33	
50	25		x = 1.76	27	
85	24		x = 1.70	29	
$Er_3Cr_{0.25}Fe_{4.75}O_{12}$	520	1129, 1130	$ErOs_2$	3	41
Er_2CrSbO_7	10	511	$ErPt_2$	4.2	1024
Er-10% Dy	31	1014		≈3	257
Er-50% Dy	~140	1014	$ErRh_2$	≈6	257
Er-90% Dy	65	1464	$ErRu_2$	8	41
$Er_{0.20}Dy_{0.80}$	~40	1320	Er_5Si_4	25	949
$ErFe_2$	596	2163	$Er_{0.90}Tb_{0.10}$	25	1320
	587	630	$Er_{0.20}Tb_{0.80}$	150	1320
	590	1608, 1944	Er-Th		711, 395
	473	542	95% Er	80	
$ErFe_3$	550	630	90% Er	74	
	455	1608	80% Er	48	
Er_2Fe_{17}	298.5	386	ErZn	50	1295
	293	630	$EuAl_4$	14	723
	310	1608	$Eu_3Al_2O_6$	10	897
Er_6Fe_{23}	491	630	$Eu_5Al_2O_8$	6	897
	495	1608	Eu_3As_2	18	1832
$Er_3Fe_5O_{12}$	551.4	239	EuB_6	8.5	1164
	556	430	EuC_{1-x}	80	2125
	556 ± 2	741	$Eu_{1-x}Ca_xO$	44	779
	542	1314	x = 0.55		
$Er_{0.95}Ho_{0.50}$	21	1579	Eu_2CrSbO_7	4	511
$Er_{0.50}Ho_{0.50}$	33	1320	$Eu_3Fe_5O_{12}$	566 ± 2	741
$ErIr_2$	3	41		563	1314
$ErMn_5$	415	875	$Eu_{0.96}Gd_{0.04}O$	128 ± 1	1846
Er_6Mn_{23}	415	204	$Eu_{≈0.96}Gd_{0.04}O$		
ErN	6	512	single crystal	137	1868
	5	355, 522	$EuGd_2S_4$	6	338
	3.4	1614	EuH_2	24	543
ErNi	10	802		17.8	2128
$ErNi_2$	14	802	EuI_2	5	17
	21	736	$Eu_{0.1}La_{0.9}Al_2$	11	1195
$ErNi_3$	62	538	$Eu_{0.4}La_{0.6}Al_2$	<4	1195
			$EuLiH_3$	38	2110

CURIE TEMPERATURE

Material	Temperature Curie (°K)	Reference	Material	Curie Temperature (°K)	Reference
			$EuPt_2$	105	975
$EuLu_2O_4$	7.5 (thermodynamic coefficients)	851	$(Eu_{1-x}RE_x)O$ $0.002 < x < 0.12$	> 130	1848
			EuS	14.5–16	1028
	9.5 (lines of equal magnetization)	851		15.8	1933
				16	371
				16.2	329
	But also see ref 346			16.3	1283, 1705
				16.4	1858
$Eu(NH_2)_2$	5.4	1969		16.4 ± 0.3	1231
$EuNb_2S_4$	~15	1662		16.5	1760
	existence questioned—transition probably to presence of EuS (see ref 1662)	2054	polycrystalline	16.5	1868
				16.5 ± 0.1	1939
				16.5 ± 0.5	532
				18	358
			EuSe	2.8	512, 994
EuO	69	17, 203, 1239		3.8	425, 1760
				4.58 ± 0.03	315
	69 ± 0.02 (Mossbauer)	1956		4.6	1290, 1933
				6	371
	69.2	1423		7	158
	69.4	1959		1.6 (>7kOe)	1260
	69.5 ± 0.5	532			
	69.594	1909	Eu_2SiO_4	7	17
	~70	1366	Eu_3SiO_5	19	897
	71	1717	$EuY_2Fe_5O_{12}$	556	1134
	72	51	Fe	1035 ± 3 (thermal expansion)	1910
single crystal	74	1868			
	≈76	2076		1040 ± 2	760
	77	359, 410, 1760, 1855		1040.2 ± 1	753
				1041	1604
	77 (photoresponse)	1895		1043	444
				1044 ± 2	319
	97(?)	53	FeAl (Fe rich)	923	713
≈5% Fe	150	2079			
7 wt.% Fe	≥180	1982	Fe_3Al	773	713
7.7% Fe	180	2079	$FeAlO_3$	<300	1121
2% Gd	~123	1772	$(Fe_{0.77}Al_{0.23})_2O_3$	810	1110
2.5 wt.% Gd	120	2109	FeB	598	302
(doped with RE_2O_3)	~135	643	$FeBO_3$	388	1178
Eu_3O_4	7.8	586		350	2144
	77	452		348 (single crystal)	1349
$EuO_{≈0.9}F_{≈0.1}$ polycrystalline	66	1868			
$(EuO)_{1-x}(RS)_x$		1143	10% Ga	272	2144
R = rare earth 0.01 < x < 0.30	73–132		Fe_3BO_6	508	1509
			Fe_5B_2P	628	1049
Eu_3P_2	25	1832	$FeBe_5$	75	1017
$Eu_3(PO_4)_2$	5	897		<273	53
$EuPd_2$	80	975			

Material	Temperature Curie (°K)	Reference
Fe_3C	483	331
FeCo	1253	1167
Fe-Co		1604
3 wt. % Co	1079 ± 9	
5 wt. % Co	1093	
5.95 wt. % Co	1105 ± 9	
9.35 wt. % Co	1124 ± 5	
10 wt. % Co	1143	
16 wt. % Co	1203	
$Fe_{1.2}Co_{0.8}$	~1200	1237
$Fe_8Co_{12}Al_3B_6$	815	916
ESD(Fe-Co-O)(35Co)	1238	745
$Fe_{2-x}Co_xP$	423	436
x = 0.3		
70Fe-30Co-P	363	787
$Fe_{2.4}Co_{0.6}P$	670	503
Fe_3Cr	893	1189
	1273	912
$Fe_{51.4}Cr_{48.6}$	9	107
$Fe_{56.5}Cr_{43.5}$	47	107
$FeCr_2O_4$	88	53
	69	598
$Fe_{3-x}Cr_xO_4$		1293
x = 0	853	
x = 0.4	~763	
x = 1.2	273	
x = 2.0	~80	
$(Fe, Cr)_3P$		1033
13.04 at. % Cr	489	
33.33 at. % Cr	99	
$FeCr_2S_4$	193	53
	180	1704
	177	1642
$Fe_{1.5}Cr_{1.5}S_4$	302	1704
$(Fe_{1-x}Cr_x)_3Se_4$		2032
x = 0.2	330	
x = 0.5	220	
$Fe_{2-x/2}Cr_xTi_{1-x/2}O_4$		2140
x = 2	80	
$Fe_{1-x}Cu_xCr_2S_4$	~355	885
x = 0.5		
$Fe_{0.5}Cu_{0.5}Cr_2S_4$	360	1377
FeF_2	363	1723
FeF_3	362	1215
$FeGa_{1.3}$	483, 697	435
$Fe_{0.69}Ga_{0.31}$	620	404
$Fe_{0.75}Ga_{0.25}$	760	404

Material	Temperature Curie (°K)	Reference
$Fe_xGa_{2-x}O_3$		1110
x = 1.4	350	
x = 0.93	262	
x = 0.82	208	
FeGe	280 ± 2	1644
(cubic B20 phase)	280 ± 2	1282
Fe-Ge	1046.2-1022.0	251
0.6-10.1 at. % Ge		
$\beta\text{-}Fe_{1.67}Ge$	485	448
Fe_2Ge	478	310
$Fe_{2-x}Ge$		256
x = 0.42	470	
x = 0.35	470	
x = 0.5	444	
Fe_3Ge	638	310, 471
	761	451
(cubic)	755	447
(hexagonal)	655	447
Fe_5Ge_3	485	448
$Fe_{2(1-s)}Ge_sCu_{1+s}O_4$		1527
s = 0.05	~719	
s = 0.10	~709	
s = 0.20	~700	
s = 0.30	~700	
$Fe_{2+x}Hf_{1-x}$		1974
at.% Hf:		
30	591	
30.3	625	
29.5	595	
Fe_2MgO_4	653	896
$Fe_xMg_{1-x}SiO_3$		1390
x = 1.0	37 ± 0.5	
x = 0.866	18 ± 1	
x = 0.758	11 ± 1	
Fe-Mn	1042.0-1017.8	251
0.5-2.4 at. % Mn		
$Fe_{1.95-x}Mn_xAs$		955
(x = 1.19)	127	
(x = 1.28)	153	
Fe-Mn-Ge	435	1857
FeMnGe	245	1859
(ferri- to ferro-)		
385		
(ferro- to para-)		
$FeMn_2Ge$	233	471
$Fe_{0.9}Mn_{0.9}Ge$	241	491
$Fe_{1.5}Mn_{1.5}Ge$	333	471
Fe_2MnGe	433	471

CURIE TEMPERATURE

Material	Temperature Curie (°K)	Reference	Material	Temperature Curie (°K)	Reference
$FeMn_2O_4$	117-120	1359	$(Fe_{1-x}Ni_x)_2P$		1612
	390	1358	$x = 0$	~267	
$Fe_{2.4}Mn_{0.6}P$	680	503	$x = 0.1$	~340	
$(Fe_{1-x}Mn_x)_{0.75}P_{0.15}C_{0.10}$		1987	$x = 0.3$	~270	
$x = 0.1$ and 0.2	> 300		$x = 0.5$	~60	
$x = 0.3$	≈210		$(Fe_{0.92}Ni_{0.08})_2P$	342	1369
$x = 0.4$	≈145		$Fe_{0.75}Ni_{2.25}P$	120	503
$x = 0.6$	≈50		$Fe_{2-x}Ni_xP$	343	436
Fe-Mo	1042.6-1032.4	251	$x = 0.1$		
0.5-3.0 at. % Mo			Fe-Ni-Sb Spinels	590-858	1149
$\gamma'\text{-}Fe_4N$	761	1202	Fe_2O_3	848	763
Fe-Ni			(α)	948	715
% Fe:			(α)	950	577
4.5	683.0	596	(α)	956 ± 2	1066
19	834.0	596	(α)	959	755
20	840	1398	(α)	963 ± 5	774
20 (γ)	200	1764	(ϵ)	483	561
23	876.0	596	(γ)	743	438
35 Fe	943	1880	Fe_3O_4	858	405
	(ordered)			848-858	683
35 Fe	863	1880	$5Fe_2O_3 \cdot 3Dy_2O_3$	551.4	239
	(disordered)		$5Fe_2O_3 \cdot 3Er_2O_3$	551.4	239
40	880	1398	$5Fe_2O_3 \cdot 3Gd_2O_3$	574.6	239
40 Fe	923	1880		~560	682
	(ordered)		$5Fe_2O_3 \cdot 3Y_2O_3$	562.7	239
40 Fe	851	1880	$5Fe_2O_3 \cdot 3Yb_2O_3$	542.6	239
	(disordered)		FeOOH		
45 Fe	948	1880	(δ)	450	214
	(ordered)		δ	460	1232
45 Fe	823	1880	δ	$C_1 = 455$	1422
	(disordered)			$C_2 = 420$	1422
50	786.1	596	FeP	215	131, 436
55 Fe	956	1880		not ferromagnetic	1475
	(ordered)				
55 Fe	710	1880	$Fe_{1.91}P$	223	1903
	(disordered)		Fe_2P	278	787
60	648	1398	(hexagonal)		
63 Fe	916	1880		306	1304
	(ordered)			225	1903
63 Fe	553	1880	Fe_3P	686	1033
	(disordered)			716	503
$FeNi_3$	865	1602	$Fe_2P_{1-x}B_x$		436
ordered	~943	1534	$x = 0.05$	291	
$Fe_{0.5}Ni_{0.5}$	790	1602	$x = 0.10$	391	
$Fe_{0.7}Ni_{0.3}$	340	1602	$x = 0.20$	512	
$Fe_8Ni_{12}Al_3B_6$	285	916	$Fe_3P_{1-x}B_x$	758	436
			$x = 0.30$		

Material	Temperature Curie (°K)	Reference	Material	Temperature Curie (°K)	Reference
$Fe_3P_xB_{1-x}$	806	436	$Fe(Rh_{1-x}Ir_x)_{1.08}$		113
$x = 0.20$			$x = 0.121$	613	
Fe_5PB_2	619–639	436	$x = 0.084$	630	
$Fe_{80}P_{12.5}C_{7.5}$	586 ± 2	457	$x = 0.056$	643	
FePd	738	436	$FeRhMo_{0.0833}$	771	364
Fe-Pd		1657	$FeRhNb_{0.0833}$	834	364
Fe at. %:			$FeRhNi_{0.0833}$	426	364
2.8 ± 0.1	95 ± 3 (hydrogen free)		$Fe(Rh_{1-x}M_x)_{1.08}$ M = Pd		113
4.0 ± 0.1	106 ± 3 (hydrogen free)		$x = 0.029$	664	
	103 ± 3 (α_{max} hydride)		$x = 0.058$	650	
	6.4 ± 5 (β_{min} hydride)		$Fe(Rh_{1-x}M_x)_{1.08}$ M = Pt $x = 0.056$	636	113
7.2 ± 0.1	168 ± 3 (hydrogen free)		$FeRhTa_{0.0833}$	773	364
	163 ± 3 (α_{max} hydride)		$FeRhW_{0.0833}$	1051	364
	54 ± 3 (β_{min} hydride)		$Fe_{0.87}S$	593	1174
			$Fe_{0.902}S$	530	408
			Fe_3S_4	580	1190
			Fe_7S_8	578	181
10.3 ± 0.1	246 ± 3 (hydrogen free)			593	1174
	230 ± 3 (α_{max} hydride)		$FeSb_{0.8}$	<80	1467
			$Fe_{1.22}Sb$	4.2	181
			Fe_3Se_4	318	2032
	155 ± 3 (β_{min} hydride)			~320	181, 1402
			Fe_7Se_8	449	892
				~450	1406
$FePd_3$	540	460		455	1437
$Fe_{0.013}Pd_{0.987}$	≅ 55	821		483	1331
$FePd_2Pt$	~360	1399		425	181
$FePd_{1.6}Pt_{1.4}$	~270	1399	Fe-Si	1043.9-1012.6	251
$Fe_7Pd_{73}Si_{20}$			0.9-7.4 at. % Si		
amorphous	~28	1658	Fe-13% Si	~953	816
	28	1942	Fe-30 At. % Si		1034
FePt	743	436	heat treatment (°C)		
Fe-Pt		1404	1100	683	
34.5% Fe	255 ± 2		1050	723	
Fe_3Pt	453	397	950	763	
Ferrites	tabulation	467	800	808	
FeRh	668	113	Fe_3Si	808	479
Fe-Rh		799	Fe_5Si_3	381	476, 477
48% Rh	773			385	971
50% Rh	678			376	1897
$Fe_{0.49}Rh_{0.51}$	667	489	$Fe_{1.3}Sn$	676	861
$FeRhCo_{0.0833}$	480	364	Fe-8% Sn	1033	1133, 2075
$FeRhCr_{0.0833}$	484	364	$Fe_{1.6}Sn$	583	331
$FeRhCu_{0.0833}$	486	364	Fe_3Sn	743	331

CURIE TEMPERATURE

Material	Temperature Curie (°K)	Reference	Material	Temperature Curie (°K)	Reference
Fe_3Sn_2	612	331	Gd	289	198, 588
Fe_5Sn_3	583	331		290	43, 444
$Fe_{1.11}Te$	479	692		290.1	1959, 1960
Fe-30 at. % Ti	~380	557		291	1836
$Fe_{2+x}Ti_{1-x}$				291.21 ± 0.02	1937
x = 0.0671	318	1687		292 ± 0.1	1628
x = 0.027	70	1947		(microhardness)	
Fe_2TiO_4	142	903		292.5	555
$(Fe_2TiO_4)_x(Fe_3O_4)_{1-x}$		1490		293.2	547
x = 1.0	853			293 (resistivity)	1505
x = 0.9	853			294 (thermal cond.)	1505
x = 0.8	850			294	201
x = 0.5	873			294 for H∥a	555
x = 0.2	853			294 ± 2	826
	(all were "heated in air")			295.5	556
$FeTi_2S_4$	60.2	1382	$GdAg_{0.3}In_{0.7}$	116	347
Fe-53.3% V	280	863	$GdAg_{0.8}In_{0.2}$	40	347
$Fe_{0.54}V_{0.46}$		1440	$GdAl_2$	171	957, 201
annealed at 650°C	~165			173	1338
quenched from 1000°C	~100			176	444
$Fe_{55.5}V_{44.5}$	~160	63		182	1190
$Fe_{61.0}V_{39.0}$	~210	63		151	1341
Fe_2VO_4	440	993, 1113		157	1972
FeV_2O_4	109	993, 1113		(elec. resist.)	
			Gd_3Al_2	282	581, 2082
$(Fe_{0.50}V_{0.50})_2O_3$	455 ± 10	1132		275	201
Fe-Zn		416	$Gd_3AlFe_4O_{12}$	435	1094
11.2 ± 0.2 at. % Zn	981.3 ± 0.5		Gd_4Bi_3	340	202, 444
20.1 ± 0.3 at. % Zn	925.0 ± 1.0		$GdBr_3$	2	1494
Fe_2Zr	628	864		1.32	1875
(36 to 25 at. % Zr)	590-780	530	GdC	313-373	1038
Garnets	tabulation	467	Gd_3C	500	33
$GaAl_2$	176	41	$\{Gd_{2-x}Ca_x\}Fe_{5-x}Ge_xO_{12}$		709
$Ga_{0.5}Fe_{2.5}O_4$	686 ± 3	424	x = 2.00	250	
$Ga_{0.7}Fe_{2.3}O_4$	620 ± 3	424	x = 2.50	50	
$Ga_{0.85}Fe_{1.15}O_3$	308	1238	$\{Gd_{2-x}Ca_x\}Fe_{5-x}Si_xO_{12}$		709
$Ga_{2-x}Fe_xO_3$			x = 2.00	250	
x = 1.08	266.5	1245	x = 2.50	60	
x ≃ 1.08	350	409	GdCd	262	347
x = 1.20	305	583	$Gd_{1.2}Cd$	~280	593
x = 0.80	205	583		265	937
x = 0.86	167.5	1245	Gd-30% Ce	160	1735
			$GdCl_3$	2.20	360
				2.218	2168

Material	Temperature Curie (°K)	Reference	Material	Temperature Curie (°K)	Reference
Gd–Co		2024	$Gd_xDy_{1-x}Ni_5$		
at.% Co:			$x = 0$	30	1812
25	180		$x = 0.2$	~26	1529
60	325		$x = 0.4$	31.5	1812
80	925		$x = 0.8$	37.5	1812
90	1225		$x = 0.8$	~37	1529
$GdCo_2$	404	417	$GdFe_2$	490	528
	408	444		782	444
	≈400	257		≈785	1944
	393	1972		793	630
	(elect. resist.)			813	542
$GdCo_3$	612	417	$GdFe_3$	728	630
$GdCo_5$	1008	442	$GdFe_5$	455	444
	1030	444	Gd_2Fe_{17}	460	630
Gd_2Co_7	775	417		459	386
	762	630		472 ± 3	1696
Gd_2Co_{17}	1209	417	Gd_6Fe_{23}	659	630
	1213	537, 759		468	1479
Gd_4Co_3	230	1151	$(Gd_3)Fe_{5-x}Al_xO_{12}$		709
$GdCo_2$-$GdAl_2$		742	$x = 1.50$	345	
10 mol % $GdAl_2$	291		$x = 2.50$	175	
50	116		$x = 3.50$	30	
70	82		$GdFe_2$-$GdAl_2$		742
100	180		50 mol % $GdAl_2$	264	
Gd_2CoMnO_6	113 (mxtl, annealed)	1343	92.5 mol % $GdAl_2$	138	
			$Gd_3Fe_5O_{12}$	564	430
	104 ± 2 (pxtl)	1343		564 ± 2	741
$Gd(Co_{1-x}Ni_x)_2$		1536		574.6	239
$0 \leq x \leq 1$	420 to 85			556	1314
$Gd_3Cr_{0.25}Fe_{4.75}O_{12}$	250	1129, 1130	$Gd_3Fe_5O_{12-x}F_x$		1835
			$x = 0$	564	
Gd_2CrSbO_7	12	511	$x = 0.40$	549	
Gd-Dy			$x = 0.60$	528	
Gd at. %: 90	285	361	$Gd_xHo_{1-x}Ni_2$		
83.4	269	1653	$x = 0$	25	1812
58.6	~234	959	$x = 0.2$	~28	1529
50	226	361	$x = 0.4$	42.5	1812
42	203	959	$x = 0.8$	70	1812
42	202	1653	$x = 0.8$	~70	1529
39	193	361	$x = 1$	85	1812
28.7	165	1653	$Gd_xHo_{1-x}Ni_5$		
12.5	120	361	$x = 0$	25	1812
$Gd_xDy_{1-x}Ni_2$			$x = 0.2$	~19	1529
$x = 0$	30	1812	$x = 0.2$	20	1812
$x = 0.2$	~42	1529	$x = 0.8$	35	1812
$x = 0.4$	52.5	1812	$x = 0.8$	~35	1529
$x = 0.8$	75	1812	GdI_3	< 1	1875
$x = 0.8$	~75	1529	$GdIr_2$	88	41
$x = 1$	85	1812		90	444

CURIE TEMPERATURE

Material	Temperature Curie (°K)	Reference	Material	Temperature Curie (°K)	Reference
Gd-30% La	185	1735	$GdNi_{2-x}Al_x$		1800
$Gd_{0.1}La_{0.9}Al_2$	24	1195	x = 0.11	92	
Gd-Lu		395	x = 1.55	68	
80.0% Gd	249		$GdNi_{1.65}Cu_{0.35}$	103 ± 1	1805
64.3	150		$GdNi_2$-$GdAl_2$		742
60.0	95		50 mol % $GdAl_2$	66	
Gd-Lu		52	92.5	181	
92-8	274-42		$GdOs_2$	66	41
$Gd_{0.69}Lu_{0.31}$	221	699	GdPd	39.5 ± 1	1187
$Gd_{0.91}Lu_{0.09}$	267	699	$GdPd_2$	335	350
$Gd_{1-x}Mg_x$		2129	Gd_5Pd_2	335	444
x = 0	293		Gd-35% Pr	165	1735
x = 0.06	233		$GdPt_2$	4.2	1024
x = 0.12	184			22	1972
$GdMn_2$	300	880		(elec. resist.)	
	575	444		≈37	257
$GdMn_5$	465	875, 444	$GdRh_2$	>77	883
	468 ± 3	197		≈72	257
Gd_6Mn_{23}	478	444		66	1972
	468	204		(elec. resist.)	
GdN	60	522	$GdRu_2$	83	41
	67.4	1614	Gd_4Sb_3	260	444, 202
	69	276, 812	$Gd_4(Sb_xBi_{1-x})_3$		362
	72	410, 512	x = 0	346	
Gd-50% Nd	145	1735	x = 0.5	290	
$Gd_xNd_{1-x}Ni_2$			x = 1	260	
x = 0	15	1812	Gd-Sc		
x = 0.2	42.5	1812	Gd %: 90	251	395
x = 0.2	~44	1529	80	212 ± 4	264
x = 0.6	65	1812	72	171	395
x = 0.8	~78	1529	69	60	395
x = 1	85	1812	69% Gd	160	264
$Gd_xNd_{1-x}Ni_5$		1529	$\{Gd_3\}[Sc_xFe_{2-x}]$		709
x = 0.2	~17		$(Fe_3)O_{12}$		
x = 0.8	~33		x = 1.00	300	
GdNi	73	802	x = 1.25	216	
$GdNi_2$	85	802, 444, 736	x = 1.50	30	
	90	698	Gd_5Si_4	336	949
	77	257	$Gd_{1-x}Th_x$		2129
	75	1524	x = 0	293	
	75	1972	x = 0.05	273	
	(elec. resist.)		x = 0.10	244	
			Gd-Y		
$GdNi_3$	116	538	Gd %: 90.0	281	395
$GdNi_5$	36	802	80	254 ± 4	264
	27	444	66.7	211	395
Gd_2Ni_7	118	954	60	84 ± 4	264
Gd_2Ni_{17}	205	712	$Gd_{0.70}Y_{0.30}$	220	699
	623	1181	$Gd_{0.85}Y_{0.15}$	268	699

CURIE TEMPERATURE

Material	Temperature Curie (°K)	Reference	Material	Temperature Curie (°K)	Reference
$Gd_xY_{1-x}Co_2$		417	Ho_2Co_{17}	1173	417
x = 0.2	60			1183	537, 759
x = 0.4	205		Ho_4Co_3	44	1151
x = 0.8	350		Ho_2CrSbO_7	10	511
$Gd_xY_{1-x}Fe_2$		1944	Ho-Er	35	756
x = 0.2	600		$Ho_{0.5}Er_{0.5}$	33	217
x = 0.5	680		$HoEr_2$	13 ± 1	1950
x = 0.9	≈775			612	1944
$Gd_{1.5}Y_{1.5}Fe_{4.5}Al_{0.5}O_{12}$	180	2103		600	1840
$Gd_{1.3}Yb_{0.7}Fe_{4.1}Ga_{0.9}O_{12}$	340	2103	$HoFe_2$	608	542
GdZn	~280	593		603	630
	270	1295	$HoFe_3$	567	630
$GdZn_2$	68	1729	Ho_2Fe_{17}	325	386
H (theoretical)		1575		319	630
	1 (at 10^4 g/cm³)		Ho_6Fe_{23}	501	630
	10 (at 10^6 g/cm³)		$HoFeO_3$	500	1840
			$Ho_3Fe_5O_{12}$	47 ± 1	1780
$HgCr_2S_4$	36	507, 510		567	430
	36.1	1392		567 ± 2	741
	60	563		548	1314
$HgCr_2Se_4$	120	563	$HoIr_2$	12	41
	106	510	$HoMn_5$	425	875
	105.5	1392		434 ± 3	197
Ho	20	295, 444, 24, 711, 1732	Ho_6Mn_{23}	434	204
	20 (resistivity and thermal cond.)	1505	HoN	19	522
				18	355
				13.3	1614
	20, 19	198		13	512
	23	40	$Ho_xNd_{1-x}Ni_2$		
HoAl	26	996, 1168	x = 0	17.5	1813
$HoAl_2$	27	695	x = 0.2	17.5	1812
	29	1338	x = 0.2	~78	1529
	42	1194	x = 0.8	22.5	1812
Ho_3Al_2	33	581	x = 0.8	~83	1529
2 at. % Ho-Au	0.001	1172	$Ho_xNd_{1-x}Ni_5$		1529
$HoC_{0.5}$	95	1038	x = 0.2	~64	
Ho_2C	≈100	2043	x = 0.8	~70	
$Ho_{1.3}Cd$	30	937	HoNi	31	802
$HoCo_2$	87, 95	306	$HoNi_2$	23	802
	95	736		22	736
	≈90	257	$HoNi_3$	66	538
$HoCo_3$	418	417	$HoNi_5$	22	742
$HoCo_5$	1025	444		10	802
	1000	442	Ho_2Ni_7	70	954
Ho_2Co_7	670	417	Ho_2Ni_{17}	162	712
	644	630		611	1181
			$HoNi_{1.09}Al_{0.91}$	27	1800

CURIE TEMPERATURE

Material	Temperature Curie (°K)	Reference	Material	Temperature Curie (°K)	Reference
$HoNi_{2-x}Al_x$		1800	$LaErO_3$	7	1073
x = 0.16	30		(meta magnetic below 4°K)		
x = 1.55	31		$LaFeFe_{11}O_{19}$	728	1077
$Ho(OH)_3$	2.55	1000	$La_{0.94}Gd_{0.06}Ru_6$	10 ± 1	1148
$HoOs_2$	9	41	$La_{0.98}Lu_{1.00}Tb_{1.00}$	≈1.5	1985
HoP	5.5	355	(supercond. at 2.582 K)		
$Ho_{5.04}Pd_{1.96}$	~10	350	$La_{0.98}Lu_{1.15}Tb_{0.85}$	≈1.3	1985
$HoPt_2$	4.2	1024	$LaMnO_3$	166	1834
	≈9	257	La_2NiMnO_6	285 quenched 1000°K	562
$HoRh_2$	≈6	257		300 slowly cooled	
Ho_5Si_4	76	949		310 annealed	
Ho-Tb		675			
20% Ho	171				
Ho-Th		395, 711	$La_{1-x}Pb_xMnO_3$		
Ho %: 90	44		x = 0.25	315	1650
85	72		x = 0.35	335	1652
Ho-Y		40	x = 0.45	350	1650
Ho at. %: 80.4	20		$(La_{0.70}Pb_{0.30})MnO_3$	361	369
50.0	24		$La_{0.65}Pb_{0.35}MnO_3$	330	2121
32.3	25		$La_{0.62}Pb_{0.38}MnO_3$	~340	1740
HoZn	80	1295	$(La_{0.45}(Pr,Nd)_{0.15}Ba_{0.40})MnO_3$	275	369
$K_2CuCl_4 2H_2O$	1	1759			
K_2CuF_4	9.5	1837	$(La_{0.15}(Pr,Nd)_{0.45}Sr_{0.40})MnO_3$	327	369
K_2IrCl_6	3.05 ± 0.04	1177			
$KMnF_3$	81.5 ?	548	$La_{1-x}RE_xRu_2$		990
$La_{0.85}Ba_{0.15}MnO_3$	230	2038	(RE = Pr-Gd)	10	
$La_{1-x}Ba_xMn_{1-x}Ti_xO_3$		2038	(RE = Gd)	9.5	
x = 0.05 to 0.10	150		(RE = Gd-Tb)	8.5	
$(La_{1-x}Ca_x)MnO_3$			$La_{0.5}Sr_{0.5}CoO_3$	~228.5	1029
x = 0.1	≈152	1976	$La_{0.5}Sr_{1.5}MnO_4$	295	814
x = 0.2	≈200	1976	$(La_{0.60}Sr_{0.30}Ba_{0.10})MnO_3$	358	369
x = 0.3	≈250	1976			
x = 0.3	260	1908	$LiCo_{0.5}Mn_{1.5}O_4$	50	209
$(La_{0.70}Cd_{0.30})MnO_3$	326	369	$Li_{0.5}CuFe_{4.5}O_8$		615
$LaCo_5$	840	1030	(slowly cooled)	818	
$LaCo_{13}$	1290	1030	(quenched)	809	
La_2Co_7	490	1030	$Li_{0.5}Fe_{2.5-x}Al_xO_4$		
$LaCo_4Cu$	690	1069	x = 0	1184	1866
La_2CoMnO_6	245 quenched 1000°C	562	x = 0.164	1124	1866
	260 slowly cooled		x = 0.330	1071	1866
			x = 0.8	≈650	1861
	270 annealed		$Li_{0.5(1-x)}Fe_{0.5(1-x)}Zn_xFe_2O_4$		1912
	280	1233	x = 0.1	873	
$LaCo_4Ni$	690	1069	x = 0.18	823	
$LaCrO_3$	300	73	x = 0.24	773	
La_2CuMnO_6	55 slowly cooled	562	$Li_{0.5}Fe_{2.5-x}Ga_xO_4$		1481
			x = 0.06	850	
			x = 0.43	705	
			x = 1.18	301	

CURIE TEMPERATURE

Material	Temperature Curie (°K)	Reference	Material	Temperature Curie (°K)	Reference
$LiFe_5O_8$	903	1135	$MgFe_2O_4$		
	893	615	(ferrimag.)	612	1810
$Li_{0.5}Fe_{2.5}O_4$	923 ± 10	278		(quenched)	
	911	1481	(ferrimag.)	679	1810
	908	1342		(slowly cooled)	
	~880	770	$Mg_{0.46}Fe_{2.54}O_4$	768 to 861	1351
$Li_{0.45}Fe_{2.52}O_4$	808.5	1255	$Mg_{0.8}Fe_2O_{3.3}F_{0.7}$	693	1119
$LiFe_2O_3F$	903	524, 776	$Mg_{1.6}Fe_{1.1}Sb_{0.3}O_4$	370	281
Existence of this compound questioned		992	$Mg_{1+2x}Fe_{2-3x}Sb_xO_4$		281
$Li_{0.5}Fe_{2.5-x}Rh_xO_4$		281	x = 0	715	
x = 0	958		x = 0.10	620	
x = 1.00	580		x = 0.30	370	
x = 2	130		$Mg_{0.5}Li_{0.5}Fe_2O_{3.5}F_{0.5}$	~768	776
$Li_{0.5+x}Fe_{2.5-2x}Sb_xO_4$		281	$Mg_{0.5}Mn_{2.5}O_4$	20	794
x = 0	955		Mg_2SnO_4-$MgFe_2O_4$	~220	1211
x = 0.25	750		55 at. % Mn-45 at. % Al	623	468
x = 0.1	885		Mn(45% Al)	~653	380
$Li_{0.6}Fe_{2.3}Sb_{0.1}O_4$	885	281	Mn_3AlC	272	1949
$LiMg_{0.5}Mn_{1.5}O_4$	38	209	Mn-Al-Co	Co rich: 370	378
$LiNi_{0.5}Mn_{1.5}O_4$	130	209		Mn rich: 466	
$Li_{0.388}Ni_{0.612}O$	210	1115	MnAlGe	518	436, 366
(Li_2O, Al_2O_3, MnO_2) ferrites	703–863	474	MnAs	305 (decreasing)	603
$LiZn_{0.5}Mn_{1.5}O_4$	22	258, 209		312	181
$LuCo_2$	< 2	2105		313	459, 1683
$LuCo_3$	362	2104, 2105		317 ↑	743
Lu_2Co_7	459	630		307 ↓	
Lu_2Co_{17}	1192	2104, 2105		313	865
	1210	537, 759		318 (increasing)	603
Lu_3Co	< 2	2105		~318	1332
Lu_4Co_3	< 2	2105	$MnAs_{0.9}P_{0.1}$	~240	970
$LuFe_2$	589	630	$MnAu_4$	360	866
	610	1945	MnB	578	302
$LuFe_3$	529	630	MnB_2	157	392
Lu_2Fe_{17}	308	386		>157 (?)	1456
	263 ± 2	1696		143	240
	235	630		140 ± 2.0	1483
Lu_6Fe_{23}	485	630	MnBi	633	181, 436, 1345
$Lu_3Fe_5O_{12}$	549	430	thin films	453	1765
	549 ± 2	741	(quenched)		
Lu_2Ni_{17}	601	1181	thin films	623	1765
$MgFeAlO_4$	273	53	(normal)		
$MgFe_2O_4$	490 (aged at 800°C)	891	Mn_2Co_2C	808	1405
	520 (aged at 700°C)	891	$MnCo_2O_4$	203	53
			$Mn_{0.8}Cr_{0.2}As$	200	1501
(continued)			$Mn_xCr_{1-x}O_2$ x = 0.5	250	1122

CURIE TEMPERATURE

Material	Temperature Curie (°K)	Reference	Material	Temperature Curie (°K)	Reference
$MnCr_2O_4$	42	962	Mn_3Ge_2	300	165
	43	1109, 1589	Mn_5Ge_3	311	907
	~43	1635		≈310	2139
	~47	1462		304	1216
	55	2001		300	2031
$MnCr_2S_4$	66	817		293	868
	103	53		≈290 (films)	2066
$Mn_{1-x}Cr_xSb$		376	Mn_2Ge_2Cu	612	514
x = 0 annealed	600		$Mn_2Ge_{0.12}Sb_{0.88}$	~340	1478
x = 0 quenched	565		Mn_3In	583	870
x = 0.51	390		$MnIn_xCr_{2-x}S_4$		
x = 0.7	255		x = 0	67	1790, 964
Mn_3CuN	149	952	x = 0.2	48	1790
$MnFe_{0.5}Cr_{1.5}O_4$	224	53	x = 0.3	36	1790, 964
$MnFe_{2-x}Cr_xO_4$		1952	$Mn_{0.5}Li_{0.5}Fe_2O_{3.5}F_{0.5}$	~703	776
x = 0.5	485		Mn_4N	745	1945
x = 1	375		$Mn_4N_{1-x}\square_x$	759	911
x = 1.5	240		(19.87 at. % N, x = 0.0099)		
$Mn_{0.1}Fe_{0.9}HoO_3$	600	1234			
$Mn_{0.7}Fe_{0.3}HoO_3$	97	1234	$Mn_4N_{1-x}C_x$	848	911
$MnFe_2O_4$	573-638	683	x = 0.303		
(ferrimag.)	573	1810	$Mn_4N_{0.75}C_{0.25}$	850	1945
	552.5	534	$Mn(NH_4)_2(SO_4)_2 \cdot 6H_2O$	0.18	1862
	555-620 (depending upon heat treatment)	772	$Mn_{1/3}NbS_2$	48	1695
			$Mn_{0.33}NbSe_2$	22	2151
			$Mn_{1/4}NbSe_2$	~25	1695
$Mn_{0.95}Fe_{2.05}O_4$	571	405	$MnNi_3$	750 (ordered)	871
$(Mn_{0.5}Fe_{0.5})_2O_3$	24	488		132 (disordered)	872
$MnFe_4Si_3$	325	1897			
$Mn_2Fe_3Si_3$	160 ('intermediate state of magnetization')	1897	$Mn_{2.94}Ni_{0.06}B_4$	146	734
			$(Mn_xNi_{4-x})(Mn_{2-x}Ni_x)O_4$ x = 0.93	116	402
			Mn_3O_4	30	525
Mn-Ga	690	336		42	794
66 at. % Mn				43	2085
ζ Mn-Ga	210	44		45	762
Mn_2Ga	690	436		46	1359
$Mn_{2.85}Ga_{1.15}$		2004	MnO-35.5%, ZnO-15%, Fe_2O_3-49.5%	467.5	398
DO_{19} phase	470 ± 10				
DO_{22} phase	> 770		MnO-26%, ZnO-24%, Fe_2O_3-50%	368.5	398
Mn_3Ga	470	867			
Mn_3GaC	246 ± 2	342	MnP	291	595
	248	1682, 1685		291.5 ± 0.2	18, 1785
	251	567		298	53
$Mn_3GaC_{0.95}$	265	1685	$Mn_{1.20}Pt_{0.80}$	540 ± 15	1264
$Mn_3GaC_{0.9}N_{0.1}$	223	567	MnSb	587	241
Mn_3GdC	150	565		556	827
Mn_3Ge	28	471	Mn_2Sb	550	873

Material	Temperature Curie (°K)	Reference	Material	Temperature Curie (°K)	Reference
$Mn_{53}Sb_{47}$	450	1205	$Nd_{1.3}Cd$	105	937
$MnSi$	38	639	$NdCl_3$	1.745	1498
	30	1009	$NdCo_2$	116	306, 736
Mn_5Si_3	60	309		≈100	257
Mn_5SiC	284	1036	$NdCo_3$	395	417
$Mn_5Si_3C_{0.22}$	≤156	1036	$NdCo_5$	910	442
Mn_9Si_2C	220	1036		913	630
$Mn_{1.5}Sn$	269	874		925	444
$Mn_{1.8}Sn$	256	874	Nd_2Co_7	609	630
Mn_2Sn	263	53	Nd_2Co_{17}	1150	417
	<300	1217		1166	537, 759
$Mn_{1/3}TaS_2$	~80	1695	Nd_2CrSbO_7	8	511
$MnTe$	260	9	$NdFe_7$	327	386
Mn_2TiO_4	~77	196		326	321, 923
α-Mn_2TiO_4	62	1154		332	349
	69	254	Nd_2Fe_{17}	327	630
β-Mn_2TiO_4	78	254	Nd_6Fe_{23}	429 ?	630
$MnVO_3$		2148	$Nd_3Fe_{5-x}Sc_xO_{12}$		1546
ilmenite	70		$x = 0.9$	~320	
perovskite	65		$x = 1.25$	~200	
MnV_2O_4	56	254, 993, 1561	$x = 1.3$	~150	
			$NdGe$	28	515
Mn_2VO_4	62 ± 3	332	$NdGe_2$	3.6	450
	45	418	Nd_5Ge_3	45	906
Mn-Zn	>450	1076		58.4	1312
(β_1)	(phase decomposes)		NdH_2	9.5	925
			$NdIr_2$	11.8	41
$MnZn_3$	250 (hcp)	877	Nd_6Mn_{23}	437	960
	>400 (hexagonal)	876		416	1479
			NdN	19	522
$Mn_xZn_{1-x}Fe_2O_4$		1489		27	266
$x = 0.2$	~228			27.6	1614
$x = 0.6$	414			32	512
$x = 0.9$	533			35	781
$x = 1.0$	563		$NdNi$	35	802
$Mn_{0.63}Zn_{0.37}Fe_{2.05}O_4$	423	1519	$NdNi_2$	20	698
$Mn_{1-x}Zn_xFe_{1.9}Ti_{0.1}O_4$		1927		16	736
				10	1615
$x = 0.2$	485		$NdNi_3$	27	538
$x = 0.35$	430		$NdNi_5$	9	802
$MoMn_2B_4$	590	956	Nd_2Ni_7	87	954
$Mo_{16}N_7$	175	1884	$NdNi_{2-x}Al_x$		1800
$Na_5Fe_3F_{14}$	80	208	$x = 0.4$	16	
	~100	900	$x = 1.73$	25	
$Na_{0.80}Fe_{2.40}O_4$	663	1127	$NdNi_{1.20}Al_{0.80}$	17	1800
$Na_{0.70}MnO_2$	≈60	2134	$NdOs_2$	23	41
Nd	29	1477	$Nd_{0.62}Pb_{0.38}MnO_3$	175	2121
$NdAl_2$	68	695	$NdPt_2$	4.2	1024
	65	412		≈3	257
	76	1338			

CURIE TEMPERATURE

Material	Temperature Curie (°K)	Reference	Material	Temperature Curie (°K)	Reference
$NdRh_2$	8.1	883	$Ni_{1-x}Cu_xMn_{0.02}Fe_{1.9}O_4$		703
	≈7	257	x = 0	565	
$NdRu_2$	28	41	x = 0.2	565	
Nd_3S_4	50	1619	x = 0.4	560	
	45.6	2159	$Ni_xCu_{1-x}MnSb$		1767
$Nd_{1-x}Sr_xMnO_3$		1690	x = 0.1	~649	
x = 0.15	130		x = 0.3	~383	
x = 0.5	261		x = 0.7	~167	
$Nd_{1-x}Sr_xMnO_3$	~260	1213	Ni–1% Fe-57	641	1843
$NdZn_2$	24	1729	$NiFe_{2-x}Al_xO_4$	860-470	1085
Ni	627	835	x = 0 – 1.5		
	625 ± 2	2050	$Ni_3\lfloor Fe(CN)_6 \rfloor_2$	19	1934
	628.3 ± 1	753	$Ni_xFe_{1-x}Cr_2O_4$		801
	≈631	2094	x = 0.1	~183	
	633	444	x = 0.2	~138	
	592 ± 3	1910	$NiFe_{0.5}Ga_{1.5}O_4$	173	53
	(thermal expansion)		Ni-Fe-Mn		1398
			78.0:19.5:2.5 at. %	832	
Ni_3Al	75	546, 915	72.0:18.0:10.0 at. %	672	
Sample homogeneity questionable			58.5:39.0:2.5 at. %	852	
			54.0:36.0:10.0 at. %	640	
75.5 at. % Ni	58.1	1364	39.0:58.5:2.5 at. %	608	
$Ni_3B_7O_{13}I$	60	1193	36.0:54.0:10.0 at. %	396	
	64	605	$NiFe_2O_4$	858	281
Ni-C	(decreases	1507	(ferrimag.)	868	1810
	~30°K/at.		$NiFe_{2-x}Rh_xO_4$		281
	% C)		x = 0	858	
$NiCo_2O_4$	350	317, 136	x = 0.5	720	
		1116,	x = 1.0	540	
	~500	1280	$Ni_{1+2x}Fe_{2-3x}Sb_xO_4$		281
	513	927	x = 0	858	
Ni-Cr	324	798	x = 0.10	820	
5.6 at. % Cr			x = 0.25	710	
$NiCrO_2$	389	1836	$Ni_{76}Ga_{24}$	10	1364
$NiCr_2O_4$	78	53	$Ni_{75-x}Ga_{25}Fe_x$		1196
	60	1461	x = 0.05	13.4 ± 0.4	
Ni-Cu			x = 0.2	41.3 ± 0.4	
40 at. % Cu	173	1016	x = 1.0	85 ± 1	
40 at. % Cu	216	798	Ni-8 at. % Ge	~348	1035
50 at. % Cu	110	798	$Ni_{1-x}Ge_xFe_{2-2x}O_4$		1158
$Ni_{0.5}Cu_{0.5}$		1407	x = 0	860	
(H/Ni = 0.086)	~50		x = 0.08	810	
Ni-47.5 at. % Cu	72 ± 2	1435	x = 0.30	680	
Ni-50 at. % Cu	35-45 ± 3	1435	x = 0.41	610	
Ni-53.6 at. % Cu	10-16 ± 1	1435	$Ni(IO_3)_2 \cdot 2H_2O$	≅ 3.2	740
Ni-57 at. % Cu	<1	1435	$Ni_3La_2(NO_3)_{12} \cdot 24H_2O$	0.393	1533
			$Ni_{0.5}Li_{0.5}Fe_2O_{3.5}F_{0.5}$	~863	776

Material	Temperature Curie (°K)	Reference	Material	Temperature Curie (°K)	Reference
Ni-Mn	500	798	Ni-8.8% Si	292	53
11.3 at. % Mn			$NiSiF_6 \cdot 6H_2O$	0.15	293
Ni_3Mn	∼610	1508	Ni-8 at. % Sn	∼348	1035
	730	1882	Ni-5.93 at. % V	308	1287
	(completely ordered)		Ni-11.31 at. % V	19	1287
			$NiV_xFe_{2-x}O_4$		281
Ni_2MnGa	379	1648	x = 0	858	
Ni_2MnIn	223–228	699	x = 0.70	670	
	323	1648	x = 1.75	445	
$NiMnO_3$	437	137	$Ni_{0.35}Zn_{0.65}Fe_2O_4$	550	497
	389	1836	$Ni_{0.78}Zn_{0.22}Fe_2O_4$	750	497
$NiMn_2O_4$	113–162 (depending upon heat treatment)	773	NpC	200	493
				220	1376
			$NpC_{0.95}$	190	659
			Np_2C_3	109	2097
Ni_2MnSb	186–198	699	$NpCl_4$	6.7 ± 0.1	2161
	334	1756	$NpFe_2$	≈600	1867
	360	1648	NpN	82	659
Ni_2MnSn	200–202	699	$(Pb_{0.95}Ba_{0.05})(Ni_{0.5}Nb_{0.5})O_3$	263	757
	342	1756			
	344	1648	$PbCo_{1/2}W_{1/2}O_3$	67	421
$NiO-Li_2O$		1573	$PbCr_2S_4$	138	1206
2 at. % Li	258		$Pb(Fe_{0.5}Ta_{0.5})O_3$	143	757
20 at. % Li	230		$Pb(Fe_{2/3}W_{1/3})O_3$	132	363
30 at. % Li	190		$(Pb_{0.95}Sr_{0.05})(Mn_{0.5}Ta_{0.5})O_3$	103	757
$NiO_{1-x}ZnO_xFe_2O_3$		1556			
x = 0	∼755		Pd-Co		
x = 0.2	∼720		at. % Co		
x = 0.4	∼625		0.07	1.55 ± 0.1	1412
x = 0.6	∼490		0.098 ± 0.005	0.79 ± 0.01	1750
Ni–P			0.19	6.5 ± 0.5	1412
5.3% P			0.20 ± 0.01	2.95 ± 0.05	1750
annealed at 373 K	≈465	1940	0.49	18.8 ± 3	1412
annealed at 673 K	≈641	1940	0.51 ± 0.02	16.2 ± 0.2	1750
12% P	370	1809	1.05 ± 0.05	44.0 ± 0.5	1750
Ni-Pd			1.91	90 ± 1	1412
% Ni:			4.5	186 ± 1	1412
20	250	1720	8	∼281 (0 pressure)	1768
50	448	1597			
50	467	1720	10	423	1850
80	583	1720	12	∼357 (0 pressure)	1768
Ni-40 at. % Pt	213	1287			
Ni-45% Pt	202	53	15	∼419 (0 pressure)	1768
Ni-70 at. % Pt	20	1287			
Ni-24 at. % Rh	273	1287	Pd-40 at. % Cr		1268
$NiRh_2O_4$	360	914	annealed at 790°K	520 (weak ferro.)	
Ni-8 at. % Si	∼453	1035			

CURIE TEMPERATURE

Material	Temperature Curie (°K)	Reference
Pd-Fe		
at. % Fe		
0.10 ± 0.005	0.78 ± 0.002	1750
0.15	2.12	977
0.16	2.8	1750
0.18	2.8	1493
0.41	13.0	1750
0.45	13.0	1493
0.5	∼23	23
0.78	32.6	1750
0.8	∼49	797
0.90	32.6	1493
1	37	128
2.5	90	128
3	114	819
7	200	819
7.5	193	128
10	237	128
Pd_3Fe	540	819
$Pd_{94}Fe_6$	173	732
Pd-8.9 at.% Gd	≈5	1885
Pd–Mn		
at.% Mn		
1	3.414	2064
1.05	3.90 ± 0.02	1584
1.35	4.477	1839
2.40	7.35 ± 0.02	1584
2.91	7.71 ± 0.02	1584
5	2	1925
PdMnSb	500	1966
Pd_2MnSb	247	882, 881
	247–251	699
Pd_2MnSn	185–189	699
	189	881, 882
Pd-1.95% Ni	>2	1720
Pd-5 at. % Ni	∼90	23
Pr	8.7	1477
PrAg	14	1274
$PrAl_2$	33	1023
	34	695, 27
	38	494
	38.5	1180
	31	1338
Pr_3Al	16	1182
$PrAl_2$-$PrCo_2$		742
0 mol % $PrCo_2$	35	
90	45	
22.5	24	
$PrAl_2$-$PrMn_2$(15 mol %)	26	742
PrBi	0.010 ? (sp. ht.)	1496
$PrCo_2$	50	736
	≈50	257
$PrCo_3$	349	417
$PrCo_5$	639	504
	875	444
	885	630
	912	442
Pr_2Co_7	574	630
Pr_2Co_{17}	1171	417
	1160	537, 759
Pr_2CrSbO_7	7	511
$Pr_{1-x}Dy_xCo$		1800
x = 0.2	61	
x = 0.6	104	
x = 0.8	137	1800
$Pr_{1-x}Dy_xNi_2$		1800
x = 0.2	22.1	
x = 0.8	26.4	
$Pr_{0.183}Er_{0.15}Mn_{0.667}$	422	2039
$PrFe_2$	501 ?	630
$PrFe_7$	283	386
	280	321, 349
Pr_2Fe_{17}	287	630
	282 ± 1	1696
Pr-Gd		1708
at. % Gd:		
92.7	254	
63.8	278	
PrGe	39	515
$PrGe_2$	19	450
PrH_x		1064
(x = 0.99-2.57)	No ordering to 4.2	
$Pr_{1-x}Ho_xNi$		1800
x = 0.2	18.7	
x = 0.8	22	
$PrIr_2$	16	41
$Pr_{0.5}La_{0.95}Al_2$	5	1180
$Pr_{0.8}La_{0.2}Al_2$	30	1180
$PrMnO_3$	80	1834
$[(Pr,Nd)_{0.60}Ba_{0.40}]MnO_3$	188	369
$Pr_{0.38}Nd_{0.36}Pb_{0.26}MnO_3$	185	2121
$[(Pr,Nd)_{0.70}Sr_{0.30}]MnO_3$	263	369
PrNi	20	802
$PrNi_2$	8	802
	50?	736
$PrNi_3$	20	538

CURIE TEMPERATURE

Material	Temperature Curie (°K)	Reference
Pr_2Ni_7	85	954
$Pr_{0.63}Pb_{0.37}MnO_3$	215	2121
$PrPt_2$	≈6	257
$PrRh_2$	8.6	883
	≈9	257
$PrRu_2$	40	41
$Pr_{0.167}Tm_{0.167}Mn_{0.667}$	455	2039
$Pr_{0.1}Y_{0.9}Al_2$	13	1180
$Pr_{0.8}Y_{0.2}Al_2$	32	1180
Pt-0.8% Co	2.8	1411
Pt-1 at. % Co	~8	23
Pt-10 at. % Co	~210	23
Pt-25 at. % Cr	~423	1532
Pt-Fe		
17.4% Fe	305	702
24.9% Fe	425	702
28% Fe	~230	1679
30% Fe	300	1679
34.3% Fe	250	702
$Pt_{0.25}Fe_{0.75}$	560	1856
$Pt_{35}Mn_{65}$	~250	688
PtMnGa	220	2112
PtMnSb	575	2112
$Pt_{1.01}Mn_{0.99}Sb_{1.00}$		1758
Slow-cooled	582	
PtMnSn	330	2112
$Pt_{1.01}Mn_{0.99}Sn_{1.00}$		1758
water-quenched	360	
slow-cooled	353	
PtNi	136 ± 2	921
(disordered, 53.3 at. % Ni)		
$PuGe_2$	34.5	878
PuP	126	1276, 1686
RE_2Co_{17}	1068-1213	537, 759
	1083-1209	417
$RE_3Fe_5O_{12}$		1540
RE = Gd, Tb, Dy, Ho, Er, Yb, Tm, Y	~530-555 (magnetocaloric effect)	
$RENi_5, RENi_2, RENi$	85–1.4	289
$RbFeCl_3$	109	2127
$RbFeCl_3(6H)$	109	2056
$RbFeF_3$	87	1370, 1427
$RbMg_{0.40}Co_{0.60}F_3$	~25	887
$RbMgF_3-RbCoF_3$		1389
>35% $RbCoF_3$	to 36	
$RbMgF_3$-50 at. % $RbCoF_3$	~36	1516
$RbNi_{0.7}Co_{0.3}F_3$	109	946

Material	Temperature Curie (°K)	Reference
$RbNi_{0.75}Co_{0.25}F_3$	115	1173
$RbNiF_3$	139	443, 946, 1363
	145	385
$RbNi_{0.67}Mg_{0.33}F_3$	~58	1005
$RbNi_{0.67}Mn_{0.33}F_3$	~103	1005
Rh_2MnGe	450	2112
RhMnSb	320	2112
RuMnGa	220	2112
Sc-24% In	6	735
Sc_3In	6.1 (initial susceptibility)	1068
	7.5 ± 0.5 (high field)	
$Sc_{3.12}In$	7	1706
$ScNi_2$	630 (cooling from 650°K; Ni particles)	1472
$SmAl_2$	122	695
SmC	~30	1038
$SmCo_2$	≈200	257
	259	736
$SmCo_5$	1020	442
	1015	444
	997	630
	747	504
Sm_2Co_7	713	630
Sm_2Co_{17}	1190	417
	1195	537, 759
Sm_2CrSbO_7	12	511
$Sm_{1.1}Dy_{1.9}Fe_{4.2}Ga_{0.8}O_{12}$	350	2103
$Sm_{0.9}Dy_{1.5}Y_{0.6}Fe_{4.1}Ga_{0.9}O_{12}$	330	2103
$SmFe_2$	675	630
	674	542
$SmFe_3$	651	630
$SmFe_7$	399	386
Sm_2Fe_{17}	395	630
$SmFeO_3$	660	238
	673	1298
	448	1973
$Sm_3Fe_5O_{12}$	578 ± 2	741
	562	379, 1314
$Sm_{0.37}Gd_2Dy_{0.63}Fe_5O_{12}$	230	2103
$Sm_{0.1}Gd_{2.24}Tb_{0.66}Fe_5O_{12}$	267	2103
$SmIr_2$	37	41
$SmMn_5$	440	875
	439 ± 3	197

CURIE TEMPERATURE

Material	Temperature Curie (°K)	Reference	Material	Temperature Curie (°K)	Reference
Sm_6Mn_{23}	439	204	$TbAl_2$	114	957
SmNi	45	802		121	1194, 695
$SmNi_2$	77	698		111	1338
	21	736, 802	Tb_3Al_2	190	581, 1214
	22	257	Tb_2C	266 ± 2	1640
$SmNi_3$	85	538	$Tb_{1.2}Cd$	185	937
$SmNi_5$	25	802	$TbCo_2$	256, 237	306
Sm_2Ni_{17}	186	712		256	736
	641	1181		≈245	257
$SmOs_2$	34	41	$TbCo_3$	506	417
$SmPt_2$	≈6	257	$TbCo_5$	980	442
$SmRh_2$	≈20	257	Tb_2Co_7	717	417
SmZn	125	1295		693	630
$SnMn_2O_4$	53	1212	Tb_2Co_{17}	1180	417
$Sn_{0.97}Te$	≅0.20	1700		1194	537, 759
$(SnTe)_{1-x}(MnTe)_x$		1537	Tb_2CrSbO_7	15	511
x = 0.0089	2.5		TbCu	80	284
x = 0.159	33.5		$TbFe_2$	695	630
Spinels	tabulation	467		696	257
$Sr_{1.8}Ca_{0.2}FeMoO_6$	413	231		711	1944
Sr_2FeMoO_6	422	231	$TbFe_3$	648	630
(ferri.)	≈450	1823	Tb_2Fe_{17}	409	630
$SrFe_{12}O_{19}$	750	1391		408	386
$Sr_{0.3}La_{0.7}MnO_3$	363–393	73	Tb_6Fe_{23}	574	630
$Sr_{2.5}La_{1.5}Mn_3O_{9.98}$	~130	1378	$Tb_3Fe_5O_{12}$	568	430, 741
$SrO \cdot (6-x)Fe_2O_3 \cdot xAl_2O_3$		1753		554.2	239
x = 1.8	543			553	1314
x = 2.2	468		TbGa	155	284
$SrO \cdot 3.6Fe_2O_3 \cdot 2.4Al_2O_3$	523	1609	Tb-Ho		
$Sr_{1-x}Pb_xRuO_3$		1667		~79	1596
x = 0	160		% Ho:		
x = 0.5	near 0		10	196	1981
$Sr_{0.5}Pb_{0.5}RuO_3$	≈0	1791	50	78	161
$SrRuO_3$	160 ± 5	997	50	80	1981
	160 ± 2 (electrical resist.)	1506	90	22	1981
			$TbIr_2$	45	41
	164	744		46 ± 1	1950
$TaMn_2B_4$	780	956	Tb-Lu	180 ± 10	24
Tb	210	444	95.3% Tb	180	395
	218	24	$TbMn_5$	445	875
	218, 223	198	TbN	38	522
	219 (resistivity and thermal cond.)	1505		40	512
				42	355
				33.7	1614
			TbNi	50	802
	222	1960	$TbNi_2$	46	802
	235	40		45	736

Material	Temperature Curie (°K)	Reference	Material	Temperature Curie (°K)	Reference
TbNi$_3$	98	538	TmCo$_2$	18	736
TbNi$_5$	27	802	TmCo$_3$	370	417
Tb$_2$Ni$_7$	101	954	TmCo$_5$	1020	442
Tb$_2$Ni$_{17}$	178	712	Tm$_2$Co$_{17}$	1185	537, 759
	615	1181		1182	417
TbNi$_{2-x}$Al$_x$		1800	Tm$_2$CrSbO$_7$	6	511
x = 0.05	53		TmFe$_2$	613	542
x = 0.09	56			610	1945
x = 1.60	59		TmFe$_3$	539	630
x = 1.79	84		Tm$_2$Fe$_{17}$	313.5	386
x = 1.92	118			271 ± 3	1696
TbNi$_{1.01}$Al$_{1.01}$	47	1800		248	630
Tb(OH)$_3$	3.72	1000	Tm$_6$Fe$_{23}$	475	630
	3.7	1388	Tm$_3$Fe$_5$O$_{12}$	549	430
TbOs$_2$	34	41		549 ± 2	741
Tb$_{5.10}$Pd$_{1.90}$	~30	350	TmIr$_2$	1	41
TbPt$_2$	4.2	1024	Tm$_{0.78}$Lu$_{0.22}$	11	1579
	≈17	257	TmMn$_2$	12 ± 2	267
TbRh$_2$	≈40	257	TmNi	4	802
Tb-Sc		395	TmNi$_2$	14	802
Tb %: 90	159			12?	736
80	<15		TmNi$_3$	43	538
Tb$_5$Si$_4$	225	949	TmNi$_5$	7	802
Tb-Y	175–99	24	Tm$_2$Ni$_{17}$	152	712
	217-156	40		603	1181
90% Tb	175	395	Tm$_3$Ni	12	506
Tb$_x$Y$_{1-x}$Fe$_2$		1944	U$_3$As$_4$	198	1039
x = 0.2	≈588			197	1881
x = 0.4	620			205 ± 50	1426
x = 0.9	≈685		UAs-USe		1153
TbZn	206	1295	8 mole % USe	184	
	160	284	35 mole % USe	~138	
ThCo$_5$	630	1030	40 mole % USe	~140	
ThCo$_3$Fe$_2$	690	1069	100% USe	171	
TiCr$_2$Te$_4$	214	465	U$_3$Bi$_4$	108	917, 1039
TiFe$_2$	323–424	653	UD	172	1415
TiFe$_x$Co$_{1-x}$	50	499	UFe$_2$	172	879, 1379
0.3 < x < 0.7			UFeO$_4$	47 ± 2	1050
TlFeF$_3$	~77	1299		42	1354
Tl$_{0.5}$La$_{0.5}$Fe$_{12}$O$_{19}$	703	1096	UGa$_2$	126	1256
TlNiF$_3$	129	1005	UGe$_2$	52	878
	150	389	U$_3$Ge$_4$	94	1156
Tm	35	246	UH$_3$		
	32	1829	β	174	1415, 1416
	22	444		185	1889
	22,38	198	U$_2$N$_3$	235 ± 5	901
TmAl$_2$	≈4	695	(α)	186	1039
TmAu$_4$	2	2146	UP	~125	1574

CURIE TEMPERATURE

Material	Temperature Curie (°K)	Reference
U_3P_4	165	901
	144 ± 5	1426
	138.3	738
	138	1039, 1881
	136.5	832
UP–US		1574
0.2 US	~90	
0.4 US	~130	
0.6 US	~159	
0.8 US	~175	
U_3P_4–U_3As_4		1881
U_3P_4 %:		
0.2	161	
0.4	176	
0.8	181	
$UP_{0.75}S_{0.25}$	100 ± 3	1278, 1544, 1731, 1896, 1915
$UP_{0.72}S_{0.28}$	109 ± 2	1278, 1544
$UP_{0.67}S_{0.33}$	118 ± 2	1278, 1544
$UP_{0.50}S_{0.50}$	150 ± 3	1278, 1544, 1896, 1924
$U_{0.54}Pt_{0.46}$	29.8 ± 0.3	2154
US	178 ± 2	901, 965
	178	2033
	180	351, 352, 782, 1297
(elect-cond., thermoelect.)	180	1207
	180.1	592
	180 ± 2	1278
	180 ± 5	965
	185	34, 1574
U_3Sb_4	148	1039
	149	1383
USe	160.5	782
	160	2033
	171	1153
	180	1297
	185	965
	210	965
USe_2	13.1	782
U_2Se_3	180	1297
U_3Se_4	160	1297
UTe	103	777, 965
	123	965
U_2Te_3	122	777
U_3Te_4	105	777
$U_{2.85}Th_{0.15}P_4$	141	1452

Material	Temperature Curie (°K)	Reference
VBO_3	32.5	1678
$VOSO_4$		
α	~4.2	1655
(α)	≈4	1833
(ferrimagnetic)		
WMn_2B_4	560	956
$Y_3AlFe_4O_{12}$	430	1093
$Y_{2.75}Bi_{0.25}Fe_{5-x}Al_xO_{12}$		325
x = 1.00	430	
x = 1.50	360	
$Y_{3-x}Bi_xFe_5O_{12}$		1160
x = 0.58	577	
$Y_2BiFe_5O_{12}$	591	1092, 1094, 1095
$(Y_{3-x}Ca_x)(Fe_{5-x}Ge_x)O_{12}$		1980
x = 0	563	
x = 0.5	478	
x = 1.5	353	
$Y_{3-2x}Ca_{2x}Fe_{5-x}V_xO_{12}$		1541
x = 1.35	494	
x = 0.3	540	
$Y_{1.8}Ca_{1.2}Fe_{4.4}V_{0.6}O_{12}$	~560	1624
$Y_{3-2x}Ca_{2x}Fe_{5-x-y}V_xIn_yO_{12}$		1585
x = 0	521	
y = 0		
x = 0.3	460	
y = 0.3		
x = 0	444	
y = 0.4		
x = 0.4	438	
y = 0.4		
x = 0	422	
y = 0.5		
x = 0.5	415	
y = 0.5		
x = 0	397	
y = 0.6		
x = 0.6	382	
y = 0.6		
(continued)		

Material	Temperature Curie (°K)	Reference	Material	Temperature Curie (°K)	Reference
$Y_{3-2x}Ca_{2x}Fe_{5-x-y}V_xIn_yO_{12}$			$Y_3Fe_5O_{12}$	545	320
$x = 0$	377			545.5	379
$y = 0.7$				546	1624
				≈549	1905
$x = 0.7$	357			547.90 ± 0.03	1931
$y = 0.7$				550	1160
YCo_2	<2	417, 2104, 2105		551	1314
				551.3	534
	291	1183	polycrystal	552 ± 1	1448
	296	742		553	566
	320	736		559	262, 822
YCo_3	301	417		560	430
	284	1445		560 ± 2	741
	310	2104, 2105		562.7	239
YCo_5	630 or 700	504	powder	570 ± 2	1448
	903	517	$Y_3Fe_5O_{12}$:Hf	541	1259
	922	630	YIG: Nd	548–568	305
	975	516	0-40% Nd_2O_3		
	977	442	$Y_3Fe_5O_{12-x}F_x$		2036
Y_2Co_7	639	417	$x = 0-1$	560–542	
Y_2Co_{17}	1213	537, 759	$Y_3Fe_{5-x}Zn_xO_{12-x}F_x$		2036
	1167	417	$x = 0-0.5$	560–541	
Y_3Co	<2	2104, 2105	$Y_3Ga_xFe_{5-x}O_{12}$	327–394	441
Y_4Co_3	13	1151	$x = 1.46$ to 1.3		
	<2	2104, 2105	$Y_3GaFe_4O_{12}$		1448
$\{Y_3\}Co_xFe_{5-2x}Ge_xO_{12}$		282	single crystal	450 ± 1	
$x = 0.80$	420		powder	458 ± 1	
$x = 2.20$	50		$Y_3Ga_{1.84}Fe_{3.16}O_{12}$		1448
Y_2CrSbO_7	15	511	single crystal	226.9 ± 1	
YFe_2	533	630	$(Y_{0.52}Gd_{0.48})_6Mn_{23}$	468	1479
	535	898	$(Y_{0.46}Gd_{0.54})_6$		1479
	545	1944, 2037	$(Mn_{0.37}Fe_{0.63})_{23}$	395	
	552	742	$Y_{0.5}Gd_{1.6}Se_{2.9}$	47	362
Y_2Fe_{17}	302.5	386	$Y_3In_xFe_{5-x}O_{12}$		1160
	301	630	$x = 0.14$	525	
	332	2037	$x = 0.50$	475	
Y_6Fe_{23}	471	630	$Y_{2.5}La_{0.5}Fe_5O_{12}$	555	379
	484	1479		558	1624
$YFe_{1-x}Al_xO_3$		1623	(Y-mischmetal)Co_5	~700	504
$x = 0.1$	590		YMn_5	490	875
$x = 0.2$	496			486 ± 3	197
$x = 0.4$	332		Y_6Mn_{23}	486	204
$YFe_{0.9}Al_{0.1}O_3$	600	958	$Y(Mn_xFe_{1-x})_2$	180	1001
$YFe_{0.7}Mn_{0.15}Al_{0.15}O_3$	440	1623	$(x = 0.5)$		
$YFe_{0.8}Mn_{0.1}Al_{0.1}O_3$	512	1623	$Y_6(Mn_xFe_{1-x})_{23}$	~380	1001
$YFe_{0.6}Mn_{0.4}O_3$	375	1623	$(x = 0.1)$		
$YFe_{0.8}Mn_{0.2}O_3$	535	1623	$YNd_2Fe_5O_{12}$	557	379
$YFe_{0.9}Mn_{0.1}O_3$	592	1623	YNi_3	33	538
$YFeO_3$	643	238	Y_2Ni_7	58	954

CURIE TEMPERATURE

Material	Temperature Curie (°K)	Reference
Y_2Ni_{17}	160	712
	~600	1530
	621	1181
$Y_2Ni_{17-x}Cu_x$		1807
x = 2	≈470	
x = 4	≈180	
$Y_2Ni_{14.5}Cu_{2.5}$	~465	1530
$3Y_2O_3 \cdot Al_2O_3 \cdot Fe_2O_3$	415	458
$3Y_2O_3 \cdot 0.25Cr_2O_3 \cdot 4.75Fe_2O_3$	515	458
$3Y_2O_3 \cdot 0.25Ga_2O_3 \cdot 4.75Fe_2O_3$	519	458
$3Y_2O_3 \cdot 0.50In_2O_3 \cdot 4.50Fe_2O_3$	444	458
$3Y_2O_3 \cdot 0.25Sc_2O_3 \cdot 4.75Fe_2O_3$	500	458
$(3-x)Y_2O_3 \cdot xSm_2O_3 \cdot 5Fe_2O_3$ x = 0 to 3	570 ± 10	374
$Y_2PrFe_5O_{12}$	553	379
$Y_3Sc_xFe_{2-x}Fe_3O_{12}$ x = 0.9	320	678
$YbCo_5$	973	444
Yb_2CrSbO_7	10	511
$Yb_3Fe_5O_{12}$	548	430
	542.6	239
	548 ± 2	741
$YbNi_3$	<20	538
$YbNi_5$	1.4	802
$Zn_2Ba_2Bi_{0.2}Fe_{12}O_{22}$	360	403
$Zn_2Ba_2Fe_{12}O_{22}$	374	403
$Zn_{1-x}Cd_xCr_2Se_4$		1360
x = 0.6	~105	
x = 0.9	~137	
$Zn_{0.65}Cd_{0.35}Cr_2Se_4$	22	507
$ZnCr_2Se_4$	129.5	507
$ZnFe_2O_4$	780	1336
$Zn_{0.5}Li_{0.5}Fe_2O_{3.5}F_{0.5}$	~593	776
$ZnMn_2As_2$	~30	1503
$Zn_{0.3}Mn_{2.7}O_4$	31	794
$Zn_{0.5}Mn_{2.5}O_4$ (or $Mg_{0.5}$)	20	794
$ZrFe_2$	610–798	653
	630	1353
	633	742
		972
$Zr(Fe_xCo_{1-x})_2$		
(x = 0.8)	~500	
(x = 1.0)	~620	

Material	Temperature Curie (°K)	Reference
$Zr(Fe_xMn_{1-x})_2$		972
(x = 0.6)	~115	
(x = 1.0)	~620	
$Zr_{1-x}Gd_xZn_2$ (x = 0.01)	7	998
$Zr_{1-x}Hf_xZn_2$		1138
x = 0.10	~11	
x = 0.12	~4	
$Zr_{0.9}Hf_{0.1}Zn_2$	10.2	1186
$(Zr_xNb_{1-x})Fe_2$		969
x = 1.00	590	
x = 0.85	490	
$Zr_{1-x}Nb_xFe_2$		1621
x = 0.15	~490	
x = 0.3	~327	
x = 0.4	~218	
$Zr_{0.6}Nb_{0.4}Fe_2$	230	1353
(magnetization)	230	1621
(permeability)	210	1621
$Zr_{0.7}Nb_{0.3}Fe_2$	~336	1353
$ZnNiMnO_4$	90	209
$Zr_{0.9}Ti_{0.1}Zn_2$	~65	1244
$Zr_{0.92}Ti_{0.08}Zn_2$	40	1186
$ZrZn_{1.9}$	27.5 ± 0.5	1420
	≈26.7	2124
	26	1775
$ZrZn_2$ (probably extrinsic ferromagnet)	<4.2	718, 8
	16	1244
25 to 500 ppm Fe	17-22	1310
	17.8	714
	18 (thermal expansion)	1773
	18 ± 3	1138
(effect of impurities)		940
	21.5 ± 0.5	1853
	24 for solid	433
	25	1627
	≈27.8 (isotope effect negligible)	1792
	32 for powder	433
	35	597
	35 ± 2	247

NÉEL TEMPERATURE

Material	Temperature Neel (°K)	Reference
$Ag_{0.5}Al_{0.5}Cr_2S_4$	7	1790, 1803
$AgCrS_2$	40	1176
$AgCrSe_2$	50	1176
AgDy	53-65 (elec. res.)	766
AgEr	10-24 (elec. res.)	766
AgF_2	163	1499, 1888
$AgFeCl_4$	13 ± 0.3	1273
	13.1 ± 0.2	1739
$Ag_{0.5}Ga_{0.5}Cr_2S_4$	≈10	1803
$Ag_{0.5}Ga_{0.5}Cr_2Se_4$	40	1790
AgGd	137-140 (elec. res.)	766
AgHo	32-42 (elec. res.)	766
$Ag_{0.5}In_{0.5}Cr_2S_4$	14	1790
	≈14	1803
Ag-Mn		
5.30% Mn	11.4	334
1.75% Mn	1.9-1.2	334
AgNd	24-28 (elec. res.)	766
AgTb	102-122 (elec. res.)	766
AgTm	9-20 (elec. res.)	766
$AlCo_2O_4$	30	317, 1116, 136
$AlCr_2$	598 ± 5	554
$Al_{0.4}Cr_{1.6}BeO_4$	20 ± 2	326
Al_2O_3: Ti	≈3	2058
Au-Cr		1995
Cr:		
18.4	230	
23.5	300	
Au_4Cr	~380	1408
$Au_{80}Cr_{20}$		1724
ordered	~400	
disordered	~270	
AuDy	24-34 (elec. res.)	766
AuEr	13-19 (elec. res.)	766
Au-Fe		2090
8% Fe	28	
2% Fe	14	
AuGd	42-50 (elec. res.)	766
AuHo	13-16 (elec. res.)	766
AuMn	513	1204
	500	1251
	493	573, 574
Au_2Mn	363	1075
Au_3Mn	140	32
Au_5Mn_2	348	1046
	353	1325
$Au_{48.1}Mn_{51.9}$	500	1665
$Au_2Mn_{2-x}Al_x$		1026
x = 0.6	415	
x = 1.0	65	
$Au_4(Mn_{1-x}Cr_x)$		2153
x = 0.8	≈250	
x = 0.85	≈340	
Au_2MnIn	233	788
$Au_2Mn_{1.25}In_{0.75}$	233	788
$Au_2Mn_{1.75}In_{0.25}$	~403	788
AuTb	48-62 (elec. res.)	766
AuTm	8-19 (elec. res.)	766
Ba_2CoF_6	250	1051
Ba_2CoWO_6	17	487, 396
$BaCo_{0.5}W_{0.5}O_3$	17	487
$Ba(Fe_xAl_{1-x})_2O_4$		1916
x = 0.5	230	
x = 0.6	455	
x = 0.75	640	
x = 1	900	
$BaFeF_4$	60 ± 2	664
$Ba_2(FeMo)O_6$	351	1060
$BaFeO_4$	10	1879
$BaFe_{12}O_{19}$	709.5 (for H ⊥ c)	558
	713.3 (for H ∥ c)	558
Ba_2FeSbO_6	12	541
$BaMnF_4$	~25	1513
$BaMnO_3$ 9H	≈350	1808
$BaNiF_4$	~150	1701
Ba_2NiF_6	200	1051
$BaNi_{0.5}W_{0.5}O_3$	55	396
$Ba_{0.4}Sr_{1.6}Zn_2Fe_{12}O_{22}$	380	1559
$Bi_{1-x}Ca_xMnO_3$	~110	708
$BiCrO_3$	123	1072

Material	Temperature Neel (°K)	Reference
BiFeO$_3$	643	225
	~643	1482
	640 ±5	2084
	645	775
	653	1631
	620 ?	399
Bi$_2$Fe$_4$O$_9$	256 ± 4	1751
BiFeO$_3$-40% LaFeO$_3$	~420	1482
BiFeO$_3$-Pb(Fe$_{0.5}$Nb$_{0.5}$)O$_3$	643	268
BiMn$_2$O$_5$	52	1081
CaCoSiO$_4$	16	520
	16 ± 2	679
CaCo$_{0.5}$W$_{0.5}$O$_3$	26	396
CaCrO$_3$	90	1057
CaCr$_2$O$_4$ (β)	80	1120
Ca$_2$FeAlO$_5$	333 ± 15	1012
Ca$_2$Fe$_{2-x}$Al$_x$O$_5$		1796
x = 0.50	568 ± 4	
x = 0.72	473 ± 5	
Ca$_2$Fe$_{0.5}$Ga$_{1.5}$O$_5$	28 ± 2	492
Ca$_2$FeGaO$_5$	315 ± 4	492
Ca$_2$Fe$_{1.5}$Ga$_{0.5}$O$_5$	560 ± 3	492
Ca$_2$Fe$_2$GaO$_5$	730 ± 2	492
Ca$_3$⌊Fe$_2$⌋Ge$_3$O$_{12}$	11.5 ± 0.2	2015
Ca$_2$(FeMo)O$_6$	398	1060
CaFeO$_3$	120	1762
CaFe$_2$O$_4$	200	938
	180	599
Ca$_2$Fe$_2$O$_5$	720	1040
	725 ± 2	1512
	730	1043
	730 ± 2	995
Ca$_2$FeSbO$_6$	31	541
CaFeSiO$_4$	<8	520
Ca$_{0.75}$La$_{0.25}$MnO$_3$	165	304
Ca$_{0.8}$La$_{0.2}$MnO$_3$	165	304
Ca$_3$⌊Mn$_2$⌋Ge$_3$O$_{12}$	13 ± 0.2	2015
CaMnO$_{2.98}$	123	944
CaMnO$_3$	~100	73
	123	211, 744
CaMn$_2$O$_4$	225 ± 5	237
Ca$_2$MnO$_4$	114	944
Ca$_3$Mn$_2$O$_7$	120	744
Ca$_4$Mn$_3$O$_{10}$	125	744
CaMnSiO$_4$	9	339, 520
Ca$_{1.5}$Pb$_{0.5}$MnO$_4$	111	2133
Ca$_{1-x}$Sr$_x$MnO$_3$		2122
x = 0	125	
x = 0.15	139	
x = 0.25	147.5	
x = 0.50	174.5	
Ca$_{0.75}$Sr$_{0.25}$MnO$_{2.98}$	150	944
CaRuO$_3$	110 ± 10	997
CdCr$_2$O$_4$	9	434
CdCr$_2$Se$_4$	55	1083
Cd and Zn manganites with $0.8 \leq x \leq 1$	200	31
Ce	13	444
CeAl	10 (2.2 kOe)	415
	9	996
CeAl$_2$	4	1194
	≈3.5	1849
CeAs	7.5	579
	8	345, 1752, 2061
CeB$_6$	$\lesssim$3.0	1164
	7	2164
CeBi	25	343, 2061
	26 ± 0.5	2131
CeC$_2$	33	422, 535
CeCl$_3$	~0.2	1498
CeCrO$_3$	257	519
CeCu	2.7	1309
Ce$_2$Fe$_{17}$	≈270	2037
CeGe	10	515
CeH$_{2.0}$	<4.3	498
CeH$_{2.35}$	5.3 ± 0.1	498
CeH$_{2.6}$	<4.3	498
CeIn$_3$	10	344
	11	1421
Ce$_2$Mg$_3$(NO$_3$)$_{12}$·24H$_2$O	~1.8 × 10^{-3}	1539
molar volume = 21 cc/mole	0.30 ± 0.21	1992
molar volume = 24 cc/mole	2.0 ± 0.26	1992
CeP	9	579
	10	345, 2061
CeS	7	406, 387, 1818
CeSb	18	345
	16.0 ± 0.2	2089
	16	2061
N$_1$	~16	1327
N$_2$	~9	1327

NÉEL TEMPERATURE

Material	Temperature Neel (°K)	Reference	Material	Temperature Neel (°K)	Reference
CeSe	9?	2074	$Co_{1-x}Cu_xRh_2S_4$		1465
	12	406, 387, 1818	x = 0.25	295	
			x = 0.1	~380	
CeTe	10	406, 387, 1818	CoF_2	37.7	55, 74, 132
Ce-Y	37	785		37-45	815
55.0 Ce-45.0 Y				39 ± 0.3	1892
CeZn	36	1295		50	645
$CeZn_2$	7	1729	CoF_3	460	541, 57
$Ce_2Zn_3(NO_3)_{12} \cdot 24H_2O$	0.0063	157	$CoFe_2O_4$		1218
$CoAl_2O_4$	6	136, 1114, 1116	quenched	792	
			slowly cooled	798	
$CoAlRhO_4$	24	317, 136, 1116	$CoGeO_3$	41	784
			$Co(H_2O)_6 \cdot SiF_3$	0.15	143
$Co_3B_2O_6$	37	252	CoI_2	12	270
	30	445		3	133
$Co_3B_7O_{13}X$		1179	$Co(IO_3)_2 \cdot 2H_2O$	0.362	740
X = Cl	15		$Co_3La_2(NO_3)_{12} \cdot 24H_2O$	0.181	1533
X = Br	20		$Co_{0.8}Mg_{0.2}O$	230	610
X = I	38		$Co_{0.67}Mg_{0.33}O$	155	609
$CoBr_2$	19	133, 135, 383	$Co_{1-x}Mn_x$		1725
			x = 0.35	13	
$CoBr_2 \cdot 2H_2O$	9.5	187	x = 0.37	≅60	
$CoBr_2 \cdot 6H_2O$	3.2, 3.07	142	$CoMn_2O_4$	70	1562
	2.91 ± 0.01	1226	$(Co_{1-x}Mn_x)_2P$		
$CoCO_3$	18.1	837	x = 0.66	230	1369
	18	55	x = 0.7	~175	1612
	17.5	1722	x = 0.8	~80	1612
$CoCl_2$	24.9	133, 134	$CoMoO_4$	5	292
	25	384	Co_3N	11	1570
$CoCl_2 \cdot 6D_2O$	2.40	1643	$Co(NC_5H_5)_2Cl_2$	3.7	139
$CoCl_2 \cdot 2H_2O$	17.5	260, 285	$Co(NH_4)_2(SO_4)_2 \cdot 6H_2O$	0.084	111
	18	139	$Co_{0.67}Ni_{0.33}O$	358	609
$CoCl_2 \cdot 6H_2O$	2.3	843	$Co_{0.9}Ni_{0.1}O$	311	608
	2.29	140, 141	CoO	271	11
	2.28	1643		289.3	1959
	2.25	1991		289.3 (ultrasonic attenuation)	1522
$CoCl_2 \cdot 6H_2O$ 92% D_2O	2.32	694			
$CoCrO_4$	14	1660		291	55, 56, 57
$CoCr_2S_4$		1429		292	541
(NiAs type)	300			~292	236
$Co_{0.25}Cr_{0.75}Sb$	<700	1467		293	26, 1444
$CoCr_2Se_4$	199?	1801		294	1836
$CoCs_3Br_5$	0.282	780		295 ± 1	1263
$CoCs_3Cl_5$	0.523	780		328.6	183
	0.53	1918	1.1% Li	284.5	1874
$Co_{0.46}Cu_{0.54}O$	285	608	CoO(I)	288	1146, 1147

Material	Temperature Neel (°K)	Reference	Material	Temperature Neel (°K)	Reference
CoO(II)	270 ± 10	1146, 1147	Cr	310 (single crystal)	1903
Co_3O_4	40	136		310 (polycrystal)	1903
	33.0 ± 1.0	1401		310	55
β-Co(OH)$_2$	12.3	724		308	790
$CoRh_{1.5}Cr_{0.5}S_4$	360 ± 10	1054		303	1659
$CoRh_2O_4$	27	138, 381, 234	N_2	122 (recalescence)	1599
$CoRh_2S_4$	?400	234			
Co_2RuO_4	~20	1654	Cr–Al at.% Al:		
CoS	358	611	0.3	300 ± 5	2081
Co_9S_8	~300	600	1.2	202 ± 5	2081
$CoSO_4$	12	55, 96	1.9	310 ± 10	2081
$Co(S_xSe_{1-x})_2$			3.7	338 ± 10	2081
x = 0.40	90	1302	6.2	440 ± 10	2081
x = 0.5	≈73	1930	15–30	800	947
x = 0.70	90	1302	CrAs	823	181
CoSb	40	181		280	1394
$CoSe_2$	93	966	Cr_2As	393	485, 244
	AF ordering questioned	1673		423	932
$CoSeO_4$	30	1634		438 ± 15	1037
$CoSiF_6 \cdot 6H_2O$	0.20	809	CrB_2	86	1577
$CoSiO_3$	50	784		85	2052, 2091
Co_2SiO_4	49	520, 290		88.0 ± 2.0	1483
$CoTa_2O_6$	16	1811	$CrBO_3$	15	1678
$Co_{0.67}Te$	115	1436	$Cr_3B_7O_{13}X$		1179
Co_2Ti	43	1199	X = Cl	25	
$CoTiO_3$	38	206	X = Br	50	
	42	292	X = I	95	
$CoUO_4$	12	97	$Cr_3B_7O_{13}I$	90	604
$CoVO_3$	142 (triclinic)	1694	Cr_2BeO_4	28	520, 1082
				28 ± 2	326
$CoWO_4$	55	292	$Cr(CH_3NH_3)\cdot(SO_4)_2 \cdot 12H_2O$	0.015-0.02	614
Cr	473	789			
	450 (powder)	1355	$CrCl_2$	20-40	384
	317	791		20	66
	312.5 ± 0.5	322		40	1157
	312	38, 1387	$CrCl_3$	16.8	68, 69, 375
	311.5	261, 820			
	311.25 ± 0.4	1983	Cr–Co		
	311	39, 783, 570, 1598	% Co:		
			0.97	300	2047, 2048
	311 (AF$_1$ phase)	219	1	300	39
			1.90	298	2047, 2048
	120 (AF$_2$ phase)	219	2.79	311	2047, 2048
			9.30	298	2047, 2048
(continued)			CrF_2	53	66

NÉEL TEMPERATURE

Material	Temperature Neel (°K)	Reference
CrF_3	80	541, 57, 67, 1005
CrFe	308	737
Cr-Fe		
0.08 at.% Fe	310	2014
0.49 at.% Fe	311–241	825
0.5 wt.% Fe	300	390
0.78 wt.% Fe	296	390
0.98 at.% Fe	300	2047, 2048
1.0 at.% Fe	300	39
2.3 wt.% Fe	260	390
3.3 at.% Fe	248	1277
3.3 at.% Fe	≈255	2014
4.7% Fe	260	769
$Cr_{2-x}Fe_xAl$		1488
x = 0	663	
x = 0.05	603	
x = 0.09	413	
x = 0.15	310	
x = 0.24	0	
Cr(0.5% Fe, 0.16% Ni)	~300	322
$Cr_{0.75}Fe_{0.25}Sb$	680	1467
CrGe	62	233
$CrK(SO_4)_2 \cdot 12H_2O$	0.004	56
Cr-Mn		
0.35% Mn	~460	934
0.12% Mn	320	420
0.44% Mn	450	420
0.5% Mn	375	219
1.0 at.% Mn	451	1903
1.03% Mn	510	420
2.1% Mn	520	219
$Cr_{0.993}Mn_{0.007}$	448	1581
$Cr_{0.982}Mn_{0.018}$	523	1581
$Cr_{0.3}Mn_{0.7}As$	190	1394
$Cr_{0.7}Mn_{0.3}As$	263	1394
Cr-Mo		
Mo %:		
0.6	≈305	1994
2	287 ± 2	407
3	273	1754
5	248 ± 3	407
5.1	≈245	1994
8.6	207	1903
16	101 ± 10	407
CrN	273	616
	275.6–284.7	1607
	277	1836
$Cr_{1.000}N_{0.997}$	286	2087
Cr-1% Ni	200	39
Cr–0.99 Ni	215	2047, 2048
Cr_2O_3	300	36
	303	1439
	306 ± 2	1170
	307	12
	308	70, 1878, 1986
	310	1836
	≈318	2057
Cr_2O_5	<100	617
Cr_2O_3–Al_2O_3		1877
90 mole % Cr_2O_3	275	
60 mole % Cr_2O_3	188	
40 mole % Cr_2O_3	50	
Cr_2O_3-0.35% Fe_2O_3	289	1042
Cr_2O_3-10 mol % Fe_2O_3	247	1439
Cr_2O_3-22 mol % Fe_2O_3	146	1439
Cr_2O_3-26 mol % Fe_2O_3	210	1439
Cr_2O_3-28 mol % Fe_2O_3	216	1439
Cr_2O_3-30 mol % Fe_2O_3	222	1439
Cr-Os		
at.% Os		
0.3	359	1711
0.3	423	2047, 2048
2.0	566	1711
3.0	568	2047, 2048
Cr–Pd		1816
at.% Cr:		
30	≈44	
32	≈57	
34	≈65	
$Cr_5Pd_{75}Si_{20}$	10	1629
Cr-2 at. % Pt	563	1532
Cr–Re		
at.% Re:		
0.1	333	2047, 2048
0.5	≈410	1355
1	473	2047, 2048
5.0	573	2047, 2048
12.5	523	2047, 2048
15	578	1903
16.8	428	2047, 2048
$Cr_{0.85}Re_{0.15}$	540	1581
Cr–Rh		
at.% Rh:		
1	548	2047, 2048
3	560	2047, 2048
(continued)		

Material	Temperature Neel (°K)	Reference	Material	Temperature Neel (°K)	Reference
CrRh			$CrSe_{1.04}$	271	847
4	553	2047, 2048	$CrSe_{1.07}$	232	847
6	560	2047, 2048	$CrSe_{1.10}$	205, 92 (annealed but 2 phases)	184
14	438	2047, 2048			
Cr-Ru					
at. % Ru			$CrSe_{1.20}$	83 ± 2	184
0.3	385	1903	Cr_3Se_4	80	848, 45, 13, 167, 2032
0.65	460	1903			
0.9	507	1711			
2.0	573	1903	Cr-0.9 at. % Si	241	523
4.8	558	1711	Cr-0.4 at. % Si	277	523
0.6	433	2047, 2048	CrTe	150	1566
6.0	568	2047, 2048	Cr_3Te_4	80	13
10.0	483	2047, 2048		85 ± 2	2080
14.0	308	2047, 2048	Cr_5Te_6	102 ± 2	2080
$Cr_{0.997}Ru_{0.003}$	377	1581	Cr_2TeO_6	105 ± 5	1163
$Cr_{0.994}Ru_{0.006}$	477	1581	$CrTe_{1-x}Sb_x$		1566
$Cr_{0.99}Ru_{0.01}$	519	1581	x = 0.25	~100	
CrS	460	408	x = 0.50	~260	
$CrS_{1.17}$	152	687	x = 0.75	500	
	153	687	$CrTe_{0.1}Sb_{0.9}$	700	618
$Cr_{0.96}S$	~450	1258	$CrTe_{0.25}Sb_{0.75}$	520	618
Cr_2S_3	15	1555	Cr-V		
	120	2135	0.45 at. % V	268	299
	122	1243	0.5 at.% V	255	1903
Cr_3S_4	215–230	1032	1% V	~113	1743
	260	13	1.0% V	210	291
	280	167, 13	1.10 at.% V	229.9	1962
Cr_5S_6	168	726	1.24 at.% V	199	1903
	158	71	1.8% V	~108	1743
	150-160	1656	1.88 at.% V	168.8	1962
Cr_7S_8	~125	1258	2.39 at.% V	135	1962
CrSb	693	303	Cr-V-Mn	433	1742
	700	618	0.59 at. % V		
	705	376	1.18 at. % Mn		
	718	181	$CrVO_4$	50	620, 621
	720	1600		51 ± 1	1672
	723	65	Cr_2WO_6	69	842
$CrSb_2$	273 ± 2	2116	$CsCoF_3$	8	1593, 1703
CrSe	279	181, 847	$Cs_4Co_3F_{10}$	33	2142
	270	619	$CsCuCl_3$	10.4	1605
	285	1350	$Cs_3Cu_2Cl_7 \cdot 2H_2O$	1.620 ± 0.005	1485
	200	1201		1.625	1913, 1932
$CrSe_{1.00}$	280 ± 5 (quenched)	184	$Cs_4Fe_3F_{10}$	≈22	2142
			$CsMnCl_3$	69 ± 3	22
	94 ± 2 (annealed)			67	2005
			Cs_2MnCl_4	55	2000

NÉEL TEMPERATURE

Material	Temperature Neel (°K)	Reference	Material	Temperature Neel (°K)	Reference
$CsMnCl_3 \cdot 2H_2O$	4.8	968, 1025	$Cu_2FeSnSe_4$	AF ordering suggested at low temperature	1999
	4.88	1484	$Cu_{0.5}Ga_{0.5}Cr_2S_4$	31	1790, 1803, 1890
	4.89	2034	$Cu_{0.5}Ga_{0.5}Cr_2Se_4$	7	1790, 1890
	~9	1617		≈7	1803
$Cs_2MnCl_4 \cdot 2H_2O$	1.86	968	CuGd	135-145 (elec. res.)	766
	1.8	1326			
$CsMnF_3$	53.5	180, 1315	$Cu(HCO_2)_2 \cdot 4H_2O$	17	110, 151
	64	307	CuHo	27 (elec. res.)	766
$CsNiCl_3$	4.5	1763, 1865	$Cu_{0.5}In_{0.5}Cr_2S_4$	26	1790, 1890
$Cs_4Ni_3F_{10}$	15	2142		≈26	1803
$Cu_{0.5}Al_{0.5}Cr_2S_4$	14	1790, 1803	$Cu_{0.5}In_{0.5}Cr_2Se_4$	14	1790, 1890
$Cu_{0.5}Al_{0.5}Cr_2Se_4$	6	1790, 1803, 1890		≈14	1803
$Cu_3B_7O_{13}X$		1179	$CuK_2(SO_4)_2 \cdot 6H_2O$	0.05	56
X = Cl	20			0.0295	1019
X = Br	24		$Cu_3La_2(NO_3)_{12} \cdot 24H_2O$	0.089	1533
$CuBr_2$	189	626	Cu-Mn		
	193	53	5.5 at. % Mn	~48	1504
$Cu(C_2H_3O_2)_2$	270	625	4.82% Mn	11.5	334
$Cu(C_{12}H_{23}O_2)_2$	230	627	4.3 at. % Mn	~60	1572
$Cu(C_2H_3O_2)_2 \cdot H_2O$	250-280	625, 626	2.0 at. % Mn	~20	1572
$Cu_3(CO_3)_2 \cdot (OH)_2$	1.86	152	1.40% Mn	4	334
$CuCl_2$	70	53	1.0 at. % Mn	~8.5	1504
$CuCl_2 \cdot 2D_2O$	4.3	690	0.1 at. % Mn	~0.9	1504
$CuCl_2 \cdot 2H_2O$	4.3	55, 259	Cu_3Mn_2Al	AF to decomposition temp. (≈800 K)	1955
	4.335	1988			
	4.357 ± 0.010	1362			
Cu_2CoGeS_4	25	1900	Cu_2MnGeS_4	≈10	1900
$CuCrO_2$	32 ± 1	1693	CuMnSb	55	1222
$CuCrS_2$	39	1176	Cu_2MnSb	38	575
$Cu_2Cs_3Cl_7 \cdot 2H_2O$	~1.62	1616	(Is *not* single phase; contains αCu solid solution)		1150
CuDy	62 (elec. res.)	766			
CuEr	10-15 (elec. res.)	766	Cu_2MnSnS_4	< 20	1900
			$Cu(NH_4)_2Br_4 \cdot 2H_2O$	1.773 ± 0.001	2062
CuF_2	68.7 ± 2	177	$Cu(NH_3)_4SO_4 \cdot H_2O$	0.42	961
$CuF_2 \cdot 2H_2O$	~10.9	271		0.43	1460
	10.9	811		0.37	283
	26	89	Cu_2NiGeS_4	36	1900
$CuFeCl_4$	21.7 ± 0.5	162	CuO	230	25
	21.8 ± 0.2	1273	$(CuO)_x(NiO)_{1-x}$		1870
	20.5 ± 0.3	1739	x = 0.30	≈523	
Cu_2FeGeS_4	12	1900	x = 0.34	≈473	
$CuFeO_2$	19	1329	$CuRh_2O_4$	25	381
	25	1693	$CuSO_4$	34.5	55, 96
$CuFe_2O_4$	780 ± 20	91		~36	1526
$CuFeS_2$	>300	123			

NÉEL TEMPERATURE

Material	Temperature Neel (°K)	Reference
$CuSO_4 \cdot 5H_2O$	0.029	283
Cu_2Sb	373	485
	Not antiferromagnetic	1246, 1247, 2107
Cu_3Se_2	Possibly at very low temp.	1458
$CuSeO_4 \cdot 5H_2O$	0.046	283
$CuSiO_3 \cdot H_2O$	21	182
CuTb	100-118 (elec. res.)	766
CuTm	10-28 (elec. res.)	766
$CuWO_4$	90	292
Dy	145	1847
	(at 10 kOe)	
	173.3	531
	177	1959, 1960
	178	444, 1941
	178.5, 179	198
	178.65 ± 0.05	2155
	179	294
	179 ± 0.2	1901
	179.9 ± 0.5	1011
	184	24
DyAg	63 ± 2	274
	55	1274
	51 ± 1	888
$DyAg_2$		1639
α	15.0	
β	9.5	
DyAl	20	996
$DyAlO_3$	3.5 ± 0.2	858
	3.5	1921
	3.4	733, 1633
$Dy_3Al_5O_{12}$	2.53	227, 1384
	2.5	373
DyAs	8.5	544
$DyAsO_4$	2.50 ± 0.10	1718
DyAu	14	1220
$DyAu_2$		1639
α	33.8	
β	25.5	
DyBi	12	343
	13 ± 1	2131
DyC_2	59	1185
Dy_3C	170	33
$DyCoO_3$	3.6	1921

Material	Temperature Neel (°K)	Reference
$DyCrO_3$	2.08	1161
(Dy)	2.16	1065
	2.1	519
(Cr)	146	1065
	146.4	519
DyCu	64 ± 2	274
	61	1309
$DyCu_2$	24	170
$DyCu_5$	7	1965
$DyFeO_3$	3.7 (sp. heat)	983
	3.7	1007
(Dy)	3.7	1289
	4.5 (Dy)	765
	36	1007
	635 (Fe)	765
	645	828
(Fe)	648	1289
$Dy_3Fe_5O_{12}$	2.5175	1675
DyGe	36	515
$DyGe_{1.67}$	12	450
$DyGe_2$	28	450
Dy_5Ge_3	40	906
Dy_5Ge_4	40	949
DyH_2	8	46
Dy-50% Ho	157	395
$Dy_{0.4}Ho_{0.6}Sb$	~6.5	1375
$Dy_{0.8}Ho_{0.2}Sb$	~8.3	1375
$DyIn_3$	23	1421
	24 ± 0.5	1275
$DyMn_{12}$	110	204
$DyMnO_3$	<30	808
	<2	1834
$DyMn_2O_5$	~8	584, 796
$Dy_2Mn_4O_9$	8	584
($DyMn_2O_5$: see 796)		
Dy_2NbO_7	1.0	1191
Dy_3Ni	35	506
Dy_2O_3	1.20	200
Dy_2O_2S	5.85	1192
$DyPO_4$	505	919
	3.40 ± 0.04	1718
	3.40 ± 0.01	1664
	(susceptibility)	
	3.391 ± 0.001	1664
	(heat capacity)	
	3.390 ± 0.002	1845
	3.39	1978

NÉEL TEMPERATURE

Material	Temperature Neel (°K)	Reference
$Dy_{5.07}Pd_{1.93}$	41	350
$DyPt_3$	13.2 ± 0.3	1275
$DySb$	9.5	844
$DySi_2$	17	580
Dy_2TiO_5	1.55	1191
$Dy_2Ti_2O_7$	1.3	1191
$DyTl_3$	11	2065
$DyVO_4$	3.0	1486, 1798
	3.05 ± 0.10	1718
	3.066	2167
	1.9 to 4.2	1777
		395
Dy-Y		
Dy %: 90.0	163	
50.0	108	
20.0	59	
20	59	24
	168-59	24
Er	88	46
	86	536
	86.5	1789
	85, 80	198
	84	444
	80	24, 296
	79	711
ErAg	21 ± 1	274
	15	1274
ErAl	13	996
	10	1230
$ErAl_3$	5	1361
Er_3Al_2	9	581
$ErAlO_3$	0.6	1921
ErAs	3.5	544
	3.03	1883
ErBi	3.9	343
$Er_{(1-x)}Bi_xFeO_3$		1236
$0 \leq x \leq 0.03$	638 ± 1	
ErC_2	14	2041
Er_3C	80	33
$ErCl_3 \cdot 6H_2O$	0.356	1844
$ErCoO_3$	< 0.3	1921
$ErCrO_3$	133.2, 16.8	519
	133.2	1707
	133, 16.8	482
	No phase change at 16.8 K	1906
	9.82 ± 0.02	2019
	No AF transition 1 to 20 K	2035
ErCu	17	1309

Material	Temperature Neel (°K)	Reference
$ErCu_2$	11	170
Er-10% Dy	93	1014
$Er_{0.90}Dy_{0.10}$		1320
N_1	~80	
N_2	~100	
$Er_{0.20}Dy_{0.80}$	~165	1320
$ErFeO_3$	636	828
	638	1236
(Fe)	643	1289
	620 (Fe)	55, 363
	5.3	1286
(Er)	4.3	126, 1289
	3.9	1288
$Er_3Ga_5O_{12}$	0.789	823
	0.8	1048
ErGe	7	515
Er_5Ge_3	31	906
Er_5Ge_4	7	949
$Er_{0.50}Ho_{0.50}$	104	1320
$Er_{0.9}Ho_{0.1}$	88	1320
$ErIn_3$	6	1421
$ErMnO_3$	79	232, 1834
Er_3Ni	9	506, 1797
Er_2O_3	3.36	495
	3.4	200, 1266
	4	67
ErP	3.1	355
ErSb	3.5	844
	3.51	1883
	3.7	355
$Er_{0.20}Tb_{0.80}$	~205	1320
Er-5% Th	80	711
$Er_2Ti_2O_7$	1.25	1618
Er-Y		395
Er %: 90.0	78	
50.0	56	
30	30 ± 5	24
50	56	
70	65	
Eu	87 ± 1	1951
	89.6 ± 0.2	2072
	91	35
	90, 96	203
	103	444
$EuAl_2$ (two points)	15, 30	1195
$EuAl_4$	13	1195
$EuCO_3$	1.05 ± 0.05	1939
EuC_2O_4	1.9 ± 0.1	1939

Material	Temperature Néel (°K)	Reference	Material	Temperature Néel (°K)	Reference
$EuC_2O_4 \cdot H_2O$	2.8 ± 0.1	1939	$EuTiO_3$	5.3	312
$EuCl_2$	1.1 ± 0.2	1939	$EuZn_2$	23	975
$EuCrO_3$	181	519		20	1729
	181 (weak ferromagnetism observed)	1307	Fe fcc	67 (neutron diffraction)	1920
			γ-Fe	8	323
$EuCu_2$	13–15	975		55-67	607
EuF_2	2	17	Fe gel	60	1744
	19.5	179, 301	Fe–Al	2.5 not AF transition	2102
	1.0 ± 0.1	1939			
$EuFeO_3$	662	700, 828	at. % Al:		
$EuFeS_2$	823	122, 123	33	100	2020
$EuGd_2O_4$	4.5	346	> 33	20	1953
$EuMnO_3$	45	808, 967, 1834	39	20	2020
			47	2	2020
Eu_2O_3	Not AF at 90°K	1569	$Fe_{1-x}Al_x$		
			x = 0.35	≈43	2118
Eu_3O_4	6.2 ± 0.2	1175	x = 0.4	≈27	2118
	5	346	x = 0.45	≈11	2118
	5.3 ± 0.2 (thermodynamic coefficients)	851	Fe_3Al	750	849
			$FeAl_2O_4$	8	1114, 136, 1116
	7.5 (lines of equal magnetization)	851	Fe_2As	368	932
				368 ± 3	1037
				353	391
	But also see ref 346			323	549
			$FeBO_3$	348.35 ± 0.2	1715
EuP	4	314		348	2008
$EuSO_4$	0.43 ± 0.05	1939	$Fe_3B_7O_{13}X$		1179
EuSe	5.8 ± 1	1142	X = Cl	11.5	
	4.7	425, 1760	X = Br	15	
	4.6	512, 2009	X = I	30	
	≅4.4	1449	$FeBr_2$	11	383, 115, 168
	11.6 ± 0.3	1231			
$EuTb_2Se_4$	3	518	$FeCO_3$	∼34	837
EuTe	6	371		20	1722
	7.8	357, 1760		20-60	81, 631, 632
	8	158		30.6	118
	8.7	1933		35	55
	9.58 ± 0.10	1647		38	1514
	9.6 (Faraday rotation)	2017		38.4 ± 0.2	1367
				38.5 ± 0.3	1418
	9.6 ± 0.2	1943	$FeC_2O_4 \cdot 2H_2O$	∼22	472
	9.6 ± 1	1592	Fe_2CaO_4	200	413
	9.64 ± 0.06	315	$FeCl_2$	23.5	559, 560
	9.7 ± 0.2	1979		23	56, 86, 87
	11	159, 173		24	383, 384, 270

NÉEL TEMPERATURE

Material	Temperature Neel (°K)	Reference
FeCl$_3$	16	384, 629
	15 ± 2	685
	10 ± 2	162
	9.75 ± 0.2	1515
	9 ± 1	2147
FeCl$_2$·2H$_2$O	~23	255
FeCl$_2$·4H$_2$O	~1	287, 169
	1.1	1787, 1869
Fe$_{0.95}$Co$_{0.05}$Sn$_2$	348	1710
FeCr$_2$O$_4$	80	1630
FeCr$_2$S$_4$		
(NiAs type)	200	1429
	180	1630
	172	1225
FeCr$_2$Se$_4$	4.2	716
(Fe$_{1-x}$Cr$_x$)$_3$Se$_4$	230	2032
$x = 0.8$		
Fe$_{0.5}$Cu$_{0.5}$Rh$_2$S$_4$	140	1783
FeF$_2$	78.3	55, 114
	78.35 ± 0.03	1689
	~78.37	1433, 1606
	78.377	2141
	78.2 ± 0.2	394
	78.26 ± 0.01	2070
	78.11 ± 0.01	505
	78	1699
	80-82	815
	90	645
FeF$_3$	394	541, 57
	364.80 ± 0.05	746
	362.4 ± 0.2	942
FeF$_3$·3H$_2$O	≈132	2003
FeGe	412 ± 2	945
	411	464
	410	308
	400	210
FeGe$_2$	190	448
	270 to 280 (thermal expansion)	1907
	~270 (ND),	701
	287 ± 2 (Mossbauer)	
	285	845
	315	19
Fe(HCOO)$_2$·2H$_2$O	3.68	178
FeI$_2$	10	115, 383
Fe$_{30}$Ir$_{70}$	~40	1755
γFe-Mn		
Mn: 13-60%	190-500	552, 839, 327, 916, 917, 918
εFe-Mn		
Mn: 17%	240	327
FeMnAs	463	229
(Fe,Mn)$_{\geq 2}$As		1400
Mn/Fe = 2.04	456	
= 1.66	452	
= 1.90	460	
Fe$_{0.10}$Mn$_{0.90}$HoO$_3$	40	729
FeMn$_2$O$_4$	55	1358
Fe$_{0.21}$Mn$_{2.76}$Pt$_{1.03}$	455 ± 10	2046
Fe$_{0.41}$Mn$_{2.52}$Pt$_{1.07}$	415 ± 10	2046
Fe$_{0.60}$Mn$_{2.36}$Pt$_{1.04}$	425 ± 10	2046
Fe$_x$Mn$_{1-x}$YO$_3$	80	729
$x = 0$ to 1		
FeMoO$_4$-II	45	1188
Fe(NH$_4$)(SO$_4$)$_2$·12H$_2$O	0.043	56
Fe$_{1/4}$NbSe$_2$	175	1695
Fe$_{0.70}$Ni$_{0.15}$Cr$_{0.15}$	21 ± 1	1815
Fe$_{0.95}$Ni$_{0.05}$Sn$_2$	362	1710
FeO	198, 187	55, 56, 57
	186	541
Fe$_{0.896}$O	209 (susceptibility)	533
	203 (Youngs modulus vs temperature)	
Fe$_{0.92}$O	198	1088
Fe$_{0.932}$O	199 (susceptibility)	533
	195 (Youngs modulus vs temperature)	
Fe$_{0.98}$O	200	1056
Fe$_2$O$_3$	259	722
	≃262	1270
(α)	998	715
(α)	963 ± 3	774
(α)	960 ± 5	2078
(α)	958	935
(α)	956 ± 2	1066
(α)	955 ± 10	288
(α)	953	633, 634, 635
(α)	263	665, 2023

(continued)

Material	Temperature Néel (°K)	Reference	Material	Temperature Néel (°K)	Reference
Fe_2O_3			$Fe_3(PO_4)_2 \cdot 8H_2O$	8.8 for long-range order	840
(δ)	950 (P → AF)	55		8.84 ± 0.05	1968
(δ)	250 (AF → AF)	55		4-11 (?)	130
FeOCl	92 ± 3	2108		12.40	833
FeOF	315	1550	Fe-Pt		1404
	317	1836	24.0% Fe	212	
$3Fe(OH)_2 \cdot 2FeSO_4$	≈8.5	1795	30.0% Fe	145 ± 1	
FeOOH			$Fe(Pt_xPd_{1-x})_3$	170	460
(α)	353	1209	x = 0 to 1		
(α)	367 ± 2	194	$FePt_{3-x}Pd_x$		215
(α)	370	829	(½ ½ 0)	160–180	
(α)	393.3	195	(½ 00)	145–160	
α	400	1661	FeRh	328	113
(α)	403 ± 2	1070		∼330	502
(β)	110–300	1125		333	1393
β	237	1661		338	799
β	∼285	1232	50% Rh	360	841
β	295 ± 4	1422	$Fe_{49}Rh_{51}$	310	1409
γ	50	1661	$Fe(Rh_{1-x}Ir_x)_{1.08}$		113
γ	73	1666	x = 0.056	449	
(γ)	<100	1125, 1099	x = 0.084	545	
			x = 0.121	585	
$(Fe_2O_3)_{1-x}(Rh_2O_3)_x$		215, 758	$Fe(Rh_{1-x}Pd_x)_{1.08}$		113
(by neutron diffraction)			x = 0.029	252	
x = 0	950 ± 10		x = 0.058	169	
x = 0.11	875 ± 15		$Fe_{48}Rh_{49}Pd_3$	210	1409
x = 0.22	760 ± 15		$Fe(Rh_{1-x}M_x)_{1.08}$	403	113
x = 0.25	715 ± 10		M = Pt		
x = 0.41	575 ± 25		x = 0.056		
x = 0.51	455 ± 15		Fe-30 at. % Ru	100	1200
x = 0.66	290 ± 10			∼15	1432
x = 0.81	cca 100		FeS	593	53, 86, 181
$(Fe_2O_3)_{1-x}(Rh_2O_3)_x$		750		∼595	1646
(by Mossbauer effect)			Fe_xS	∼599	1636
x = 0	958 ± 3		$Fe_{0.902}S$	600	408
x = 0.11	823 ± 10		Fe_3S_4	570 ± 8	1998
x = 0.22	805 ± 10		Fe_7S_8	455	1967
x = 0.25	783 ± 10		$FeSO_4$	21	55, 96
x = 0.41	628 ± 10			23	472
x = 0.81	<82		$FeSb_2$	773	117
FeP	"antiferromagnetic at 120°K"	1548	$FeSb_2O_4$	46 ± 1	1305
			FeSe	847	112
	126	1688	Fe_7Se_8	475	693
FeP_2	250	131	FeSi	443 (but see also ref. 807)	611
$FePO_4$	25	79			
$Fe_3(PO_4)_2 \cdot 4H_2O$	∼15	129			
	15.3	1152			
	20	67			

NÉEL TEMPERATURE

Material	Temperature Neel (°K)	Reference	Material	Temperature Neel (°K)	Reference
Fe_2SiO_4	~30	803	GdAl	42	996
	20	509	$GdAlO_3$	4.0	689
	65	53, 104, 520, 706		3.95	1921
FeSn	373	890		3.89	813
	368	2075		3.87	2051
	367	1133		3.69	1003
	365	1710	GdAs	25	544
$FeSn_2$	384	677		19	1380
	380	116, 1710, 2075	GdAu	31 (19.3 kOe) 37 (5.5 kOe)	415
	377	1133	GdB_6	18	1164
$FeTa_2O_6$	14	1811		16	2164
FeTe	70	53	GdBi	28	343
	63	636		32	1380
$FeTe_{1.95}$	~85	637, 638	$GdBr_3$	2	1528
$FeTe_2$	83	638	Gd-30% Ce	110	1735
	85	637, 638	$GdCl_{1.6}$	~50	591
$FeTe_{2.10}$	~85	637, 638	$GdCl_3 \cdot 6H_2O$	0.185 ± 0.001	1052
Fe_2TeO_6	218.5 ± 0.5	1162		0.183	283
	213	1836	Gd_2CoMnO_6	~115	1343
	201	1281	$GdCoO_3$	2.9	1921
$Fe_{2+x}Ti_{1+x}$		1687	$GdCrO_3$	170	519
x = 0.0049	276			~4	519
x = −0.0291	282			2.42	1161
Fe-38 at. % Ti	~260	557	GdCu	2.28	2168
$FeTiO_3$	55-68	120, 94, 207		41	593
				41 (?)	1309
	68	55, 119, 120, 207		45	347
				~143.3	1736
			$GdCu_2$	41	170
$FeTi_2Se_4$	134	1318	Gd-Dy		
$FeVO_4$	22 ± 1	2126	Gd at. %:		
FeV_2S_4	131	1318	42	~218	959
FeV_2Se_4	94.5	1318	42	217	1653
$FeWO_4$	66	850	39	217	361
	75.9	1518	28.7	197.5	1653
$Fe_{0.762}Zn_{0.238}F_2$	58.4 ± 0.3	394	12.5	190	361
$Fe_{0.259}Zn_{0.741}F_2$	19.3 ± 1.0	394	$GdFeO_3$	558	55
Ga	(no transition 2.3 to 1.1)	941		657	700, 828
				678 (Fe^{3+})	640
$Ga_{0.92}Fe_{1.08}O_3$	335	1321	N_2	1.47	1721
$Ga_{1.2}Fe_{0.8}O_3$	210	1321	(Gd)	2.16	1288
GdAg	145	593, 347		2.5 (Gd)	127
	150 ± 3	274	$Gd_3Fe_5O_{12}$		1425
	~137.4	1736	N_1	90	
	138	1274	N_2	328	
$GdAg_{0.5}Au_{0.5}$	138	347	GdGe	62	515
$GdAg_{0.2}In_{0.8}$	77	347	$GdGe_{1.67}$	22	450
$GdAg_{0.1}In_{0.9}$	155	347	$GeGe_2$	28	450

Material	Temperature Neel (°K)	Reference	Material	Temperature Neel (°K)	Reference
Gd$_5$Ge$_3$	48	906	HFeO$_2$	330	929
Gd$_5$Ge$_4$	15	949	H$_2$IrCl$_6$·6H$_2$O	2.32 ± 2	1779
	45	949	He-3	0.00198	1946
GdH$_2$	21	920	HgCr$_2$S$_4$		
GdIn	28	593, 347	(meta)	25	510
GdIn$_3$	45	1421		60	1545
Gd-30% La	120	1735	Ho	133	4, 24, 444, 295, 711
Gd-Lu					
Gd %: 60.0	192			132	540, 40
20.0	100			132, 133	198
55 to 90% Lu	≈0.1	2021		132.1	1959, 1960
	320-80	52		∼131	1505
GdMnO$_3$	>30	808		125.4	1087
	21	1834	HoAg	33 ± 1	274
GdNbO$_4$	1.67	1191		32	1274
Gd$_2$NbO$_7$	<1.0	1191	HoAg$_2$	5.7	1638
Gd-50% Nd	100	1735	Ho$_3$Al$_5$O$_{12}$	0.8	1048
Gd$_3$Ni	100	506		∼0.95	1340
Gd$_2$O$_3$	∼1.6	1564	HoAs	4.8	544
Gd(OH)$_3$	2.0	1000	HoAu	10 (2.3 kOe)	415
Gd$_2$O$_2$S	5.7	1192	HoC$_2$	26	423, 535
GdP	15	314, 173	HoCl$_3$	2.15	1324
GdPO$_4$	225	449	HoCoO$_3$	2.4	1921
Gd-35% Pr	110	1735	HoCrO$_3$	141.3	519
Gd$_{70.5}$Pr$_{29.5}$	55	2053		∼12	519
GdS	50	1380	HoCu	28	1309
Gd$_2$(SO$_4$)$_3$·8H$_2$O	0.182 ± 0.001	1052	HoCu$_2$	9	170
GdSb	28	844, 1380	HoD$_2$	8	1954
Gd-Sc			Ho$_{0.5}$Er$_{0.5}$	104	217
Gd %: 69	150	395	HoFeO$_3$	639	828
50	103 ± 4	264	(Fe)	647	1289
35	60	395		700 (Fe)	126
25	46 ± 4	264	(Ho)	6.5	169, 1289
15	<4.2	395		4.5 ± 0.5	858
GdSe	60	1380		<1.2	1288
	68	2074	HoGe	18	515
Gd$_2$Se$_3$	6	362	HoGe$_2$	11	450
GdSi$_2$	27	580	Ho$_5$Ge$_3$	10	906
GdTe	80	1380	Ho$_5$Ge$_4$	21	949
GdVO$_3$	7.5 (Gd)	127	HoH$_2$	8	924, 1591
GdVO$_4$	2.5	1486	HoIn$_3$	11	1421
Gd-Y				11.5 ± 0.5	1996
10% Gd	59 ± 4	264	HoMnO$_3$	76	232, 1834
50% Gd	182	395	HoMn$_2$O$_5$	46	1081
60% Gd	196 ± 4	264	Ho$_3$Ni	20	506
GeCo$_2$O$_4$	20	317	Ho$_2$O$_2$S	2.5	1192, 1784
GeFe$_2$O$_4$	10	317	Ho$_{5.04}$Pd$_{1.96}$	27	350
GeNi$_2$O$_4$	16	317	HoSb	9	355
	15	341, 167		5.5	844

NÉEL TEMPERATURE

Material	Temperature Neel (°K)	Reference	Material	Temperature Neel (°K)	Reference
$HoSi_2$	18	580	$KMnCl_3 \cdot 2H_2O$	2.74	968
$Ho_{0.5}Sm_{0.5}FeO_3$	6.5 (Ho)	1431		2.70 ± 0.01	1417
Ho-Tb	213	675	$KMnF_3$	88.3	99, 455, 548, 1139, 1315
20% Ho					
Ho-Th	90	711, 395			
90% Ho				88.06 ± 0.02	6
$Ho_2Ti_2O_7$	1.3	1191		88	47, 10, 541, 704, 705
Ho-Y					
Ho: 10% to 100%	26–133	395, 24		89–95	622, 623, 666
20% to 100%	23–132	40			
32.3% Ho	57	40	powder	95	1716
$HoZn_2$	12	1729	single crystal	90	1716
IrMnGa	60	2112	K_2MnF_4	<75	419
$KCoF_3$	144	99		45.0 ± 1	1670
	135	400	K_3MoCl_6	5	1128
	114	10		4.7	1348
	109.5	1159, 1061		6.6	1348
			$KNiF_3$	280	400
	109	10		275	541, 818
$KCo^{2+}F_3$	114	541		275, 253	99, 10
K_2CoF_4	125	1123		253.5	1139
	110	1582		250 (optical scattering)	1374
	107	1649			
$KCrF_3$	<77	622, 623	$KNiF_3$	235.3(DTA)	1919
	40	311	K_2NiF_4	~200	564
$KCuCl_3$	No magnetic ordering above 1.3 K	1958		~110 (∥ to c axis)	564
				190	148
$KCuF_3$	220	10		102	1582
	233.2	1139		97.23	1714
	243	99, 541		97.1	1356, 1677, 1841
D_{4h}^{18} polytype	38 ± 1	1730			
D_{4h}^{5} polytype	22 ± 4	1730	$KNi_{0.073}Mn_{0.027}F_3$	~300	1480
$K_3Fe(CN)_6$	0.129	283	KO_2	7	216
$KFeF_3$	115	622, 623	K_2ReBr_6	15.3	466
	113	99, 10, 277, 541	K_2ReCl_6	12.3 ± 0.5	1177
				11.9	466
	112	400	$La_{1-x}Ca_xMnO_3$		1632
$KFeF_4$	137	2114	x = 0	140	1908
$KFeO_2$	983 ± 10	1989	x < 0.1	141	1632, 1976
K_2FeO_4	3.6	1455	x = 0.4	140	1908
	5	1879	x = 0.175	185	1908
$KFe_{11}O_{17}$	803	793	$La_2Co_3(NO_3)_{12} \cdot 24H_2O$	0.181	893
$K_4Fe_3(OH)_6(SO_4)_2$	60	1339	$LaCoO_3$	80	541
K_2IrCl_6	3.05	895, 2022		No AF order at 90°K	1430
	3.08, 3.05	160			
$KMnCl_3$	100 ± 3	22		(no measurable transition)	72, 1495
	100	1285			

Material	Temperature Neel (°K)	Reference	Material	Temperature Neel (°K)	Reference
$LaCrO_3$	~4.2	967	$Li_{0.1}Mn_{0.9}Se$	70	1413
	282	519, 174, 967, 1063	$LiNiPO_4$	23	520
				23 ± 2	846
			$Li_xV_{2-x}O_2$	463	1257
	291	1836	$LuCrO_3$	~4.2	967
	295	73		112	967
	320	541, 213, 72		112.4	519, 174, 1063
$LaErO_3$	2.4	684, 1073, 1626	$Lu_{25}Er_{75}$	68.4	1789
			$Lu_{50}Er_{50}$	49.2	1789
$LaFeO_3$	740	828, 700	$Lu_{75}Er_{25}$	27.4	1789
	738	125	$LuFeO_3$	623	828
	750	72, 218, 541	$LuMnO_3$	91	232, 1834
			$MgCr_2O_4$	15	317, 1097
La-Gd	155	785		$\cong 14$	1319
30.1 La-69.9 Gd				<77	624
$La_2Mn_3(NO_3)_{12} \cdot 24H_2O$	0.230	893		$N_1 \approx 16$	1891
$LaMnO_3$	150	2038		$N_2 \approx 13.5$	1891
	141	1975	MgO: 9 cat. % Fe	>8	1549
$LaMnO_3$	131	967	$Mg(V_2)O_4$	45	317
	130	1834	Mn	~100	55, 1659
	100	72, 73, 10, 541, 808		100	1333
			α	96	1031
			α	95	1728, 64
$La_2Ni_3(NO_3)_{12} \cdot 24H_2O$	0.393	893, 2162	$Mn_{1.8}Al_{0.2}O_3$	70	228
$LaVO_3$	137	213	$MnAl_2O_4$	4	1114, 136, 1116
Li_2CoGeO_4	13.5	2068			
$LiCoPO_4$	23	520	MnAs	400	181
	23 ± 2	846	N_2	388 (AF questioned)	1683
$LiCuCl_3 \cdot 2H_2O$	5.9, 4.4	98			
	4.5	1137	Mn_2As	573	263, 932
	4.4	2101		573 ± 15	1037
$LiFeF_6$	105 ± 5	1419		583	222
$LiFeO_2$	295	7	$MnAs_{0.9}P_{0.1}$	~120 (meta)	1311
	300	2016	MnAu	370	852
	42	1055		520	866
(α and β)	42	7	$MnAu_2$	373	1045
(γ-tetragonal)	315	124, 356		370	1681
(cubic)	90	356		363	88
$LiFePO_4$	50	520, 710		≈ 363	2045
$Li_{0.5}Ga_{0.5}Cr_2S_4$	14	1790	$MnAu_3$	145	866
	≈ 14	1803		150	853
$Li_{0.5}In_{0.5}Cr_2S_4$	27	1790, 1803	MnB_2	760	1936
$LiMn_{0.7}Fe_{0.3}PO_4$	42	67	Mn_3B_4	393	426
$LiMnPO_4$	35	520		392	956
	34.85	484, 67		$N_1 \approx 210$	1872
	34.6 ± 0.4	1463		$N_2 \approx 410$	1872
	38.4	473	$Mn_3B_2O_6$	35	445
	42	483			

NÉEL TEMPERATURE

Material	Temperature Neel (°K)	Reference	Material	Temperature Neel (°K)	Reference
$MnBr_2$	2.16	86, 87, 383	$Mn_{1-x}Cr_xSb$		376
	2.158	1838	x = 1.0	705	
$MnBr_2 \cdot 4H_2O$	2.136	95, 109	x = 0.7	500	
	2.13	1645	$Mn_{2-x}Cr_xSb$	260 to 230	1141
	≈2.125 (hydrated)	1961	x = 0.06	(increasing field)	
	2.124	1770	$Mn_{1.97}Cr_{0.03}Sb$	104	50
	≈2.102 (deuterated)	1961	$Mn_{1.90}Cr_{0.10}Sb$	289 ± 2 (decreasing)	603
$Mn(C_5H_5)_2$	134	641		248 ± 2 (increasing)	603
$Mn(CH_3OO)4H_2O$	3.2	110	$Mn_{0.1}Cr_{0.9}Sb$	673	654
$MnCO_3$	32.5	837	$Mn_{0.2}Cr_{0.8}Sb$	588	654
	32.55 (initial susceptibility)	1876	$Mn_{0.25}Cr_{0.75}Sb$	570	654
	32.69 (spontaneous magnetization)	1876	$Mn_{1.9}Cr_{0.1}Sb_{0.95}In_{0.05}$	100-350	655
			MnCu	~300	53, 77
	32.4	55	90 at. % Mn (γ)	400	1749
	32.2	1469	$Mn(DCOO)_2 \cdot 2D_2O$	3.7	1335
	31.5	642, 631, 644	MnF_2	72, 66.5	55, 56, 74
	30	1722		72-75	81, 645
	29.4	92		67.5	1738
$MnCl_2$	1.96	383, 85, 86		67.43 ± 0.1	1517
$MnCl_2 \cdot 4D_2O$	1.59	1641		67.37	725
	1.588 ± 0.002	1410, 1671		67.34	1769
				67.33 ± 0.03	1891
				67.31 ± 0.02	1799
$MnCl_2 \cdot 2H_2O$	6.8	1851		67.3	64, 1680, 1959
$MnCl_2 \cdot 4H_2O$	1.62	108, 1594			
	1.622 ± 0.001	388		~67.3 (ultrasonic attenuation)	1523
	1.623	1770			
	1.626 ± 0.002	1410, 1671		67	720, 721, 47
$Mn_{1-x}Co_xO$		2119			
x = 0.05	125			66.2 ± 0.5	795
x = 0.01	118		MnF_3	43-47	89, 646
$Mn_{0.5}Co_{0.5}O$	210	648		45	541
$Mn_{0.25}Co_{0.75}O$	251	648		47 ± 2	57, 89
$Mn_{0.75}Co_{0.25}O$	165	648	$MnF_2 \cdot 10\% ZnF_2$	61	815
Mn-2% Cr	73.2	275	$MnF_2 \cdot 56\% ZnF_2$	18	815
$MnCr_2O_4$	43	1590	$Mn_{0.95}Fe_{0.05}CO_3$	35	1469, 1845
$(Mn_{1-x}Cr_x)_2O_3$			$Mn_{0.70}Fe_{0.30}GeO_3$	125	1118
x = 0.01	N_1 80		$Mn_{0.7}Fe_{0.3}HoO_3$	9.6	1234
	N_2 31		$Mn_{0.1}Fe_{0.9}HoO_3$	8.8	1234
x = 0.075	N_1 72	1917	$MnFeO_3$	58	490
	N_2 57			78	228
$Mn_{0.33}Cr_{0.67}S$	240	727			
$MnCr_2S_4$		1429			
(NiAs type)	70				

Material	Temperature Neel (°K)	Reference	Material	Temperature Neel (°K)	Reference
$(Mn_{1-x}Fe_x)_2O_3$			$(Mn_{1-x}Ga_x)_2O_3$		
x = 0.05	65	1917	x = 0.01	N_1 78	1917
x = 0.10	28	1917		N_2 26	
x = 0.2	35	1022	x = 0.06	N_1 55	1917
x = 0.35	30	1917	x = 0.27	N_1 16	1917
$Mn_{2-x}Fe_xO_3$		1567	Mn_3Ge	350	854
x = 0			Mn_3Ge	360	1560
N_1	90			164	165, 189
N_2	50		$Mn_{3.25}Ge$	139	647
x = 0.04			$MnGeO_3$	10	784
N_1	70 ± 30			16	2158
N_2	26 ± 4		Mn_2GeO_4	24 ± 2	1938
x = 0.2			$Mn_2Ge_{0.12}Sb_{0.88}$	244	1478
N_1	40 ± 2		$Mn_2Ge_{0.2}Sb_{0.8}$	~77	1620
N_2	34 ± 3		$Mn(HCOO)_2 \cdot 2H_2O$	3.65	1748
$(Mn_{0.5}Fe_{0.5})_2O_3$	38	488		3.686	1825
$(Mn_{0.9}Fe_{0.1})_2O_3$	~28	1262		3.70	186
$Mn_{2.76}Fe_{0.21}Pt_{1.03}$	455 ± 10	1264	MnHg	198	80, 29
$Mn_2Fe_3Si_3$	160 ('intermediate state of magnetization')	1897		200	28
				>360	30
				460	590, 29
$Mn_3Fe_2Si_3$	162	1897	MnI_2	3.40	370, 383
Mn_4FeSi_3	99	1897	$MnIn_xCr_{2-x}S_4$		
$(Mn_{1-x},Fe_x)WO_4$		1583	x = 0.4	18	1790, 964
x = 0	14.4		x = 0.6	17	1790
x = 0.2			x = 0.8	16	1790
N_1	22 (neutron diffraction)		x = 1	13	1790, 964
			$Mn_3La_2(NO_3)_{12} \cdot 24H_2O$	0.230	1533
N_2	8.5 (neutron diffraction)		$Mn_{0.9}Li_{0.1}Se$	71	624, 658
			$Mn_{0.93}Li_{0.07}Se$	73	624, 658
x = 0.5	~45 (susceptibility)		$Mn_{0.95}Li_{0.05}Se$	83	624, 658
			$Mn_{0.97}Li_{0.03}Se$	90	658
x = 0.7			$Mn_{0.85}Mg_{0.15}(HCOO)_2 \cdot 2H_2O$	~3.1	1748
N_1	60.6 (neutron diffraction)		$MnMoO_4$	13	292
			Mn_2N	301	1197
N_2	58.8 (susceptibility)		$Mn(NH_3)_6Br_2$	0.35	2067
			$Mn(NH_3)_6Cl_2$	0.60	2067
x = 1	75.9		$Mn(NH_4)_2(SO_4)_2 \cdot 6H_2O$	0.176	1018
$Mn_{2.85}Ga_{1.15}$	460 ± 10	2004		0.14	111
Mn_3GaC	~150	342	MnNi	1140	76
	150	650		1073 ± 40	973, 974
	153	567	Mn-30% Ni	400	1726
Mn_3GaC	164	1685	$Mn_{0.91}Ni_{1.09}$	1033 ± 30	973, 974
Mn_3GaN	298	568, 650	$Mn_{1.07}Ni_{0.93}$	1013 ± 40	974
$MnGa_2O_4$	33	582, 313	$Mn_{2.94}Ni_{0.06}B_4$	370	734

NÉEL TEMPERATURE

Material	Temperature Neel (°K)	Reference	Material	Temperature Neel (°K)	Reference
MnO	118 ± 2	905	$Mn_{30}Pd_{70}$	235 ± 5	1702
	118	1372		220 ± 5 (decreased order)	1702
	117.9	578			
	117.7	100			
	117 (heat capacity)	1668	MnPt	973 ± 10	480, 973
			$Mn_{0.82}Pt_{1.18}$	815 ± 10	2046
	116	541	$Mn_{0.96}Pt_{1.04}$	953 ± 10	973
	113	728	$Mn_{0.9}Pt_{1.1}$	903 ± 20	480
	122, 115.9	55, 57, 74	$Mn_{1.04}Pt_{0.96}$	883 ± 20	973
	120	1741	$Mn_{2.93}Pt_{1.07}$	460 ± 10	501
	125	1441	$Mn_{1+x}Pt_{1-x}$		1264
	125 (neutron diffraction)	1668	x = 0.04	885 ± 20	
			x = 0.32	485 ± 10	
MnO_2	84	630, 649	$Mn_{3+x}Pt_{1-x}$		
	92	57, 81	x = 0.17	430 ± 10	1264
	94	1451	x = 0.18	510 ± 10	2046
Mn_2O_3	~82	488	x = 0.28	525 ± 10	2046
	80	189	Mn_3Pt	475 ± 10	501
(α)	79	1022	(ordered)	473 ± 10	221
alpha N_1	90	1567	(disordered)	310 ± 10	427
alpha N_2	50	1567	$Mn_{75}Pt_{25}$	310 ± 10	427
Mn_2O_3-2 mole % Fe_2O_3	~40	488	$Mn_{73.5}Pt_{26.5}$	270 ± 10	427
MnOOH	40	1101	$Mn_{78}Pt_{22}$	350 ± 10	427
MnP	50	75, 1228	Mn_3PtN_x		2071
	47	1249, 1977	x = 0	≈490	
Mn_2P	103	1159	x = 0.03	≈530	
	110	1291	x = 0.045	≈555	
Mn_3P	115	503	$Mn_3Pt_{1-y}Rh_y$		1264
$Mn_2P_2O_7$	14	1271, 1323	(y = 0.05)	465 ± 10	
	12 to 14	1774	$Mn_3Pt_{0.5}Rh_{0.5}$	685 ± 10	501
MnPd	813 ± 10	973	$Mn_3Pt_{0.5}Rh_{0.5}$ (disordered)	>450	427
	813	462			
	825 ± 10	1015	$Mn_3Pt_{0.8}Rh_{0.2}$	390 ± 10	427
Mn-Pd		1525	$Mn_3Pt_{0.9}Rh_{0.1}$	370 ± 10	427
19.5 at. % Mn	(no magnetic order at 77°K)		$Mn_3Pt_{0.95}Rh_{0.05}$	465 ± 10	501
			MnRh	~170	78
			Mn_3Rh	855 ± 10	501
22.7 at. % Mn	205 ± 15		(ordered)	853 ± 10	221
25.3 at. % Mn	220 ± 10		(disordered)	>450	427
30.4 at. % Mn	235 ± 10		$MnRh_2O_4$	15	381
$Mn_{0.38}Pd_{0.62}$	493	855	$Mn_3Rh_{0.5}Pt_{0.5}$ (ordered)	683 ± 10	221
$Mn_{0.80}Pd_{1.20}$	613 ± 10	973			
Mn-66 at. % Pd	290 ± 10	1015	MnS	165	56, 57
Mn_2Pd_3	653	1551		156.5	1520
N_1	643 ± 10	1552	(α)	152 ± 0.8	431
N_2	613 ± 10	1552		152	1372
Mn_2Pd_3	613 ± 10	973		150	2026
ζ phase	623	1603		149.5	1317
μ phase	433	1603	(continued)		

Material	Temperature Neel (°K)	Reference	Material	Temperature Neel (°K)	Reference
MnS			MnZn$_3$		
(α)	130	651, 652	α	150	1554
(β) (cubic)	160	652	α'	~155 (neutron diffraction)	1771
(β) (hexagonal)	110	652			
			Mn$_3$ZnN	190	952
MnS$_{0.98}$	154 ± 0.5	1241	MoF$_3$	185	156
MnS$_{1.07}$	156 ± 0.5	1241	[(NH$_2$)$_2$CS]$_4$NiBr$_2$	2.28	1078, 1089
MnS$_2$	<77	82			
	48.2	676	NH$_4$FeF$_3$	98	1235, 1521
	47.93	1886	NH$_4$Fe$_3$(OH)$_6$(SO$_4$)$_2$	~55	1339
MnSO$_4$	11.5	910, 55, 96	(NH$_4$)$_2$IrCl$_6$	2.16, 2.15	160
			NH$_4$MnCl$_3$	110 ± 15	2010
	15	620, 660	NH$_4$MnF$_3$	84	307
MnSO$_4$·H$_2$O	16	910	NaCa$_2$[Co$_2$]V$_3$O$_{12}$	8.1	2015
MnSc$_2$S$_4$	<4.2	1395	NaCa$_2$[Ni$_2$]V$_3$O$_{12}$	8.9	2015
(Mn$_{1-x}$Sc$_x$)$_2$O$_3$	N$_1$ 79	1917	NaCoF$_3$	~74 (ND)	1165
	N$_2$ 23			≈74	2029
MnTe	302 (alloyed)	2063		~78 (VSM)	1165
				~35	526
	305.5 (stoichiometric)	2063	Na$_2$CoGeO$_4$	≈3.5	2068
			Na$_2$CoSiO$_4$	≈4	2068
	306.6	602	NaCrS$_2$	18	1208, 1663
	306.7 ± 0.2	696		19	1176
	307	56, 1788	NaCrSe$_2$	40	1176
	310	20, 9, 1520	NaFeO$_2$		
			α	11 ± 1	1970
	310-323	656, 661, 662	β	723	771
			Na$_3$Fe$_5$O$_9$	375	752
	320	463	Na$_4$Fe$_3$(OH)$_6$(SO$_4$)$_2$	53	1339
	323	1334	Na$_2$GaSbO$_7$	1.158 ± 0.002	1618
MnTe$_2$	80	82	Na$_2$IrCl$_6$·6H$_2$O	≈4.1	1779
	83.0	1886	NaMnF$_3$	60	307
	83.8	1576	Na$_{0.40}$MnO$_2$	≈25	2134
	87	348	Na$_2$Mn$_2$Si$_2$O$_7$	26	1027
	87.2	676	Na$_{0.5}$Nd$_{0.5}$WO$_4$	20	899
MnTiO$_3$	41	93, 94, 95	NaNiF$_3$	149	147
	60	185		149 (neutron diffraction)	1242
	62.3 ± 0.1	1470			
	~63	576		156	707
	64	1676		165	953
MnTiO$_3$ II	24 ± 1	1492	NaNiO$_2$	20	226
MnUO$_4$	12	97	NaO$_2$	well above 193	834
MnV$_2$O$_4$	52	1561			
MnWO$_4$	16	292, 1322	Na$_{0.5}$Tb$_{0.5}$WO$_4$	39	899
	14.4	1518			

NÉEL TEMPERATURE

Material	Temperature Neel (°K)	Reference	Material	Temperature Neel (°K)	Reference
$Na_x V_x^{IV} V_{2-x}^{V} O_5$		1447	NdSb	16	844
x = 0.8-0.9	220-240		NdSe	14	406
x = 0.5-0.7	~160			11	2074
$Nb_2Co_4O_9$	>4.2, <30	612, 613		10.6	1692
	~30	105	$NdSn_3$	4.7	344, 1713
$Nb_2Mn_4O_9$	125	105	$Nd_2Sn_2O_7$	0.91	1618
$NbSe_2$	40	1782	NdTe	13	406
	No magnetic ordering	2099		10.2	1692
Nd	12	444	Nd-Tm	28.5	785
	18	1466	63.0 Nd-37.0 Tm		
	19	1563	Nd-Y	34.5	785
NdAg	22	1274	70.0 Nd-30.0 Y		
NdAl	29	996	NdZn	148	1295
	25 (19.3 kOe)	415	$Ni_3B_2O_6$	49	445
NdB_6	8.45	1171	$Ni_3B_7O_{13}Cl$	15	1179
	8.4	2164	$Ni_3B_7O_{13}I$	120	605, 604
	8.6	1164		120 NOT an AF transition	1144
NdBi	24	343			
	2.5 ± 0.5	2131	$NiBr_2$	60	383
NdC_2	29	422, 535	$NiBr_2 \cdot 6H_2O$	6.5	149
$NdCl_3$	1.035	1498	$NiBr_2 \cdot 6NH_3$	0.61	283
$NdCrO_3$	224	519, 482		0.7	1625
	~10	519	$Ni(CN)_2 \cdot NH_3 \cdot C_6H_6$	2.37 ± 0.01	1253
$Nd_{0.2}Dy_{0.8}Sb$	~12	1375	$NiCO_3$	~25	837, 686
$Nd_{6.5}Dy_{3.5}Sb$	~5	1375		25.2	1854
$NdFe_7$	573	923		30	55
$NdFeO_3$	760	126	$NiCl_2$	50	384
	687	828, 700		52	134
$Nd_3Ga_5O_{12}$	0.516	823	$NiCl_2 \cdot 6H_2O$	5.34	140
Nd_2GaSbO_7	1.16	1618		5.34-5.8	668, 669, 670
$NdIn_3$	7	1421			
$NdMn_2$	86	960		≈5.347	1990
	~86	749	$NiCl_2 \cdot 6NH_3$	1.45	283
$NdMn_{12}$	135 ± 2	960	$Ni_xCo_{1-x}Cl_2 \cdot 6H_2O$		1997
	~135 ± 2	749	x = 0.208	2.58	
$NdMnO_3$			x = 0.473	3.08	
N_1	85	808, 1834	x = 0.857	4.71	
N_2	1.50	808	$NiCrO_3$	~250	1316
$NdNbO_4$	25	768	$NiCrO_4$	23	1660
Nd_3Ni	15	506	$NiCr_2S_4$		1429
$Nd(OH)_3$	1.7	1000	(NiAs type)	200	
NdS	8	980	$Ni_{17}Cu$	≈600	1807
	8.5 ± 0.5 (heat capacity)	1692	NiF_2	~73.2	54, 337, 15, 16, 55
	8.6 (electrical resistance)	1692		73	1699
				78.5-83	610, 645, 671

Material	Temperature Neel (°K)	Reference	Material	Temperature Neel (°K)	Reference
$(Ni,Fe)_3Mn$		786	Ni_2SiO_4	34	520, 265
50% Fe	~205		$NiSO_4 \cdot 6H_2O$	2	1098, 1131
75% Fe	~425		$NiSO_4 \cdot 7H_2O$	2	1131
$Ni(HCOO)_2 \cdot 2H_2O$	15.7	178	$NiTa_2O_6$	26	1811
NiI_2	75	270	$NiTiO_3$	25	292
$NiI_2 \cdot 6NH_3$	0.31	283		23	146
				<77	94
$NiI_2 \cdot [(NH_2)_2CS]_6$	1.77	922	$NiVO_3$	153 ± 5 (ilmenite phase)	2092
$Ni(IO_3)_2 \cdot 2H_2O$	3.08	193	$NiWO_4$	67	292
β	3.04	1669	NpC	310	1376
	2.4	740	$N_pC_{0.95}$	317	659
			NpO_2	25	475, 493 1454
Ni_3Mn	≈110	2025			
$NiMn_2O_4$	70	1565		No antiferro. reflections at 5.4°K	1428
$NiMoO_4$	19	292			
$Ni(NH_3)_2 \cdot Ni(CN)_4 \cdot 2C_6H_6$	2.37	1757			
$Ni(NO_3)_2 \cdot 2H_2O$	4.20	150	O_2	24	1357
	≈4.2	1820	(α)	>20	1601
NiO	647, 523	57, 74, 144		35 (calc.)	2152
	533–650	672, 674	$Pb(Co_{0.5}Ta_{0.5})O_3$-	153	757
	526	1836	1 mol % La_2O_3		
	525.5 ±1 (high temp sample)	1874	$PbCo_{0.5}W_{0.5}O_3$	9	396
			$PbCrO_3$	240	508
				250	697
	525	377, 1547	$PbFe_{0.5}Nb_{0.5}O_3$	78	551
	524	1573	$Pb(Fe_{0.5}Ta_{0.5})O_3$	180	1169
	523	279, 541, 1778	$Pb(Fe_{2/3}W_{1/3})O_3$	383	363
			Pd	~90	1578
	515 ±5 (low temp sample)	1874	Pd–Mn Mn:		1925
			10%	4.5	
$Ni(OH)_2$	30	724, 529	15%	12.5	
	28	2060	25%	25	
$NiRh_2O_4$	18	138, 381	Pd_3Mn	170	49
NiS	263	145	(ζ phase)	653	1062
	265	1414	Pd_3Mn_2	593	90
	208	2117	Pd_2MnAl	240	882
	200	2069	Pd_2MnIn	142	882
	150	673	Pr	23 (probably not an AF transition)	1466
$Ni_{1-x}S$		1826			
x = 0.005	≈217			No AF ordering to 4.2 K in monocrystal	1971
x = 0.008	200				
x = 0.015	≈152				
NiS_2	≈40	1957	PrAl	20	996
$NiSO_4$	37	55, 96		19 (19.3 kOe)	415
$NiSeO_4$	27	1634			
			Pr_2Al	18	1182

NÉEL TEMPERATURE

Material	Temperature Neel (°K)	Reference	Material	Temperature Neel (°K)	Reference
PrAl$_4$	12 ?	1182	Pu$_2$C$_3$	120	1487
PrAl$_2$-PrAg$_2$(15 mol %)	12	742	PuN	13	1450
PrAl$_2$-PrCo$_2$ 15 mol % PrCo$_2$	6.5	742		120	1487
			PuO$_{1.52}$	16	1459
PrAl$_2$-PrCu$_2$ (15 mol %)	13	742	PuS	4.5	1459
PrAl$_2$-PrFe$_2$ (15 mol %)	13	742	PuS$_2$	15	1459
PrAl$_2$-PrNi$_2$ (15 mol %)	14.5	742	Pu$_2$S$_3$		1459
PrB$_6$	6.92	1164	α	7	
	6.9 (specific heat)	2055	Pu$_3$S$_4$	10	1459
			RECrO$_3$	282–112.4	1063
	7	2164	RE$_2$O$_2$S		1595
PrC$_2$	15	422, 535	(RE = Gd, Tb, Dy, Ho, Yb)	2.5 to 7.7	
Pr$_{1.3}$Cd	~50	937			
PrCl$_3$	0.4	978	Rb$_3$CoCl$_5$	1.14	1531
	0.428	1511	RbCoF$_3$	150	1103
	0.42 not an AF transition	2115		32	919
			Rb$_2$CoF$_4$	130	1117
PrCrO$_3$	238.7	519		101	1649
PrFeO$_3$	707	828, 700	RbFeF$_3$	75	224
Pr-Gd		1708		101	1427
at. % Gd:				102	1370
70.5	55			~102	1473
Pr$_5$Ge$_3$	12	906	RbFeF$_4$	133.4	2100
PrMnO$_3$			Rb$_2$FeF$_4$	50-60	572
N$_1$	91	808, 1834		56.3	1841
N$_2$	1.50	808	RbMnCl$_3$	~88	836
PrMn$_2$O$_5$	46	1081		86 ± 6	22
PrNi$_5$	(appears to be antiferromagnetic at low temperatures)	802		85	1286
				92	22
				≈94	2006
			Rb$_2$MnCl$_4$	≈57	2002
Pr$_3$Ni	2	506		55	2000
PrO$_2$	14	804	Rb$_2$MnCl$_4$·2H$_2$O	2.24	928, 982
PrP$_{0.86}$	~10	1368	RbMnF$_3$	54.5	101
PrS	16	406		66	102
PrSn$_3$	8.6	344, 1713		82	307, 21
	8.5	1497		82.5	1699
Pt-Fe			RbMnF$_3$	82.6	1315
28% Fe	<130	1679		83	220, 1959
28 at.% Fe	155	1852		83 (elastic constants)	1571
Pt-26.7 at. % Fe		702			
(½ ½ 0)	~165			≈83	2106
(½ 0 0)	~125			~100	205
Pt$_3$Fe	170	702	Rb$_2$MnF$_4$	38.4	1344
Pt$_{35}$Mn$_{65}$	~710	688		38.5 ± 1	1670
Pt$_2$MnAl	190	2112	RbNiCl$_3$	11	1763, 1865
Pt$_2$MnGa	75	2112		10 ± 2	2011
Pu	No T$_N$ to 16	1468	Rb$_2$NiCl$_4$·2H$_2$O	4.60 ± 0.02	191
PuC$_{1-x}$	~100	943, 1963	Rb$_2$NiF$_4$	<190	194

Material	Temperature Neel (°K)	Reference	Material	Temperature Neel (°K)	Reference
$RuCl_3$		1586	$Sr_2(FeMo)O_6$	450	1060
(alpha)	∼30		$Sr(FeMo_{0.3}W_{0.7})O_6$	∼300	1622
Sc-Er	1.3	1074	$SrFeO_3$	<100	218
(Er 39 at. %)				130	5
Sc-Ho	1.3	1074	$SrFeO_4$	<2	1879
(Ho 18 at. %)			$Sr_2Fe_2O_5$	700	1438
$ScMnO_3$	120 (uncertain because of sample purity)	232	$Sr_3Fe_2O_{6.90}$	130	432
			Sr_2FeSbO_6	21	541
			$SrFeO_3$-$SrTiO_3$	<60	353
			Sr_2FeWO_6	16	541
Sc-Tb	1.3	1074	$SrGd_2O_4$	2.8	913
Tb (25 at. %)			$Sr_{1.5}La_{0.5}MnO_{3.99}$	∼16	1378
Sm	106 ± 1	454	$SrMnO_3$		1808
	109 (probably not an AF transition)	1466	6H	≈90	
			4H	≈350	
			Sr_2MnSbO_6	∼10	541
	15	444	Sr_2MnWO_6	10	541
	14	1466	$Sr_2(NiMo)O_6$	71	437
$SmAlO_3$	1.30	1284	$Sr_2(NiW)O_6$	59	909
SmB_6	No magnetic ordering to 0.35°K	1381	Sr_2NiWO_6	54	541
			$SrTb_2O_4$	5.9	913
			$SrTb_2S_4$	6	518
SmBi	<4?	343	$SrTi_{1-x}Fe_xO_{3-x/2}\phi_{x/2}$		989
$Sm_{(1-x)}Bi_xFeO_3$		1236	(ϕ = oxygen vacancy)		
x = 0.099	672 ± 1		x = 0.9	75	
Sm_3C	30	33	x = 0.95	110	
$SmCrO_3$	192.5	519	$TaSe_2$	130	589
SmCu	40	1309		110	1782
$Sm_{0.2}Dy_{0.8}FeO_3$	<30	1385		No magnetic ordering	2099
$SmFeO_3$	674	828, 700			
	672	1236	Tb	227.6	1960
	∼450	1385		228	24
$Sm_3Ga_5O_{12}$	0.967	823		229	1386
SmGe	40	515		230	198, 444
$SmIn_3$	16	1421		∼230	1505
$SmMnO_3$				234	719
N_1	60	808, 1834	TbAg	100	284
SmN	13	781		105	1274
	15	717		106 ± 2	274
	18.2	1614	$TbAg_2$	35	1221
SmSe	52.5	2074	TbAl	72	996, 1041
$SmSn_3$	12	344	$TbAlO_3$	3.8	1013, 1793
$SmZn_2$	45	1729		3.95	1161
$SrCo_{0.5}Mn_{0.5}O_3$	34	487		3.95 ± 0.10	667
Sr_2CoSbO_6	∼35	541		4	1004, 223, 1921
Sr_2CoUO_6	7	487			
Sr_2CoWO_6	22	541, 396		5	1044
Sr_2CrSbO_6	9	541	(Tb in)		
$SrFe_{0.5}Cr_{0.5}O_{3-y}$	230	1397			

NÉEL TEMPERATURE

Material	Temperature Neel (°K)	Reference	Material	Temperature Neel (°K)	Reference
$Tb_3Al_5O_{12}$	1.5	731	Tb–Ho		
	1.35	1048, 1340	at.% Ho:		
TbAs	10.5	544	10	218	1981
	12	355	35	196	395
TbAu	40 (19.3 kOe)	415	50	183	461
	43 (2.3 kOe)		50	184.5 ±1	1981
$TbAu_2$	42.5	1145	90	144 ±1	1981
$TbAu_2$	55	1313	TbIn	190	284
(α)	55	948	$TbIn_3$	36	1421
(β)	42.5	948		37 ±1	1996
TbB_6	~22	1164	$Tb_{0.4}La_{0.6}$	68	1008
TbBi	17	343	Tb–Lu	228–80	24, 395
	18 ±0.5	2131	Tb: 100% to 23%		
TbC_2	66	423, 535	$TbMn_2$	41	171
Tb_2C_3	20	904	$TbMnO_3$		
$TbCoO_3$	3.31	1004	N_1	<30	808
	3.35	1161	N_2	4.2	808
	3.3	1921		<2	1834
$TbCrO_3$	117.6	519		<13.8	808
N_1	158	1004	$TbNbO_4$	1.82	1191
N_2	4	1004	Tb_3Ni	62	506
N_3	3.05	1004	$TbO_{1.82}$	6	1107
TbCu	115	284	TbO_2	3	243
	117	1300, 1309	Tb_2O_3	2.0	1107
$TbCu_2$	54	170	Tb_2O_2S	7.7	1192, 1784
$TbCu_5$	15	1965	TbP	8	314
TbD_2	40	1954		9	355
$TbFe_{1-x}Co_xO_3$		2123	$TbPO_4$	~415	298
x = 0.1	570			≈4.2	1777
x = 0.2	470		$Tb_{5.10}Pd_{1.90}$	62	350
$TbFeO_3$	3.1	545, 1921	$TbPt_3$	20.5 ±0.5	1996
N_3	3.1	1004	TbSb	16.5	844
(Tb)	3.1	1289		~15.2	1842
	3.2	1161		15	1442
	8.4	545		14.6	1535
N_2	8.4	1004		14	355
	647	828, 700	Tb–Sc		395
	681	545	Tb %: 90	200	
N_1	681	1004	80	174	
(Fe)	651	1289	51	104	
TbGe	48	515	25	<1.3	
$TbGe_2$	42	450	TbSi	57	1691
Tb_5Ge_3	85	906	$TbSi_2$	17	580
Tb_5Ge_4	30	949	Tb–22% Tm	205	395
$TbH_{1.98}$	40	920	Tb–Y		
TbH_2	40	924	Tb %: 5–100	25–228	395, 24
TbHo	~182	1596	10–70	50–188	40

Material	Temperature Neel (°K)	Reference
$Tb_xY_{1-x}Sb$		
$x = 0.928$	13	1535
$x = 0.8$	≈9.9	1842
$x = 0.635$	8.8	1535
$x = 0.467$	no transition from 1.6 to 4.9°K	1535
$x = 0.403$	< 1.5	1842
$TbZn_2$	55	1729
	25	
Th 36 at. % Er	3.5 ± 0.5	1010
Th 80 at. % Pr	<1.3	1010
$TiCl_3$	~100	58
	217 ± 2	730
	217 (elec. conductivity)	2018
Ti-Fe 34.6 at. % Ti	281	792
$TiFe_2$	280	898, 742
TiH_2	~300	153
$TiO_{1.75}, TiO_{1.80}, TiO_{1.83}$	~403	486
Ti_2O_3	no transition	991
	not AF	1474
	630	1136
Ti_2O_3 (probably due to Ti_3O_5-ref. 902)	248	154, 155
Ti_2O_3 (probably due to Ti_3O_5-ref. 902)	660	368
Ti_3O_5	450	991
during heating	462	1272
during cooling	432	1272
	448 (no AF ordering)	2136
$Ti_4O_7, Ti_5O_9, Ti_6O_{11}$	~130	889
$TlCoF_3$	125	1103
Tl_2CoF_4	130	1117
$TlCuF_3$	235	1103
$TlFeCl_4$	11.2 ± 0.2	1739
$TlFeF_3$	100	1071
	78	1235
$TlMnCl_3$	118 (weak ferromag. at 108 K)	2007
$TlMnF_3$	76 ± 1	500
	85	513
Tl_2NiF_4	~190	190

Material	Temperature Neel (°K)	Reference
Tm	53	246
	51	444
	56	297, 24
	60, 56	198
	~57	1254
	58	382
TmAg	9.5 ± 1	274
	10	1274
TmAl	10	996
Tm_3Al_2	3	581
$TmAu_2$	3.2 ± 0.2	1902
TmC_2	no ordering to 1.7 K	2041
$TmCrO_3$	123.7	519
	~4	519
TmCu	11	1309
$TmFeO_3$	632	828
$Tm_{0.53}Lu_{0.47}$	38	1579
$Tm_{0.78}Lu_{0.22}$	49	1579
$TmMnO_3$	86 (uncertain because of sample purity)	232
	86	1834
TmTe	0.21	2088
Tm-Y		24, 395
Tm %: 55	37	
14.9	20 ± 8	
U	36 *not* antiferro. transition	1269
(α)	43	2073
(α)	≈43	1993
(α)	~60	1265
UAs	128	965, 253, 1039
UAs_2	283	1020, 1039
UAs-USe		1153
0 mole % USe	128	
35 mole % USe	132	
40 mole % USe	134	
UBi	290	830
	285 ± 5	1457
	No magnetic structure to 4°K	1383

NÉEL TEMPERATURE

Material	Temperature Néel (°K)	Reference	Material	Temperature Néel (°K)	Reference
UBi_2	183	917, 1020, 1039	$UP_{0.95}S_{0.05}$	N_1 122 ± 3, N_2 27–35	1278, 1306, 1544
U_2C_3	59	761, 1450	$UP_{0.92}S_{0.08}$	N_1 110 ± 3, N_2 32–42	1278, 1544
	50	1487	$UP_{0.90}S_{0.10}$	N_1 102 ± 3	1278, 1544, 1896
$UC_{1-x}N_x$ $x = 0.5$	40–50	1898		N_2 79–87	1278, 1544
				N_3 41–53	1278, 1544
$UC_{0.5}N_{0.95}$	31	2095	$UP_{0.85}S_{0.15}$	95 ± 2	1278, 1544
UCl_3	≈23	2044	$UP_{0.80}S_{0.20}$	92 ± 3	1278
$UFeO_4$	55	1354	$UP_{0.75}S_{0.25}$	N_1 65 ± 5, N_2 20 ± 3	1731
UGe_2	260	857			
UI_3	2.6	1250	UP-50 at. % ThP	115	764
UN	53 ± 2	965	UP-US		1574
	53	550	0.2 US	~140	
	52	1450, 2095	0.4 US	~180	
	50	1487	0.6 US	~200	
	~50-60	1674	0.8 US	~210-15	
	~45	901	US	~220	1574
β-U_2N_2	96	1039	USb	213	965, 1039
UO_2	28.7	161			
	~30	340		217 ± 4	1457, 1929
	30.44	1864		246	1383
	30.6	527	USb_2	206	918, 1020, 1039
	30.8	248			
	~31	869			
U_3O_8	No magnetic reflection at low temp	1229	UTe_2	<77	751
			UTe_3	<77	751
UOS	55	1020, 1079	$U_{1-x}Th_xO_2$		987
			(x = 0)	~29	
UOSe	72	805	(x = 0.2)	~16	
	<90	1227	(x = 0.4)	~2.4	
UOTe	162	354	V	(no intrinsic transition; anomaly due to H content)	931
	~157	805			
UP	130	901, 965			
	129	764			
	125 ± 2	1278, 1544, 1928	VCl_3	30	384
			$V_{1-x}Cr_xB_2$		2052
	~125	401, 1924	x = 0.8	14	
	123	965, 1039	x = 0.9	48	
	121	832	x = 1	85	
	~60-70	1574	$(V_{1-x}Cr_x)_2O_3$		2057
	N_2 22.9	1278, 1544	x = 0.3	≈75	
UP_2	206 ± 2	901	x = 0.5	≈163	
	203	439	x = 0.8	≈273	
$UP_{0.98}S_{0.02}$	N_1 124 ± 2	1278	x = 1	≈318	
	N_2 23–25		VF_2	7	1124

Material	Temperature Neel (°K)	Reference	Material	Temperature Neel (°K)	Reference
V oxides V_2O_3 VO_2 VO	Magnetic anomaly comes from crystalline distortion at the transition temp.	886	V-Te 64.3 to 66 at. % Te	410-480	539
			WV_2O_6	370 ± 30	1021
			$YCrO_3$	140.6	519
				141	411
				142	440
$VO_{1.147}$	4.6	470	$Y_{25}Er_{75}$	70.6	1789
$VO_{1.257}$	7.0	470	$Y_{50}Er_{50}$	54.1	1789
$VO_{1.50}$	170	1053	$Y_{75}Er_{25}$	31.5	1789
$VO_{2.16}$	154	1102	$YFe_{0.9}Al_{0.1}O_3$	602	958
VO_2	343	53, 60	$YFeO_3$	643	238, 125
	345 ± 3	3		640	828
	338	1538		650 ± 3 (neutron diffraction)	1794
V_2O_3	Structure agrees with Goodenough's theory (see 886)	862	$(Y_3Fe_5O_{12})_{1-x}(Ca_3Fe_2Si_3O_{12})_x$		747
	AF	1926	x = 0.60	302	
	150	1224	x = 0.70	234	
(monoclinic)	~154	1698	x = 0.82	130	
	160	2111	x = 0	553	748
	No antiferromagnetism at 168°K	62	x = 0.60	305	
			x = 0.70	233	
	168	3, 42	x = 0.82	112	
	170	61, 62	$YMnO_3$	46	167, 324, 762
$V_2O_{3.06}$	No transition	1813		80	1834
V_3O_5	235 ± 3	3		~80	762
	253 ± 3	42, 3	YVO_3	110	213
V_4O_7	130 ± 3	42, 3	YbC_2	no magnetic ordering to 2 K	2042
V_5O_9	162	42, 3	$YbCrO_3$	158	519
	140	1786		3.05	519
V_6O_{13}	154	42, 3	$YbFeO_3$	627	828
$(VO_2)_x(NbVO_4)_{1-x}$		1538		625	126
x = 0.8	213			632 ± 3 (neutron diffraction)	1794
x = 0.9	241				
x = 0.95	298				
x = 0.97	313		$YbMnO_3$	87.3	1923
$(VO_2)_x(TaVO_4)_{1-x}$		1538		86	1834
x = 0.8	217		$YbMn_2O_5$	46	1031
x = 0.9	241		YbN	3.8	1614
x = 0.95	277		Yb_2O_3	2.25	495
$VS_{1.33}$	473	1346		2.3	200, 1266
$VOSO_4$		1655, 1833	Yb_2O_2S	~3 (measurement not described)	1510
β	25				
VSe	~400	1542		≈3	1784
$VSe_{0.90}$	~300	1557			

NÉEL TEMPERATURE

Material	Temperature Neel (°K)	Reference	Material	Temperature Neel (°K)	Reference
$Zn_{1-x}Cd_xCr_2Se_4$			$ZnFe_2O_4$	15	230
x = 0 to 0.35	~20	507		10.6	1887
x = 0.1	~20	1360		10	1831
x = 0.5	~12	1360		9	53
$Zn(Cr_{1.96}Fe_{0.04})O_4$	~12.5	1543	$ZnMn_2As_2$	~138 ?	1503
$ZnCr_2O_4$	15	53	$Zn_{0.87}Mn_{0.13}Fe_2O_4$	11 ± 0.5	1737
	16	434	$ZnMnO_3$	18	1830
	≅11.5	1319	$ZnMn_2O_4$	330	770
$ZnCr_2S_4$	18	434, 962	$(Zr_xNb_{1-x})Fe_2$		969
	<20	563	x = 0.25	300	
$ZnCr_2Se_4$	~20	606	$(Zr_{1-x}Nb_x)Fe_2$		1621
	20	434	x = 0.45	~300	
$ZnFe_{2-x}Cr_xO_4$		2132			
x = 1.9	16				

ADDENDUM (CURIE TEMPERATURE)

Material	Temperature Curie (°K)	Reference	Material	Temperature Curie (°K)	Reference
AgCl	5.7 (optically induced magnetic order)	2343	$CuCr_2S_4$ (powder) (crystal)	~360 ~335	2266
$Au_{3.9}Ag_{0.1}V$	28	2449	$CuCr_2Se_{3.5}Br_{0.5}$	~370	2236
Au_2Mn	≅360 (21 kO_e field)	2322	$CuCr_2Se_{3.5}Cl_{0.5}$	~385	2236
Au_4V	52 ± 1	2375	$Cu_{1+x}Cr_2Te_4$		2347
	53	2449	x = 0.0	~360	
$BaFe_{12-x}Cr_xO_{12}$		2296	x = 0.5	~260	
x = 1	~670		x = 0.9	~170	
x = 2	~630		x = 1.0	~160	
x = 3	~600		$Cu_{1-x}Mn_x$		2409
Ba_2FeReO_6	334	2473	x = 0.05	5.9	
Ba_2MnReO_6	105	2473	x = 0.1	15	
Ba_2NiReO_6	18	2473	x = 0.2	26	
$BaVO_3$	<77	2208	γ-Cu-40 at. % Mn	130	2320
$CdCr_2Se_4$	~130	2235	$Cu_2Mn_{1.02}Al_{0.07}$		2305
$CdCr_2Se_4$	131 ± 1	2241	(annealed)	561	
$(Cd_{1-x}Fe_x)Cr_2S_4$		2362	(quenched)	613	
x = 0.0	86		$Cu(NH_4)_2Br_4 \cdot 2H_2O$	1.828	2401
x = 0.1	108		Dy	85	2311
x = 0.5	140		Dy_3Al_2	76	2425
x = 0.9	165			100	2397
x = 1.0	170		DyAs	2	2451
CdO	80 (optically induced magnetic order)	2343	$Dy_3Co_xFe_{5-2x}Ge_xO_{12}$ x = 0.0	570	2368
γ-Ce	8	2286	x = 0.4	472	
$CeCo_5$	673	2413	x = 1.0	350	
Co	1365 (thermoelec.)	2460	x = 1.5	225	
$Co_{50.9}Al_{49.1}$	13	2474	x = 2.0	101	
CoB	Not ferromag. >77	2472	$DyCu_{0.4}Zn_{0.6}$	58	2317
$CoF_2 5HF \cdot 6H_2O$	246 ± 2	2478	$DyFe_3$	605	2346
$CoFe_2Se_4$	125	2448	$DyFe_{0.8}Ni_{0.2}O_{2.8}F_{0.2}$	630	2214
Co_2MnAl	693 ± 3	2212	DyGa	116	2384
Co_2MnGa	694 ± 3	2212	75% Dy − 25% Ho	50	2220
Co_2MnGe	905 ± 3	2212	DyN	17	2451
$Co_{1.5}MnSb$	600 ± 10	2212	DyNi	62	2437
Co_2MnSi	985 ± 5	2212	DyP	4.5	2451
Co_2MnSn	829 ± 4	2212	DyRh	3.8	2317
CoS_2	120.5	2402	DyZn	137	2317
$CoSiF_6 \cdot 6H_2O$	246 ± 2	2478	$ErAl_2$	12.5 ± 0.5	2201
$CrFe_2Se_4$	320	2448	$ErAl_{2-x}Ni_x$		2207
$CrMn_2O_4$	65	2218	x = 0.03	22	
CrNiAs	193	2360	x = 0.10	33	
CrNiP	140	2360	x = 0.30	28.5	
Cr_5S_6	305	2191	$ErFe_3$	555	2346
$CrS_{1.2}$	305	2383	$Er_{0.333}Fe_{0.667-x}Al_x$		2302
$(Cr_{1.01}Sb)_{0.4}(Fe_{1.22}Sb)_{0.6}$	90	2424	x = 0.067	>297	
$Cr_{1.96}Te_3$	167	2344	x = 0.333	56	
$Cr_{1.98}Te_3$	170	2344	x = 0.600	16	
Cr_7Te_8		2170	x = 0.667	20.5	
ordered	332		$ErFe_{0.8}Ni_{0.2}O_{2.8}F_{0.2}$	620	2214
disordered	343		ErGa	32	2384
$CuAg_{0.25}Cr_2Te_4$	330	2347	$ErMn_{0.2}Fe_{0.8}O_{2.8}F_{0.2}$	595	2214
			ErN	6	2451
			ErNi	13	2437
			$Er_{0.8}Y_{0.2}Al_2$	13	957

ADDENDUM (CURIE TEMPERATURE)

Material	Temperature Curie (°K)	Reference
Eu	~85	2227
Eu_3As_2	18	2239
$EuFe_{0.8}Ni_{0.2}O_{2.8}F_{0.2}$	645	2214
$Eu_{0.7}Gd_{0.3}S$	25	2399
$Eu_{0.1}La_{0.9}Al_2$	11	2216
$Eu_{0.2}La_{0.8}Al_2$	6	2216
$Eu_{0.4}La_{0.6}Al_2$	<4	2216
$EuLiH_3$	37.6	2334
EuO	69.2	2172
	70	2404
EuS	16.4	2339
	16.6	2399
EuS – 1.1 mole % $EuBr_2$	39	2274
EuSe	~2.5	2382
	4.6	2255
EuSe – (Cl doped)	18	2274
EuSe – 0.48 mole % $EuBr_2$	12–15	2274
Eu_2SiO_4	5.2 ± 0.5	2263
Eu_2SiO_4 (monoclinic)	5.40	2452
Eu_3SiO_5	11.5 ± 0.5	2263
FeCoAs	300	2360
FeCoP	425	2360
$FeCr_2S_4$	195	2292
(powder)	~185	2266
(crystal)	~160	
$Fe_{0.5}Cu_{0.5}Cr_2S_4$		2266
(powder)	~340	
(crystal)	~315	
Fe-Ni		
at. % Ni		
28.8	373	2381
29.82	348	2189
43.2	713	2381
44.43	723	2189
FeNiP	95	2360
$Fe_{1.91}P$	223	2174
Fe_2P	225	2174
$Fe_{44}Pd_{36}P_{20}$	319	2432
Fe_3Pt	~425	2243
	430	2341
$Fe_{65}(Pt_xIr_{1-x})_{35}$		2369
x = 0.8	295	
x = 0.9	393	
x = 1.0	460	
Fe_5Sn_3		2431
γ_2	588	
γ_3	632	
$FeTiO_3(Fe_2O_3)_4$	~765	2470
Fe_2TiO_4	142	2379
FeV_2O_4	109	2434
Gd	290	2295
	290	2301
	289	2311
α-Gd	297	2231
$GdAl_{2-x}Ni_x$		2207
x = 0.1	191	
x = 0.45	68	
GdCd	264	2231
α-$GdCd_2$	129	2231
$GdCd_3$	12	2231
$GdCd_6$	3	2231
$GdCl_3$	2.22 ± 0.02	2429
$GdCo_2$	416	2210
$Gd_3Co_xFe_{5-2x}Ge_xO_{12}$		2368
x = 0.4	470	
x = 1.0	350	
x = 1.5	221	
x = 2.0	102	
$GdCu_{0.2}Fe_{0.8}O_{2.8}F_{0.2}$	660	2214
$GdFe_3$	725	2346
$GdFe_{0.8}Ni_{0.2}O_{2.8}F_{0.2}$	645	2214
GdGa	183	2384
	210	2346
$GdGa_{0.7}Cu_{0.3}$	190	2346
$Gd_{0.1}La_{0.9}Al_2$	24	2216
$Gd_6(Mn_{0.6}Fe_{0.4})_{23}$	120	2284
$GdMn_{0.2}Fe_{0.8}O_{2.8}F_{0.2}$	625	2214
GdN	60	2451
GdNi	71	2437
$GdNi_5$	28	2428
	33	2392
$Gd_{0.6}Y_{0.4}Al_2$	100	957
$GdZn_2$	68	2283
$HgCr_{2-x}In_xSe_4$		2238
x = 0	~110	
x = 0.1	~96	
x = 0.2	~87	
x = 0.3	~73	
$HoAl_{2-x}Ni_x$		2207
x = 0.08	46	
x = 0.45	31	
$HoCu_{0.2}Fe_{0.8}O_{2.8}F_{0.2}$	640	2214
$HoFe_2$	600	2194
	610	2411
$HoFe_3$	575	2346
Ho_6Fe_{23}	500	2194
$HoFe_{0.8}Ni_{0.2}O_{2.8}F_{0.2}$	630	2214
HoGa	63	2384
$HoMn_{0.2}Fe_{0.8}O_{2.8}F_{0.2}$	615	2214
HoN	12.8	2451
HoNi	37	2437
$HoNi_5$	23	2392
HoP	4.8	2451
$LaF_{0.8}Ni_{0.2}O_{2.8}F_{0.2}$	720	2214
$La_{0.1}Sr_{0.9}RuO_3$	133	2350
$La_{0.25}Sr_{0.75}RuO_3$	79	2350
MnAs		2246
free film	311	
clamped film	339	
	318	2183
MnCoAs	350	2360
MnCoP	583	2360
$Mn_{0.8}Fe_{0.2}B$	740	2441

ADDENDUM (CURIE TEMPERATURE)

Material	Temperature Curie (°K)	Reference	Material	Temperature Curie (°K)	Reference
$Mn_{5-x}Fe_xGe_3$		2356	$Pr_2(Co_{1-x}Fe_x)_{17}$		2290
x = 0.0	304		x = 0.0	1163	
x = 0.5	318		x = 0.3	1090	
x = 1.0	322		x = 0.75	739	
x = 1.5	332		$PrFe_{0.8}Ni_{0.2}O_{2.8}F_{0.2}$	650	2214
$MnFe_{1.3}Sc_{0.7}O_4$	443	2327	PrGa	32	2384
Mn_3Ge_2		2325	$Pr_{0.28}Gd_{0.053}Mn_{0.667}$	450	2205
Tc_1	158		$Pr_{0.233}Ho_{0.10}Mn_{0.667}$	450	2205
Tc_2	283		$PrMn_2$	448	2205
Mn_5Ge_3	294.5	2250	PrNi	22	2437
Mn_3O_4	43	2309	$Pr_{0.22}Tb_{0.113}Mn_{0.667}$	445	2205
$MoCl_5$	22	2187	$RbNiF_3$	138.4	2374
$MoCl_5 \cdot C$ (graphite layers)	15	2187	$SmCo_5$	984	2413
$NdAl_2$	77.2	2219	$Sm_2(Co_{1-x}Fe_x)_{17}$		2290
$NdCo_2$	42	2364	x = 0.0	1199	
	113	2210	x = 0.4	1079	
$NdCo_5$	913	2413	x = 0.8	541	
$Nd_2(Co_{1-x}Fe_x)_{17}$		2290	x = 1.0	394	
x = 0.0	1290		$SmCu_{0.2}Fe_{0.8}O_{2.8}F_{0.2}$	675	2214
x = 0.3	1090		$SmFe_2$	675	2277
x = 0.75	754			676	2470
$NdFe_{0.8}Ni_{0.2}O_{2.8}F_{0.2}$	675	2214	$SmFe_3$	650	2346
NdGa	44	2384		650	2277
NdNi	28	2437	Sm_2Fe_{17}	385	2277
$NdNi_5$	13	2392	$SmFe_{0.8}Ni_{0.2}O_{2.8}F_{0.2}$	655	2214
$NdS_{1.33}$ (Nd_3S_4)	~42	2217	SmGa	108	2384
$NdS_{1.42}$	~10	2217	SmNi	45	2437
Ni	633	2312	$(SrRuO_3)_{0.6}(LaRhO_3)_{0.4}$	50	2418
	~640	2228	Tb_3Al_2	190	2425
Ni (in SiO_2)		2228		190	2397
Vol % Ni			$TbAl_{2-x}Ni_x$		2207
48	188		x = 0.08	118	
52	295		x = 0.40	59	
$Ni_{75+x}Al_{25-x}$		1364	$Tb_3Co_xFe_{5-2x}Ge_xO_{12}$		2368
x = 0	41.5		x = 0.0	565	
x = 0.5	58.1		x = 0.4	465	
x = 1.0	71.5		x = 1.0	348	
Ni-Cr		2282	x = 1.5	228	
at. % Cr			x = 2.0	103	
2.6	518		TbF_3	3.95 ± 0.05	2285
5.1	390		$TbFe_3$	655	2346
7.7	235		TbFeAl	185	2331
9.4	130		$TbFe_{0.8}Ni_{0.2}O_{2.8}F_{0.2}$	640	2214
12.5	11		TbGa	158	2384
$NiFe_2Se_4$	67	2448	$TbMn_{0.2}Fe_{0.8}O_{2.8}F_{0.2}$	565	2214
$Ni_{85}P_{15}$ (amorphous)	~400	2337	TbN	39	2451
$NiSc_{0.15}Fe_{1.85}O_4$	830 ± 5	2324	TbNi	52	2437
$NiSc_{0.4}Fe_{1.6}O_4$	740 ± 5	2324	$Tb(OH)_3$	3.7	2338
$NpCl_4$	6.7 ± 0.1	2446		3.71	2232
$PaCl_4$	182 ± 2	2446	$Tb_{0.6}Y_{0.4}Al_2$	51	957
	182 ± 2	2353	$ThCo_x$		2206
Pd – 1.35 at. % Mn	4.5	2456	x = 4.8	~410	
β-Pr	9	2286	x = 5.0	~420	
$PrAl_2$	31.8	2219	x = 5.2	~560	
$PrCo_5$	921	2413	x = 5.6	~735	
			$ThCo_5$	415	2202

ADDENDUM (CURIE TEMPERATURE)

Material	Temperature Curie (°K)	Reference	Material	Temperature Curie (°K)	Reference
$ThCo_{5-5x}Fe_{5x}$		2415	$Y_2(Co_{1-x}Fe_x)_{17}$		2290
$x = 0.2$	585		$x = 0.0$	1189	
$x = 0.4$	660		$x = 0.3$	1120	
$x = 0.8$	695		$x = 0.6$	915	
$ThNi_{5-5x}Fe_{5x}$		2415	$x = 1.0$	304	
$x = 0.2$	170		$Y_3Co_xFe_{5-2x}Ge_xO_{12}$		2248
$x = 0.6$	420		$x = 0.2$	514	
$x = 0.8$	525		$x = 2.2$	60	
$TiFe_{0.5}Co_{0.5}$	~45	2209	$Y_2(Fe_{1-x}Co_x)_{17}$		2366
$TiFe_2Se_4$	240	2448	$x = 0.0$	~300	
Tm		2308	$x = 0.1$	~500	
(ferri-)	38		$x = 0.35$	~810	
(ferro-)	22		$x = 0.7$	~1080	
	~22	2204	$x = 1.0$	~1180	
$TmFe_3$	535	2346	$YFe_{0.8}Ni_{0.2}O_{2.8}F_{0.2}$	625	2214
$TmFe_{0.8}Ni_{0.2}O_{2.8}F_{0.2}$	615	2214	$YMn_{0.2}Fe_{0.8}O_{2.8}F_{0.2}$	620	2214
TmNi	8	2437	$YbFe_{0.8}Ni_{0.2}O_{2.8}F_{0.2}$	615	2214
$TmNi_5$	22	2392	U_3Bi_4	112	2351
UAsS	128	2457	UGa_2	126	2351
UAsSe	118	2457	$Zn_xFe_{3-x}O_{4-x}F_x$		2348
$UAs_{0.4}Se_{0.6}$	170	2445	$x = 0.1$	~545	
UAsTe	<77	2457	$x = 0.25$	~470	
UGa_2	133	2458	$x = 0.4$	~400	
US	180	2476	$x = 0.5$	~365	
USe	160 ± 5	2476	$ZrFe_{0.8}Co_{0.2}$	~500	2380
USi	120	2297	$ZrFe_{0.8}Mn_{0.2}$	~465	2380
$V_xCr_{1-2x}Mn_xTe$		2306	$(Zr_{1-x}Hf_x)Zn_2$		2365
$x = 0$	~335		$x = 0.1$	~10	
$x = 0.1$	~330		$x = 0.05$	~13	
$x = 0.2$	~320		$(Zr_{1-x}Ti_x)Zn_2$		2365
$x = 0.3$	~270		$x = 0.05$	~34	
VFe_2Se_4	155	2448	$x = 0.08$	~40	
YCo_3	310	2202	$x = 0.20$	~49	
YCo_5	978	2413	$ZrZn_2$	~18	2365
	995	2202		22.2−1.8P (P in kbar)	2421

ADDENDUM (NÉEL TEMPERATURE)

Material	Temperature Neel (°K)	Reference
Au – 10% Cr	102	2226
Au – 15.7% Cr	183	2226
Au_2Mn	366	2322
Au_5Mn_2	353	2321
$BaCoF_4$	~95	2222
$BaCoF_4$		2398
$\quad T_{N_2}$	65	
Ba_2CoReO_6	40	2473
$BaMnO_3(2H)$	<2.4	2358
$Bi_2Fe_3GaO_9$	<77	2367
$Bi_2Fe_{3.4}In_{0.6}O_9$	237 ± 10	2367
$Bi_2Fe_4O_9$	265 ± 10	2367
$Bi_2O_3\text{-}2Fe_2O_3$	265	2294
$Bi_{1-x}RE_xO_{1.5}$		2395
$\quad (x \geqslant 0.4)$	4.3	
$(C_6H_5CH_2NH_3)_2FeCl_4$	72	2252
$(CH_3NH_3)_2FeCl_4$	96	2252
$(C_3H_7NH_3)_2FeCl_4$	90	2252
$Ca_{1.7}Bi_{0.3}MnO_4$	~104	2229
$CaCrO_3$	325	2269
	90	
$Ca_3Mn_2Ge_3O_{12}$	13	2335
α-Ce	23	2286
β-Ce	14	2286
CeAl	~9.4	2315
	9.5	2230
CeBi		2289
$\quad T_{N_1}$	25.2	
$\quad T_{N_2}$	12.5	
$CeIn_3$	10.4 ± 0.5	2215
	11	2430
Ce – 9% La	10	2354
$Ce_2Mg_3(NO_3)_{12} \cdot 24H_2O$	1.8×10^{-3}	2179
$CePt_2$	1.6 ± 0.1	2442
$CoCl_2 \cdot 2H_2O$	17.3	2422
$CoCl_2 \cdot 6H_2O$	2.29	2328
	2.29	2273
	~2.3	2272
$CoCl_2 \cdot 2NC_5H_5$	3.17 ± 0.02	2371
CoF_2	37.5	2477
$Co(HCOO)_2 \cdot 2H_2O$	5.1	2376
	5.10 ± 0.05	2370
		2373
$(CoMn)_{1-x}Fe_x$		
$\quad x = 0.02$	≅375	
$\quad x = 0.2$	≅345	
$\quad x = 0.4$	≅275	
CoO	289	2466
$CoSO_4$	15.5	2329
$Co_{1-x}Zn_xFe_2O_4$		2298
$\quad x = 0.4$	513 ± 5	
$\quad x = 0.6$	322 ± 5	
Cr	311	2256
	311.5	2227
	~311	2182
	312	2221
	311.7	2186
	311.5	2340

Material	Temperature Neel (°K)	Reference
CrAs	248	2303
	265 (sp. heat)	2342
CrB_2	85	2453
	85	2396
	86 ± 2	2440
$Cr_{0.8}B_2V_{0.2}$	16	2453
$CrB_2 – 25\%\ VB_2$	10	2440
Cr – 9.35 at. % Fe	181	2439
CrH_x		2203
$\quad x = 0.84$	66	
$\quad x = 0.91$	>300	
$\quad x = 0.94$	218	
Cr – 9.35 at. % Mo	197	2439
CrN	281.5	2471
	286	2169
	287	2407
$(Cr_{1.01}Sb)_{1-x}(Co_{1.13}Sb)_x$		2424
$\quad x = 0.6$	150	
$\quad x = 0.4$	450	
$\quad x = 0.2$	660	
$\quad x = 0$	713	
$(Cr_{1.01}Sb)_{0.9}(Fe_{1.22}Sb)_{0.1}$	625	2424
CrSi	300	2249
Cr – 0.17 at. % Ti	276	2420
$Cr_{0.8}Ti_{0.2}O_3$	~260	2268
$Cr_{0.6}V_{0.4}O_3$	~170	2268
$CsCoCl_3 \cdot 2H_2O$	3.4	2193
$\alpha\text{-}Cs_2MnCl_4$	0.935	2262
Cs_3MnCl_5	0.601	2262
$CsMnCl_3 \cdot 2H_2O$	4.88	2475
	4.88	2258
	~4.9	2278
	4.89	2244
$Cs_2MnCl_4 \cdot 2H_2O$	1.81	2262
$CsNiCl_3$	4.3	2253
	4.84 ± 0.03	2438
$CuCl_2 \cdot 2NC_5H_5$	<1.3	2371
$CuK_2(SO_4)_2 \cdot 6H_2O$	0.029	2184
$Cu(OH)_2$	21	2469
$CuSO_4$	35	2329
Dy	179	2311
DyAg	63	2317
$DyAlO_3$	3.5	2410
	3.525 ± 0.006	2316
$DyAsO_4$	2.44	2175
DyCu	64	2317
$DyCu_{0.6}Zn_{0.4}$	45	2317
$Dy_3Ga_5O_{12}$	0.37	2245
75% Dy – 25% Ho	167	2220
DyOCl	9	2390
DyOOH	9	2357
$DyPO_4$	3.390	2300
	3.4	2435
DySb	9	2451
$DyVO_4$	3.1	2389
	3	2386

ADDENDUM (NÉEL TEMPERATURE)

Material	Temperature Néel (°K)	Reference	Material	Temperature Néel (°K)	Reference
Er	84	2196	$LiCr_{1-x}Fe_xO_2$		2359
$ErAlO_3$	~0.6	2363	x = 0.0	~15	
$ErCl_3$	0.307	2257	x = 0.1	~56	
$ErFe_2$	600	2287	x = 0.2	~40	
	600	2352	x = 0.3	~35	
ErOOH	7.2	2357	x = 0.4	~50	
ErSb	3.6	2436	x = 0.5	~55	
ErSi	10	2195	x = 0.6	~45	
Eu	~90	2227	x = 0.7	~40	
$EuAl_2$	15	2216	$LiCrO_2$	200	2349
$Eu_{0.8}La_{0.2}Al_2$	14	2216	$LiFe_{0.5}Ta_{0.5}O_2F$	883	2444
$EuPb_3$	22	2279	$MgCr_2O_4$	~10	2234
EuSe		2427	γ-Mn	540	2361
T_{N_2}	1.8		$MnAs_{1-x}P_x$		2443
EuTe	9.6 ± 0.1	2463	x = 0.03	205	
	9.6 ± 0.2	2465	x = 0.05	240	
	9.7	2464	Mn_3B_4	392	2423
(calcn.)	9.76 ± 0.41	2281		413	2459
(exptl.)	9.81			226	2423, 2276
Fe	~80	2213	$MnBr_2 \cdot 4H_2O$	2.13	2273
$FeAsO_4 \cdot 2H_2O$	16	2414		2.12	2192
$FeC_2O_4 \cdot 2H_2O$	~25	2275	$MnCO_3$	34.5 ± 0.1	2433
$FeCl_2$	23.55	2462	α-$MnCl_2 \cdot 4H_2O$	1.62	2400
FeF_2	78.2	2477	Mn-Cr		2299
	78.377 ± 0.050	2405	at. % Mn = 2	~500	
FeF_3	363.2	2293	at. % Mn = 6	~670	
FeGe	~411	2176	Mn-Cu		2313
$Fe(HCOO)_2 \cdot 2H_2O$	3.75 ± 0.05	2370	at. % Mn = 78	~220	
	3.7	2336	at. % Mn = 84	~340	
Fe_3Mn_2	449	2345	at. % Mn = 90	~410	
$Fe_{51}Mn_{34}Ni_{15}$	311	2345	Mn-Cu		2330
$Fe_{57}Mn_{38}Ni_5$	403	2345	Mn at. % = 85	368	
FeOCl	68 ± 2	2259	Mn at. % = 90	468	
β-FeOOH	283 ± 3	2326	Mn at. % = 95	578	
FeP	125 ± 1	2173	Mn – 10 at. % Cu	410	2378
FeRh	~338	2171	MnF_2	66.8	2477
$FeSO_4$	21	2329		67.3	2265
$Fe_{1.11}Te$	68	2291		67.3	2199
$FeVO_4$	22 ± 1	2355		67.46	2251
$GdO_{1.493}$	34	2247	MnFe	~490	2313
$GdO_{1.500}$	3.9	2247	γ-MnFe	500	2417
$Gd(OH)_3$	0.94	2232	Mn – 10 at. % Ge	460	2378
	0.9	2338	$Mn(HCOO)_2 \cdot 2H_2O$	3.72	2271
$GdPb_3$	17	2279		3.7	2270
^3He	0.002	2387	Mn – 15 at. % Ir	650	2416
Ho	132.5	2196	Mn – 18.7 at. % Ni	442	2314
HoAs	6	2451	Mn – 20 at. % Ni	438	2378
$KFeF_4$	138	2307		453	2288
$KFe_4(OH)_4(AsO_4)_3 \cdot 8H_2O$	~6	2414	Mn_3NiN	266	2403
K_4MnCl_6	0.439	2263	MnO_2	93	2391
$KMnF_3$	88	2393	$Mn_2P_2O_7$	13 ± 3	2177
$KMn_{1-x}Mg_xF_3$		2408	Mn – 13 at. % Pd	~395	2314
x = 0.05	77		$MnPd_2$	415 ± 10	2455
x = 0.10	81		$MnSO_4$	11.5	2454
x = 0.15	84		$MnSO_4 \cdot H_2O$	15.8	2454

ADDENDUM (NÉEL TEMPERATURE)

Material	Temperature Neel (°K)	Reference	Material	Temperature Neel (°K)	Reference
α-MnSe	247	2318	TbAs	7	2451
MnSi	~100	2249	TbCo$_{0.3}$Al$_{1.7}$	26	2237
MnTe	307	2406	Tb – 50% Ho	183	2224
Mn – 20 at. % Zn	280	2378	TbOCl	4	2390
Mo-Cr		2299	TbOOH	10	2357
at. % Mo = 4	~250		TbPO$_4$	2.17	2223
at. % Mo = 8	~175			2.05 ± 0.01	2242
MoF$_3$	172	2394	TbRhO$_3$	~1.85	2363
NH$_4$MnCl$_3$	~105	2185	TbVO$_4$	No magnetic ordering to 1.35	2188
NaMnF$_3$	66.7 ± 0.1	2468			
NiO	522 (elastic constants)	2332			
	525	2310	Tb$_{85}$Y$_{15}$	204	2254
NiAs$_2$O$_4$	53.5 ± 0.5	2225	α-TiCl$_3$	265	2190
NiCl$_2 \cdot$2H$_2$O	7.258	2419	TlMnF$_3$	76.5	2388
NiCl$_2 \cdot$6H$_2$O	5.34	2181	Tm	56	2308
NiF$_2$	73.2	2477	UAs$_{0.30}$P$_{0.70}$		2240
	73.2 ± 0.1	2433	(neutron diffraction)	116 ± 1	
Ni(HCOO)$_2 \cdot$2H$_2$O	15.5 ± 0.2	2370	(mag. susceptibility)	122	
Ni(NH$_3$)$_2$Ni(CN)$_4 \cdot$2C$_6$H$_6$	2.39 ± 0.01	2372	UAs$_{0.35}$P$_{0.65}$		2240
Ni(NO$_3$)$_2 \cdot$2H$_2$O	4.105 ± 0.005	2264	(neutron diffraction)	117 ± 1	
	~4.2	2260	(mag. susceptibility)	121	
NiO	525	2377	UAs$_{0.69}$Se$_{0.31}$	135	2445
NiO:0.9% Li	521.2 ± 1	2200	UCl$_3$	22 ± 1	2353
NiS	210–230	2461	UHg$_2$	70	2458
	≅1000	2319	UO$_2$	30.44	2180
	264	2267	V-Cr		2299
NiSO$_4$	37	2329	at. % V = 2	~140	
NiSb$_2$O$_4$	47.0 ± 0.5	2225	V$_x$Cr$_{1-2x}$Mn$_x$Te		2306
NpPd$_3$	53	2280	x = 0	~160	
Pd$_3$Gd	7.5	2385	x = 0.1	~180	
Pd$_3$Tb	2.0	2385	x = 0.2	~185	
α-Pr	22	2286	x = 0.3	~200	
γ-Pr	18	2286	V$_{0.1}$Cr$_{0.9}$N	225	2407
PrO$_2$	13	2426	(V$_{0.96}$Cr$_{0.04}$)$_2$O$_3$	~170	2304
α-Pu	~60	2178	V$_2$O$_3$	160	2204
	No mag. order to 4.6	2447	V$_2$O$_5$ – 90 mole % MoO$_3$	110 (V$_3$O$_5$ impurity?)	2333
PuPd$_3$	25	2280	V$_2$O$_5$–P$_2$O$_5$ (glass)	203	2467
PuCl$_3$	4.5 ± 1	2353	(V$_{0.96}$Ti$_{0.04}$)$_2$O$_3$	~85	2304
RbCoF$_3$	~100	2233	Yb$_2$O$_2$S	2.65	2450
RbCrF$_3$	50	2261	Yb$_2$O$_2$Se	2.65	2450
RbMnF$_3$	83	2323	ZnCr$_2$O$_4$	9.5 ± 0.2	2198
	83.0	2374	ZnCe$_2$O$_4$	~10	2234
SmSn$_2$	9	2412	ZnMn$_3$N	4.2	2197
Sm$_2$Sn$_3$	11	2412			
Sr$_3$U$_{1-x}$W$_x$Fe$_2$O$_9$		2211			
x = 1	317				
x = 0.9	305				
x = 0.7	331				

References

1. "Magnetocaloric Effects in Dysprosium," A. C. Hudgins, Jr., and A. S. Pavlovic, *J. Appl. Phys.* **36**, 3628 (1965).
2. "Observation of a High-Lying Far-Infrared Absorption Line in Antiferromagnetic FeF_2," I. F. Silvera and M. Tinkham, *Bull. Am. Phys. Soc.* **9**, 714A (1964).
3. "Crystal Chemistry Studies," R. Roy, S. Kachi, G. J. McCarthy, O. Muller, and W. B. White, Pennsylvania State University, University Park, Quarterly Report, Aug. 13, 1965–Nov. 13, 1965, Contract DA 28-043 AMC-01304(E), January 1966.
4. "Observation of an Infrared Absorption Band from s-f Exchange Interaction in Antiferromagnetic Holmium," C. Chr. Schuler, *Phys. Letters* **12**, 84 (1964).
5. "Electric and Magnetic Properties of the Strontium Ferrates," J. B. MacChesney, R. C. Sherwood, and J. F. Potter, *J. Chem. Phys.* **43**, 1907 (1965).
6. "Critical Magnetic Scattering from $KMnF_3$," M. J. Cooper and R. Nathans, BNL-9615; also in *J. Appl. Phys.* **37**, 1041 (1966).
7. "The Magnetic Susceptibilities of $LiFeO_2$," J. C. Anderson, S. K. Dey, and V. Halpern, *J. Phys. Chem. Solids* **26**, 1555 (1965).
8. "Magnetic Moment of $ZrZn_2$ up to 150 kG: Is "Pure" $ZrZn_2$ Ferromagnetic?" S. Foner, E. J. McNiff, Jr., and V. Sadagopan, *Phys. Rev. Letters* **19**, 1233 (1967).
9. "Evidence of Weak Ferromagnetism in MnTe from Galvanomagnetic Measurements," J. D. Wasscher, *Solid State Commun.* **3**, 169 (1965).
10. K. Hirakawa, K. Hirakawa, and T. Hashimoto, *J. Phys. Soc. Japan* **15**, 2063 (1960).
11. "Magnetic Anisotropy Measurements of CoO Single Crystal," E. Uchida, N. Fukuoka, H. Kondoh, T. Takeda, Y. Nakazumi, and T. Nagamiya, *J. Phys. Soc. Japan* **19**, 2088 (1964).
12. "Antiferromagnetic Domain Switching in Cr_2O_3," T. J. Martin, *Phys. Letters* **17**, 83 (1965).
13. "Structures Magnetiques de Cr_3X_4 (X = S, Se, Te)," E. F. Bertaut, G. Roult, R. Aleonard, R. Pauthenet, M. Chevreton, and R. Jansen, *J. Phys.* **25**, 582 (1964).
14. "New Magnetic Perovskites $BiMnO_3$ and $BiCrO_3$," F. Sugawara, S. Iida, Y. Syono, and S. Akimoto, *J. Phys. Soc. Japan* **20**, 1529 (1965).
15. "The Magnetic Properties of NiF_2," A. H. Cooke, K. A. Gehring, and R. Lazenby, *Proc. Phys. Soc. (London)* **85**, 967 (1965).
16. "Far-Infrared Magnetic Resonance in NiF_2," P. L. Richards, *Phys. Rev.* **138**, A1769 (1965).
17. "Ferromagnetic Europium Compounds," T. R. McGuire and M. W. Shafer, *J. Appl. Phys.* **35**, 984 (1964).
18. "Magnetic Properties of a Single Crystal of Manganese Phosphide," E. E. Huber, Jr., and D. H. Ridgley, *Phys. Rev.* **135**, A1033 (1964).
19. "Magnetic Structures in the Iron-Germanium System," N. S. Satya Murthy, R. J. Begum, D. S. Somanathan, and M. R. L. N. Murthy, *Solid State Commun.* **3**, 113 (1965).
20. "Spin-Disorder Scattering, Anomalous Behaviour of the Hall Coefficient and Magnon-Drag in Antiferromagnetic MnTe," J. D. Wasscher, A. M. J. H. Seuter, and C. Haas, Article M-22 in *Proc. Cong. Intern. Phys. Semiconductors*, 1964.
21. "Magnetic Ordering and Fluorescence in Manganese Salts," E. W. Prohofsky, W. W. Holoway, Jr., and M. Kestigan, *J. Appl. Phys.* **36**, 1041 (1965).
22. "Resonance Observation of Antiferromagnetic Ordering in $RbMnCl_3$, $CsMnCl_3$, $KMnCl_3$," R. W. Kedzie, J. R. Shane, M. Kestigan, and W. J. Croft, *J. Appl. Phys.* **36**, 1195 (1965).
23. "The Intrinsic Magnetic Properties of Transition Elements and Their Alloys," J. Crangle, p. 53 in *Electronic Structures and Alloy Chemistry of the Transition Elements*, ed. by Paul Adams Beck, Interscience, New York, 1963.

24. "The Magnetic Properties of Heavy Rare Earths Diluted by Yttrium and Lutetium," H. R. Child, W. C. Koehler, E. O. Wollan, and J. W. Cable, *J. Appl. Phys.* **34**, 1335 (1963).
25. "The Magnetic Susceptibility of Cupric Oxide," M. O'Keeffe and F. S. Stone, *J. Phys. Chem. Solids* **23**, 261 (1962).
26. "Observation of Antiferromagnetic Domains in Cobaltous Oxide by Means of the Electron Microscope," G. Remault, P. Delavignette, A. Lagasse, and S. Amelinckx, *J. Appl. Phys.* **35**, 1351 (1964).
27. "Magnetic Structures of $PrAl_2$ and $ErAl_2$," M. Nereson, C. Olsen, and G. Arnold, *J. Appl. Phys.* **39**, 4605–09 (1968).
28. "The Magnetic and Thermal Properties of the Intermetallic Compound MnHg," M. Ohashi, *J. Phys. Soc. Japan* **20**, 911 (1965).
29. "Magnetic Structure of MnHg," A. Oles, *Phys. Stat. Sol.* **8**, K167 (1965).
30. "Magnetic Structure of the Intermetallic Compound MnHg," Y. Nakagawa, H. Watanabe, and T. Hori, *J. Phys. Soc. Japan* **19**, 2078 (1964).
31. "On the Magnetic Susceptibility of $Zn_xMn_{3-x}O_4$ and $Cd_xMn_{3-x}O_4$ Systems," M. Rosenberg and I. Nicholae, *Phys. Stat. Sol.* **5**, K127 (1964).
32. "Magnetic Properties of Au_3Mn Alloys," K. Sato, T. Hirone, H. Watanabe, S. Maeda, and K. Adachi, *J. Phys. Soc. Japan* **17**, Suppl. B-I, 160 (1962).
33. "Resistivity and Thermoelectric Power at Low Temperatures of the 'Tri-Rare Earth Carbides' of Sm, Gd, Er, and Dy," Robert Lallement, CONF-405-27, 4th Rare Earth Research Conference, Phoenix, Ariz., April 1964, 11p. (Also published by Gordon and Breach, New York, Leroy Eyring, editor.)
34. "Ferromagnetic Uranium Monosulfide," W. Trzebiatowski and W. Suski, *Roczniki Chem.* **37**, 117 (1963).
35. "Magnetic Structure of Europium," N. G. Nereson, C. E. Olsen, and G. P. Arnold, LADC-5892. (27) Also in *Phys. Rev.* **135**, A176 (1964).
36. "Direct Cation-Cation Interactions in Several Oxides," J. B. Goodenough, *Phys. Rev.* **117**, 1442 (1960).
37. "Magnetization and Electrical Resistivity of Gadolinium Single Crystals," H. E. Nigh, thesis, Iowa State University, Ames, Iowa, 1963, 144p.
38. "Magnetic Anisotropy in Antiferromagnetic Chromium," R. A. Montalvo and J. A. Marcus, *Bull. Am. Phys. Soc.* **9**, 114A (1964).
39. "The Transition in Chromium and in Some Alloys of Chromium with Small Amounts of Other Transition Elements," G. DeVries, Natuurkundig Lab. der Univ. van Amsterdam, Netherlands, preprint.
40. "Magnetic Characteristics of Tb-Y and Ho-Y Solid Solutions," S. Weinstein, R. S. Craig, and W. E. Wallace, *J. Chem. Phys.* **39**, 1449 (1963).
41. R. M. Bozorth, B. T. Matthias, H. Suhl, E. Corenzwit, and D. D. Davis, *Phys. Rev.* **115**, 1595 (1959).
42. "Electrical Conductivity of Vanadium Oxides," S. Kachi, T. Takada, and K. Kosuge, *J. Phys. Soc. Japan* **18**, 1839 (1963).
43. "Neutron Diffraction Investigation of a Gadolinium Single Crystal," G. Will, R. Nathans, and H. A. Alperin, *J. Appl. Phys.* **35**, 1045 (1964).
44. "Magnetic Properties of ζ Phase in Mn-Ga System," I. Tsuboya and M. Sugihara, *J. Phys. Soc. Japan* **18**, 1096 (1963).
45. "Etudes Magnetiques de Cr_3Se_4 et Cr_3Te_4," Chap. 6, p. 89 in *Etude Structurale des Seleniures et Tellurures de Metaux de Transition, en Particulier de Chrome*, M. Chevreton, thesis, Univ. de Lyon, France, 1964.
46. "Magnetic Characteristics of Dysprosium, Erbium, and Thulium Hydrides," Y. Kubota and W. E. Wallace, *J. Chem. Phys.* **39**, 1285 (1963).
47. "Neutron Inelastic Scattering Measurements of Antiferromagnetic Excitations in MnF_2 at 4.2°K and at Temperatures Up to the Neel Point," A. Okazaki, K. C. Turberfield, and R. W. H. Stevenson, *Phys. Letters* **8**, 9 (1964).
48. "The Ferromagnetism of the Heusler Alloy Au_2MnAl," D. P. Morris and C. D. Price, *Proc. Phys. Soc. (London)* **81**, 1074 (1963).

REFERENCES

49. "Antiferromagnetism of the Antiphase Domain Structure of Pd_3Mn," J. W. Cable, E. O. Wollan, W. C. Koehler, and H. R. Child, *Phys. Rev.* **128**, 2118 (1962).
50. "Magnetic Structures of Chromium-Modified Mn_2Sb," A. E. Austin, E. Adelson, and W. H. Cloud, *Phys. Rev.* **131**, 1511 (1963).
51. "The Effect of Pressure on the Ferromagnetic Phase Transition of EuO," R. Stevenson and M. C. Robinson, *Can. J. Phys.* **43**, 1744 (1965).
52. "The Magnetic Susceptibility of the Gadolinium-Lutetium Alloy System," J. Popplewell, A. M. Harris, and R. S. Tebble, *Proc. Phys. Soc. (London)* **85**, 347 (1965).
53. American Institute of Physics Handbook, 2d ed. (McGraw-Hill Book Co., Inc., N.Y., 1963), p. 5–200.
54. "Effects of Atomic Ordering on the Magnetic Structure of Pt_3Fe," E. Kren, P. Szabo, and T. Tarnoczi, *Solid State Commun.* **4**, 31 (1966).
55. A. S. Borovik-Romanov, *Itogi Nauki: Fiz.-Mat. Nauki, Akad. Nauk SSSR* **4** (1962).
56. P. Handler and C. A. Hutchison, Jr., *J. Chem. Phys.* **25**, 1210-1213 (1956).
57. D. Bacon, *Usp. Fiz. Nauk* **8**, 335 (1963).
58. U. Kittel, *Fiz. Tverd. Tela* (1963).
59. I. Tsubokawa, *J. Phys. Soc. Japan* **14**, 196 (1959).
60. C. M. Arya and G. Grossman, *Fiz. Tverd. Tela* **2**, 1283 (1960).
61. G. Goodman, *Phys. Rev. Letters* **9**, 305 (1962).
62. A. Paoletti and S. J. Pickart, *J. Chem. Phys.* **32**, 308 (1960).
63. M. Mori and T. Mitsui, *J. Phys. Soc. Japan* **22**, 931 (1967).
64. "Ultrasonic Attenuation in MnF_2," R. G. Evans, *Phys. Letters* **27A**, 451-52 (1968).
65. A. I. Snow, *Phys. Rev.* **85**, 365 (1952).
66. J. W. Cable et al., *Phys. Rev.* **118**, 950 (1960).
67. R. M. Bozorth and V. Kramer, *J. Phys. Radium* **20**, 393 (1959).
68. S. S. Shalit, *JETP* **8**, 518 (1939).
69. H. Bizette and C. Terrier, *J. Phys. Radium* **23**, 486 (1962).
70. S. Foner, *Phys. Rev.* **130**, 183 (1963).
71. K. Dwight et al., *J. Appl. Phys.* **33**, 1341 (1962).
72. W. C. Koehler and E. O. Wollan, *J. Phys. Chem. Solids* **2**, 100 (1957).
73. G. H. Jonker, *Physica* **22**, 707 (1956).
74. P. L. Richards, *J. Appl. Phys.* **34**, 1237 (1963).
75. R. De Blois and D. Rodbell, *J. Appl. Phys.* **34**, 1099 (1963).
76. J. S. Kasper and J. S. Kouvel, *J. Phys. Chem. Solids* **11**, 231 (1959).
77. R. Street and J. Smith, *J. Phys. Radium* **20**, 82 (1962).
78. J. S. Kouvel et al., *J. Appl. Phys.* **34**, 1095 (1963).
79. "The Measurement of the Anisotropy Constant of Antiferromagnetic $FePO_4$ by Means of the Mossbauer Effect," V. Beckmann, W. Bruckner, W. Fuchs, G. Ritter, and H. Wegener, *Phys. Stat. Sol.* **29**, 781 (1968).
80. Y. Nakagawa and T. Hori, *J. Phys. Soc. Japan* **16**, 1470 (1961).
81. H. Bizette, *J. Phys.* **12**, 161 (1951).
82. J. Hastings et al., *Phys. Rev.* **115**, 13 (1959).
83. L. Corliss and J. Hastings, *J. Appl. Phys.* **34**, 1192 (1963).
84. J. Kouvel and C. Hartelius, *Phys. Rev.* **123**, 124 (1961).
85. D. Douglass and M. Strandberg, *Physica* **27**, 1 (1961).
86. T. Nagamiya, *J. Phys. Radium* **20**, 70 (1959).
87. W. C. Koehler, M. K. Wilkinson, J. W. Cable, and E. D. Wollan, *J. Phys. Radium* **20**, 180 (1959).
88. P. G. Asch, *J. Phys. Radium* **20**, 349 (1959).
89. R. Bozorth and J. Nielsen, *Phys. Rev.* **110**, 879 (1959).
90. J. Burger et al., *J. Phys. Radium* **20**, 427 (1959).
91. "Mossbauer Resonance of ^{57}Fe in Oxidic Spinels Containing Cu and Fe," B. J. Evans and S. S. Hafner, *J. Phys. Chem. Solids* **29**, 1573-88 (1968).
92. H. Fink and D. Shaltiel, *Bull. Am. Phys. Soc., SII*, **8**, 214 (1963).

REFERENCES

93. G. Shirane et al., *J. Phys. Soc. Japan* **14**, 1352 (1959).
94. Y. Ishikawa and S. Akimoto, *J. Phys. Soc. Japan* **13**, 1298 (1958).
95. R. J. Joenk, *J. Appl. Phys.* **34**, 1097 (1963).
96. N. M. Kreines, Thesis, IFP, Moscow (1963).
97. "Etude des Uranates de Cobalt et de Manganese," E. F. Bertaut, A. Delapalme, F. Forrat, and R. Pauthenet, *J. Phys. Radium* **23**, 477 (1960).
98. P. H. Vossoc, *J. Chem. Phys.* **32**, 1590 (1960).
99. A. Okazaki and Y. Suemune, *J. Phys. Soc. Japan* **16**, 671 (1961).
100. "Magnetic Susceptibility of MnO in the Neighborhood of the Neel Temperature," R. Lindsay and F. H. Michelsohn, *Bull. Am. Phys. Soc.* **11**, 108 (1966).
101. D. Teaney et al., *Phys. Rev. Letters* **9**, 212 (1962).
102. T. R. McGuire, *Bull. Am. Phys. Soc.*, SII, **8**, 55 (1963).
103. S. Chiba, *J. Phys. Soc. Japan* **15**, 581 (1960).
104. H. Kondo and S. Miyahara, *J. Phys. Soc. Japan* **18**, 305 (1963).
105. E. Bertaut et al., *J. Phys. Chem. Solids* **21**, 234 (1961).
106. M. Wilkinson et al., *J. Phys. Chem. Solids* **2**, 289 (1957).
107. "Evidence of Itinerant Electron Ferromagnetism in Sigma Phase Alloys," D. A. Read, E. H. Thomas, and J. B. Forsythe, *J. Phys. Chem. Solids* **29**, 1569-72 (1968).
108. M. A. Lasheen et al., *Physica* **24**, 1061 (1958).
109. J. Schelleng and S. Friedberg, *J. Appl. Phys.* **34**, 1087 (1963).
110. S. Friedberg and R. Flippen, *Proc. Intern. Conf. Low Temperature Physics, 7th*, Toronto, 1960, p. 122 (1961).
111. N. Uryu, *J. Phys. Soc. Japan* **16**, 2140 (1961).
112. I. Maxim, *Studii Cercetari Fiz.* **9**, 323 (1959).
113. J. S. Kouvel, *J. Appl. Phys.* **37**, 1257 (1966).
114. R. Ohlman and M. Tinkham, *Phys. Rev.* **123**, 425 (1961).
115. H. Bizette et al., *Compt. Rend.* **245**, 507 (1957).
116. Kanematsu et al., *J. Phys. Soc. Japan* **15**, 2358 (1960).
117. T. Rosenquist, *Acta Met.* **1**, 761 (1953).
118. H. Bizette, *J. Phys. Radium* **12**, 13 (1951).
119. Y. Ishikawa and S. Akimoto, *J. Phys. Soc. Japan* **12**, 1083 (1957).
120. H. Bizette and B. Tsai, *Compt. Rend.* **242**, 2124 (1956).
121. C. Berry and C. Combs, *J. Appl. Phys.* **31**, 1130 (1960).
122. T. Teranishi, *J. Phys. Soc. Japan* **17**, 263S, B-1 (1962).
123. G. Donnay et al., *Phys. Rev.* **112**, 1917 (1958).
124. D. Cox et al., *Phys. Rev. Letters* **11**, A6 (1963).
125. D. Treves, *Phys. Rev.* **125**, 1843 (1962).
126. W. Koehler et al., *Phys. Rev.* **118**, 58 (1960).
127. R. Bozorth, H. J. Williams, and D. E. Walsh, *Phys. Rev.* **103**, 572 (1956).
128. "Measurement of the Resistive Anomaly in FePd Alloys Near the Curie Point," G. Longworth and C. C. Tsuei, *Phys. Letters* **27A**, 258-59 (1968).
129. T. W. J. Van Agt et al., *Proc. Intern. Congr. Low Temperature Physics, 8th* (Preprint) 381 (1962).
130. N. Poulis and W. v. d. Lugt, *J. Phys. Soc. Japan* **17**, S.B-1, 505 (1962).
131. A. Meyer and M. Cadeville, *J. Phys. Soc. Japan* **17**, S.B-1, 223 (1962).
132. V. I. Okogni, *JETP* **45**, 1687 (1963).
133. H. Bizette et al., *J. Phys. Radium* **20**, 421 (1959).
134. H. Bizette, *Compt. Rend.* **243**, 1295 (1956).
135. H. Bizette et al., *Compt. Rend.* **246**, 251 (1958).
136. W. L. Roth, *Bull. Am. Phys. Soc.* **8**(II), 213 (1963); "The Magnetic Structure of Co_3O_4," W. L. Roth, *J. Phys. Chem. Solids* **25**, 1–10 (1964).
137. T. Swoboda, R. C. Toole, and J. D. Vaughn, *J. Phys. Chem. Solids* **5**, 293 (1958).
138. G. Blasse and D. Schipper, *Phys. Letters* **5**, 300 (1963).
139. T. Haseda et al., *Proc. Intern. Congr. Low Temperature Physics, 8th* (Preprint), 48 (1962).

REFERENCES

140. W. K. Robinson and S. A. Friedberg, *Phys. Rev.* **117**, 402 (1960).
141. T. Haseda and E. Kanda, *J. Phys. Soc. Japan* **12**, 1051 (1957).
142. H. Forstat et al., *Bull. Am. Phys. Soc.*, SII, **5**, 59 (1960).
143. A. Ohtsubo and E. Kanda, *J. Phys. Soc. Japan* **17**, S.B-I, 497 (1962).
144. M. Trombe, *J. Phys.* **12**, 170 (1951).
145. Neutron Diffraction Study of NiS, J. T. Sparks and T. Komoto, *J. Appl. Phys.* **34**, 1191 (1963).
146. G. Heller et al., *J. Appl. Phys.* **34**, 1033 (1963).
147. S. Ogawa, *J. Phys. Soc. Japan* **15**, 2361 (1960).
148. R. Plumer and E. Legrand, *J. Phys. Radium* **23**, 474 (1962).
149. R. D. Spence et al., *J. Chem. Phys.* **31**, 555 (1959).
150. L. Berger and S. Friedberg, *Phys. Rev.* **136**, 158 (1964).
151. H. Kobayashi and T. Haseda, *J. Phys. Soc. Japan* **18**, 541 (1963).
152. E. Frikee and J. Van den Handel, *Physica* **28**, 269 (1962).
153. E. P. Wohlfarth, *Acta Met.* **4**, 225 (1956).
154. S. F. Adler and P. W. Selwood, *J. Am. Chem. Soc.* **76**, 346 (1954).
155. B. I. Alshin and D. N. Astrov, *JETP* **44**, 1195 (1963).
156. M. K. Wilkinson et al., *Phys. Rev.* **121**, 74 (1961).
157. *Proc. Intern. Conf. Low Temperature Physics*, 7th, Toronto 1960 (1961).
158. *Philips Res. Rept.* **17**, 451 (1962).
159. G. Will, *Bull. Am. Phys. Soc.*, SII, **8**, 213 (1963).
160. J. Owen, *J. Phys. Radium* **20**, 138 (1959).
161. D. G. Henshaw and B. N. Brockhouse, *Bull. Am. Phys. Soc.*, SII, **2**, 9 (1957).
162. "Low Temperature Magnetic Susceptibility of $CuFeCl_4$ and $FeCl_3$," Oliver B. Morton and Edwin R. Jones, Jr. (University of South Carolina), EA9, Am. Phys. Soc., Southeastern Section, 35th Mtg., Oct. 9–11, 1968, University of Georgia.
163. R. M. Bozorth, *Phys. Rev. Letters* **1**, 362 (1958).
164. E. A. Tyrov, *Physics of Magnetic Crystals* (1963).
165. E. G. Fakidov and V. N. Tsiovkin, *Fiz. Metal. i Metalloved.* **7**, 685 (1957); *Phys. Metals Metallog. (USSR) (English Transl.)* **7**, 5, 47 (1959).
166. K. Lee, A. M. Portis, and G. L. Witt, *Phys. Rev.* **132**, 144 (1963).
167. E. F. Bertaut et al., *J. Appl. Phys.* **35**, 952 (1964).
168. I. S. Jacobs and P. E. Lawrence, *J. Appl. Phys.* **35**, 996 (1964).
169. S. A. Friedberg and J. T. Schriempf, *J. Appl. Phys.* **35**, 1000 (1964).
170. R. C. Sherwood, H. J. Williams, and J. H. Wernick, *J. Appl. Phys.* **35**, 1049 (1964).
171. E. A. Nesbitt, H. J. Williams, J. H. Wernick, and R. C. Sherwood, *J. Appl. Phys.* **34**, 1347 (1963).
172. J. W. Cable et al., *Phys. Rev. Letters* **12**, 553 (1964).
173. G. Busch et al., *Phys. Letters* **9**, 7 (1964).
174. "Some Neutron Diffraction Investigation at the Nuclear Center of Grenoble," F. Bertaut et al., *J. Appl. Phys.* **37**, 1038 (1966).
175. L. Berger et al., *Phys. Rev.* **132**, 1057 (1963).
176. W. P. Wolf and A. F. G. Wyatt, *Phys. Rev. Letters* **13**, 368 (1964).
177. R. J. Joenk and R. M. Bozorth, *J. Appl. Phys.* **36**, 1167 (1965).
178. S. A. Friedberg et al., *J. Appl. Phys.* **36**(1965). p. 936.
179. K. Lee and H. Muir, *J. Appl. Phys.* **36** Suppl. (1965).
180. K. Lee, A.M. Portis, and G. L. Witt, *Phys. Rev.* **132**, 144 (1963).
181. K. Adachi, *J. Phys. Soc. Japan* **16**, 2187 (1961).
182. W. R. Eisenberg and H. Forstat, *J. Phys. Soc. Japan* **19**, 406 (1964).
183. M. Tachiki, *J. Phys. Soc. Japan* **19**, 454 (1964).
184. "Magnetic Susceptibilities of Chromium Selenides," Kanichi Masumoto and Tahahiko Kamigaichi, *J. Sci. Hiroshima Univ.*, Ser. A-II, **28**, 47 (1965).
185. G. S. Heller, *Proc. Intern. Conf. Magnetism*, Nottingham (1964).
186. H. Abe and M. Matsuura, *J. Phys. Soc. Japan* **19**, 1867 (1964).

187. A. Narath, *J. Phys. Soc. Japan* **19**, 2244 (1964).
188. G. Gorodetsky and D. Treves, *Phys. Rev.* **135A**, 97 (1964).
189. R. C. Meisenheimer and D. L. Cook, *J. Chem. Phys.* **30**, 605 (1959).
190. W. Rudorff, J. Kandler, and D. Babel, *Z. Anorg. Allgem. Chem.* **317**, 261 (1962).
191. "Specific Heat of Antiferromagnetic $Rb_2NiCl_4 \cdot 2H_2O$," H. Forstat, J. N. McElearney, P. T. Bailey, and J. R. Ricks, *Bull. Am. Phys. Soc.* **14**, 79 (1969).
192. B. C. Frazen and P. J. Brown, *Phys. Rev.* **125**, 1283 (1962).
193. J. S. Burgiel et al., *Phys. Rev.* **122**, 429 (1961).
194. "On the Antiferromagnetism of α-FeOOH," I. Dezsi and M. Fodor, *Phys. Stat. Sol.* **15**, 247 (1966).
195. "Mossbauer Effect in α-FeOOH," F. van der Woude and A. J. Dekker, *Phys. Stat. Sol.* **13**, 181 (1966).
196. "Sur Deux Groupes de Nouveaux Composés Ferrimagnetiques," A. Lecerf, M. Rault, J. Portier, and G. Villers, *Bull. Soc. Chim. France*, 1208 (1965).
197. "Ferrimagnetische Seltene Erdmetall-Mangan-Verbindungen I," H. Rudolf Kirchmayr and K.-H. Schindl, *Z. Angew. Phys.* **19**, 517 (1965).
198. "Magnetic Structures of Heavy Rare-Earth Metals," K. Yoshida in *Progress in Low Temperature Physics*, vol IV, North-Holland, Amsterdam, 1964, pp. 274–75.
199. "Table of Antiferromagnetics," *Inst. Atom. Energ., Gosudarstvennyi Komitet po Ispol'zovaniyu Atomnoi Energii SSSR, Moscow*, IAE-942 (1965). 26p
200. "Antiferromagnetische Umwandlung von Dy_2O_3, Er_2O_3 und Yb_2O_3 im Temperaturbereich von 1.1 bis 4.2°K," H. Bonrath, K. H. Hellwege, K. Nicolay, and G. Weber, *Phys. Kondens. Materie* **4**, 382 (1966).
201. "Magnetic Properties of Gadolinium-Aluminium Intermetallic Compounds," B. Stalinski and S. Pokrzywnicki, *Phys. Stat. Sol.* **14**, K157 (1966).
202. "Rare Earth Compounds with the Th_3P_4-Type Structure," F. Holtzberg and S. Methfessel, *IBM Res. Note*, NC 551 (1965). [Also in *J. Appl. Phys.* **37**, 1433 (1966)].
203. "Magnetic and Structural Properties of Europium Metal and Europium Monoxide at High Pressure," D. B. McWhan, P. C. Souers, and G. Jura, *Phys. Rev.* **143**, 385 (1966).
204. "Magnetic Properties of Rare Earth-Manganese Compounds," H. R. Kirchmayr, *IEEE Trans.-Magnetics* **2**, 493 (1966).
205. "Magnetic Studies of the Antiferromagnet $RbMnF_3$," W. J. Ince, Lincoln Lab., MIT, Lexington, Mass., Tech. Rept. 384, Oct. 14, 1965.
206. "Crystal Structure and Magnetic Properties of $CoTiO_3$," R. E. Newnham, J. H. Fang, and R. P. Santoro, *Acta Cryst.* **17**, 240 (1964).
207. "Neutron-Diffraction Study of Antiferromagnetic $FeTiO_3$ and Its Solid Solutions with α-Fe_2O_3," G. Shirane, S. J. Pickart, R. Nathans, and Y. Ishikawa, *J. Phys. Chem. Solids* **10**, 35 (1959).
208. "Ferrimagnetic Fluoride-$Na_5Fe_3F_{14}$," K. Knox and S. Geller, *Phys. Rev.* **110**, 771 (1958).
209. "Ferromagnetism and Ferrimagnetism of Oxygen Spinels Containing Tetravalent Manganese," G. Blasse, *J. Phys. Chem. Solids* **27**, 383 (1966).
210. "The Mossbauer Study of FeGe," S. Tomiyoshi, H. Yamamoto, and H. Watanabe, *J. Phys. Soc. Japan* **21**, 709 (1966).
211. "Weak Ferromagnetism of $CaMnO_3$," V. M. Yudin, A. I. Gavrilishina, M. V. Artem'eva, and M. F. Bryzhina, *Fiz. Tverd. Tela* **7**, 2292 (1965); *Soviet Phys.-Solid State* **7**, No. 8, 1856-60 (1966).
212. "Hydrostatic Pressure Effect on the Magnetic Moment and the Curie Temperature of a Cu-Ni Alloy," H. Tange, *J. Sci. Hiroshima Univ.*, A-2 **29**, 17 (1965).
213. "Single-Crystal Growth and Properties of the Perovskites $LaVO_3$ and YVO_3," D. B. Rogers, A. Ferretti, D. H. Ridgeley, R. J. Arnott, and J. B. Goodenough, *J. Appl. Phys.* **37**, 1431 (1966).
214. "Structure of δ-FeOOH," Shoichi Okamoto, *J. Am. Ceram. Soc.* **51**, 594-99 (1968).

REFERENCES

215. "Neutron Diffraction Studies in Hungary," E. Kren, L. Pal, P. Szabo, and B. Szigeti, Hungarian Acad. Sci., Central Res. Inst. for Phys., Budapest, KFKI, January 1966.
216. "Antiferromagnetism in Potassium Superoxide KO_2," H. G. Smith, R. M. Nicklow, L. J. Raubenheimer, and M. K. Wilkinson, *J. Appl. Phys.* **37**, 1047 (1966).
217. "Neutron Scattering—Neutron-Diffraction Studies of Magnetic Structures in Ho-Er Alloys," L. M. Corliss, G. Shirane, and S. J. Pickart, *J. Appl. Phys.* **37**, 1032 (1966).
218. "Mossbauer Studies on Iron in the Perovskites $La_{1-x}Sr_xFeO_3 (0 \leq x \leq 1)$," U. Shimony and J. M. Knudsen, *Phys. Rev.* **144**, 361 (1966).
219. "Neutron-Diffraction Investigation of Chromium with Small Additions of Manganese and Vanadium," Y. Hamaguchi, E. O. Wollan, and W. C. Koehler, *Phys. Rev.* **138**, A737 (1965).
220. "Critical Point of the Cubic Antiferromagnet $RbMnF_3$," D. T. Teaney, V. L. Moruzzi, and B. E. Argyle, *J. Appl. Phys.* **37**, 1122 (1966).
221. "Magnetic Structures and Magnetic Transformations in Ordered Mn_3(Rh, Pt) Alloys," E. Kren, G. Kadar, L. Pal, J. Solyom, and P. Szabo, *Phys. Letters* **20**, 331 (1966).
222. "Magnetic Properties of Single Crystal Mn_2As," Y. Ishizawa and E. Hirahara, *J. Phys. Soc. Japan* **21**, 189 (1966).
223. Die Magnetische Struktur der in dem Perovskittyp Kristallisierenden Verbindung: $TbAlO_3$, J. Bielen, J. Mareschal, and J. Sivardiere, *Z. angew. Phys.* **23**, 243 (1967).
224. "Magnetic Properties of $RbFeF_3$," F. F. Y. Wang and M. Kestigian, *J. Appl. Phys.* **37**, 975 (1966).
225. "On the Nature of the Dielectric and Magnetic Properties of $BiFeO_3$," Yu. E. Roginskaya, Yu. Ya. Tomashpol'skii, Yu. N. Venevtsev, V. M. Petrov, and G. S. Zhdanov, *Zh. Eksperim. i Teor. Fiz. SSSR* **50**, 69 (1966).
226. "Metamagnetism of $NaNiO_2$," P. F. Bongers and U. Enz, *Solid State Commun.* **4**, 153 (1966).
227. "Antiferromagnetism II—First- and Higher-Order Magnetic Phase Transitions in Dysprosium Aluminum Garnet," B. E. Keen, D. Landau, B. Schneider, and W. P. Wolf, *J. Appl. Phys.* **37**, 1120 (1966).
228. "Growth of Crystals and Magnetic Properties of β-Mn_2O_3 Solid Solutions," E. Banks and E. Kostiner, *J. Appl. Phys.* **37**, 1423 (1966).
229. "Crystal Structure and Magnetic Property of FeMnAs," S. Yoshii and H. Katsuraki, *J. Phys. Soc. Japan* **21**, 205 (1966).
230. "The Influence of Fe^{3+} Ions at Tetrahedral Sites on the Magnetic Properties of $ZnFe_2O_4$," F. K. Lotgering, *J. Phys. Chem. Solids* **27**, 139 (1966).
231. "Relationship Between Magnetic Curie Points and Cell Sizes of Solid Solutions with the Ordered Perovskite Structure," F. S. Galasso, F. C. Douglas, and R. J. Kasper, *J. Chem. Phys.* **44**, 1672 (1966).
232. "The Magnetic Structures of Rare Earth Manganites $RMnO_3$," W. C. Koehler, H. L. Yakel, E. O. Wollan, and J. W. Cable, *Proc. Rare Earth Research Conf., Phoenix, 4th*, April 1964, CONF-405-34. [Also in *Phys. Letters (Neth.)* **9**, 93 (1964)].
233. "Antiferromagnetism of CrGe," K. Yasukochi, K. Yamagiwa, Y. Kuwasawa, and K. Sekizawa, *J. Phys. Soc. Japan* **21**, 557 (1966).
234. "Antiferromagnetism of $CoRh_2S_4$," G. Blasse, *Phys. Letters* **19**, 110 (1965).
235. "Tabulation of First Order Magnetic Phase Transitions," C. S. Naiman and R. Gilmore, Mitras, Inc., Cambridge, Mass., MC-62-49-R3, AFCRL-64-464, AD-604424, October 1963.
236. "Electrical Properties of Cobalt Monoxide," B. Fisher and D. S. Tannhauser, *J. Chem. Phys.* **44**, 1663 (1966).
237. "Etude par Mesures Magnetiques et Neutrocristallographiques de l'Antiferromagnetique $CaMn_2O_4$," Y. Allain and B. Boucher, *J. Phys.* **26**, 789 (1965).
238. M. Eibschutz, G. Gorodetsky, S. Shtrikman, and D. Treves, *J. Appl. Phys.* **35**, 1071 (1964).
239. "Variation Avec la Pression des Points de Curie de Quelques Ferrites d'Yttrium et de Terres Rares à Structure Grenat," D. Bloch, F. Chaisse, and R. Pauthenet, *Compt. Rend.* **262**, 404 (1966).

240. "Ferromagnetic MnB_2," L. Andersson, B. Dellby, and H. P. Myers, *Solid State Commun.* **4**, 77 (1966).
241. T. Suzuoka, *J. Phys. Soc. Japan* **12**, 1344 (1957).
242. "Determination of the Curie Temperature and the Thermoremanent Magnetization of Very Small Amounts of Ferromagnetic Material," K. J. Keller and F. Schurink, *Appl. Sci. Res. Sec. B*, **12**, 218 (1965).
243. "Magnetic Properties of the Terbium Oxides at Temperatures Between 1.4 and 300°K," J. B. MacChesney, H. J. Williams, R. C. Sherwood, and J. F. Potter, *J. Chem. Phys.* **44**, 596 (1966).
244. "On the Neutron Diffraction Study of Cr_2As," H. Watanabe, Y. Nakagawa, and K. Sato, *J. Phys. Soc. Japan* **20**, 2244 (1965).
245. "Initial Susceptibility Investigation of Magnetic Transitions in Several Rare Earth Metals: Thermal Hysteresis in Ferromagnetic Transitions," F. J. Jelinek, E. D. Hill, and B. C. Gerstein, *J. Phys. Chem. Solids* **26**, 1475 (1965).
246. "Propriétés Magnetiques du Thulium à l'Etat de Sesquioxyde et de Metal Entre 0 et 110 kOe et de 2 à 1560°K," N. Perakis and F. Kern, *Phys. Kondens. Materie* **4**, 247 (1965).
247. "Ferromagnetic Resonance in $ZrZn_2$," S. Ogawa, *J. Phys. Soc. Japan* **20**, 2296 (1965).
248. "Neutron-Diffraction Study of Antiferromagnetism in UO_2," B. C. Frazer, G. Shirane, D. E. Cox, and C. E. Olson, *Phys. Rev.* **140**, A1448 (1965).
249. "Microwave Absorption and Magnetic Transition in $GdCl_3$," E. Becker, H. Bonrath, K. H. Hellwege, F. Kuch, M. Schinkmann, and B. Unger, *Phys. Letters* **19**, 86 (1965).
250. "Magnetic Properties of Lutetium," V. I. Chechernikov and I. Pop, *Soviet Phys. JETP* **19**, 831 (1964).
251. "Ferromagnetic Curie Temperatures of Iron Solid Solutions with Germanium, Silicon, Molybdenum, and Manganese," S. Arajs, *Phys. Stat. Sol.* **11**, 121 (1965).
252. "Neutron-Diffraction Study of $Co_3B_2O_6$," R. E. Newnham, M. J. Redman, and R. P. Santoro, *Z. Krist.* **121**, 6 (1965).
253. "Neutron Diffraction Study of the Antiferromagnetism of Uranium Monoarsenide," J. M. Williams, L. Heaton, and F. Campos, *J. Phys. Chem. Solids* **29**, 1702-03 (1968).
254. "Preparation, Propriétés Cristallines et Magnetiques des Spinelles de Composition: $Mn^{2+}_{(1+x)}V^{3+}_{2(1-x)}Ti^{4+}_x O_4$, pour des Valeurs de x Comprises Entre 0 et 1," G. Villers, A. Lecerf, and M. Rault, *Compt. Rend.* **260**, 3017 (1965).
255. "Magnetic Properties of $FeCl_2 \cdot FeCl_2 \cdot 2H_2O$," A. Narath, *Phys. Rev.* **139**, A1221 (1965).
256. "Propriétés Magnetiques des Alliages $Fe_{2-x}Ge$ et $FeGe_2$," G. Airoldi and R. Pauthenet, *Compt. Rend.* **258**, 3994 (1964).
257. "The Magnetization of Ferromagnetic Cubic Laves Phase Compounds of Rare Earths with Transition Elements," J. Crangle and J. W. Ross, *Proc. Intern. Conf. Magnetism, London*, 1965. pp. 240-3.
258. "Ferromagnetism and a New Type of Ferrimagnetism in Oxygen Spinels," G. Blasse, *Solid State Commun.* **3**, 67 (1965).
259. "Magnetic Structure of $CuCl_2 \cdot 2H_2O$," G. Shirane and B. C. Frazer, *Phys. Letters* **17**, 95 (1965).
260. H. Kobayashi and T. Haseda, *J. Phys. Soc. Japan* **19**, 765 (1964).
261. "First-Order Magnetic Phase Change in Chromium at 38.5°C," A. Arrott and S. A. Werner, *Phys. Rev. Letters* **14**, 1022 (1965).
262. E. E. Anderson, *Phys. Rev.* **134**, A1581 (1964).
263. M. Yuzuri and M. Yamada, *J. Phys. Soc. Japan* **15**, 1845 (1960).
264. Magnetic Structure Properties of Gd-Y and Gd-Sc Alloys, H. R. Child and J. W. Cable, *J. Appl. Phys.* **40**, 1003 (1969).
265. "Antiferromagnetism in Nickel Orthosilicate," R. Newnham, R. Santoro, J. Fang, and S. Nomura, *Acta Cryst.* **19**, 147 (1965).
266. "Specific Heats of Lanthanum Nitride and Neodymium Nitride," J. J. Veyssie, J. Chaussy, and A. Berton, *Phys. Letters* **13**, 29 (1964).

REFERENCES

267. "Investigation of the Magnetic Structure of $ErMn_2$, $TmMn_2$, $TbNi_2$ by Neutron Diffraction," G. P. Felcher, L. M. Corliss, and J. M. Hastings, *J. Appl. Phys.* **36**, 1001 (1965).
268. "X-Ray Diffractometric Study of the Phase Transitions in the System $BiFeO_3$-$Pb(Fe_{0.5}Nb_{0.5})O_3$," I. G. Ismailzade, *Soviet Phys. Cryst.* **13**, No. 3, 350-53 (1968).
269. "Magnetic Properties of Alloys of Gadolinium with Iron, Cerium, and Yttrium," V. I. Checherrnikov, I. Pop, and I. V. Burov, JPRS-28849, pp. 64–70, trans. from Voprosy Teorii i Primeneniya Redkozemel'nykh Metallov, pp. 55–9.
270. "Graphical Correlation of the Néel Temperatures of Chlorides, Bromides, and Iodides of Divalent 3d Transition Metal Ions," L. G. Van Uitert, H. J. Williams, R. C. Sherwood, and J. J. Rubin, *J. Appl. Phys.* **36**, 1029 (1965).
271. "Low Temperature Susceptibility of $CaF_2 2H_2O$ Single Crystal," S. Tazawa, K. Nagata, and M. Date, *J. Phys. Soc. Japan* **20**, 181 (1965).
272. "Magnetic and Other Studies of a Reversible Phase Change in $CoSiF_6 \cdot 6H_2O$," M. Majumdar and S. K. Datta, *J. Chem. Phys.* **42**, 418 (1965).
273. "The Antiferromagnetic Phase Transition in Mixtures of $CuCl_2$ and $CuBr_2$," C. F. Wong and R. Stevenson, in *Absorption Spectra of Solids Exhibiting Antiferromagnetic Phase Transitions*, August 1964, pp. 29–31.
274. "Propriétés Magnetiques des Composés Intermetalliques Terre Rare-Argent, RAg," J. Pierre and R. Pauthenet, *Compt. Rend.* **260**, 2739 (1965).
275. "Nuclear Magnetic Relaxation in Mn-Cr Alloys near the Critical Temperature," Y. Masuda and T. Taki, *J. Phys. Soc. Japan* **20**, 175 (1965).
276. "Proprietes Magnetiques du Nitrure de Gadolinium, NGd," J.-P. Rebouillat and J.-J. Veyssie, *Compt. Rend.* **259**, 4239 (1964).
277. "The Anomalous Thermal Conduction in $KFeF_3$ Single Crystal at Low Temperatures," Y. Suemune, *J. Phys. Soc. Japan* **19**, 2234 (1964).
278. "Properties of Single-Crystal Lithium Ferrite Grown in the Ordered State," J. P. Remeika and R. L. Comstock, *J. Appl. Phys.* **35**, 3320 (1964).
279. "Magnetic Anisotropy Measurements of NiO Single Crystal," H. Kondoh, E. Uchida, Y. Nakazumi, and T. Nagamiya, *J. Phys. Soc. Japan* **13**, 579 (1958).
280. T. Watanabe, *J. Phys. Soc. Japan* **16**, 1131 (1961).
281. "Crystal Chemistry and Some Magnetic Properties of Mixed Metal Oxides with Spinel Structure," G. Blasse, *Philips Res. Rept., Suppl. III* (1964).
282. "Substitution of Divalent Cobalt in Yttrium Iron Garnet," S. Geller, H. J. Williams, G. P. Espinosa, and R. C. Sherwood, *Phys. Rev.* **136**, A1650 (1964).
283. "Antiferromagnetic Spin Ordering Below 1°K," W. T. Duffy, Jr., J. Lubbers, H. Van Kempen, T. Haseda, and A. R. Miedema, *Proc. Intern. Conf. Low Temperature Physics, London, 8th,* 1962 (Butterworth, London, 1964), p. 245.
284. "Magnetic Order in Rare-Earth Intermetallic Compounds," J. W. Cable, W. C. Koehler, and E. O. Wollan, *Phys. Rev.* **136**, A240 (1964).
285. "Antiferromagnetism in $CoCl_2 \cdot 2H_2O$. I. Magnetic Structure," A. Narath, *Phys. Rev.* **136**, A766 (1964).
286. "Magnetic Characteristics of $CeFe_2$," J. Farrell and W. E. Wallace, *J. Chem. Phys.* **41**, 1524 (1964).
287. "Magnetic Susceptibility of $FeCl_2 \cdot 4H_2O$ from 0.35 to 4.2°K," J. T. Schriempf and S. A. Freidberg, *Phys. Rev.* **136**, A518 (1964).
288. "On the Curie Temperature of α-Fe_2O_3," P. Gilad, M. Greenshpan, P. Hillman, and H. Shechter, TN-3, AD-602466, Contract AF 61(052)-621, Nov. 20, 1963; also *Phys. Lett.* **7**, 239 (1963).
289. "The Crystal Structure and Magnetic Properties of the Rare-Earth Nickel (RNi) Compounds," S. C. Abrahams, J. L. Bernstein, R. C. Sherwood, J. H. Wernick, and H. J. Williams, *J. Phys. Chem. Solids* **25**, 1069 (1964).
290. "Antiferromagnetism in Cobalt Orthosilicate," S. Nomura, R. Santoro, J. Fang, and R. Newnham, *J. Phys. Chem. Solids* **25**, 901 (1964).

291. "The Electrical Resistivity of Dilute Solutions of Transition-Metals in Chromium," M. A. Taylor, *J. Less-Common Metals* **4**, 476 (1962).
292. "Magnetic Properties of a Number of Divalent Transition Metal Tungstates, Molybdates and Titanates," L. G. Van Uitert, R. C. Sherwood, H. J. Williams, J. J. Rubin, and W. A. Bonner, *J. Phys. Chem. Solids* **25**, 1447 (1964).
293. "Magnetic and Thermal Properties of Nickel and Copper Fluosilicates Below 1°K," A. Ohtsubo, *J. Phys. Soc. Japan* **20**, 76 (1965).
294. M. K. Wilkinson, W. C. Koehler, E. O. Wollan, and J. W. Cable, *J. Appl. Phys.* **32**, 48S (1961).
295. "Recent Progress in Magnetic Structure Determinations of Rare Earth Metals," W. C. Koehler, J. W. Cable, E. O. Wollan, and M. K. Wilkinson, *J. Phys. Soc. Japan, S-III*, **17**, 32 (1962).
296. J. W. Cable, E. O. Wollan, W. C. Koehler, and M. K. Wilkinson, *J. Appl. Phys.* **32**, 49S (1961).
297. "Magnetic Structures of Thulium," W. C. Koehler, J. W. Cable, E. O. Wollan, and M. K. Wilkinson, *Phys. Rev.* **126**, 1672 (1962).
298. "The Magnetic Susceptibility of $GdPO_4$, $TbPO_4$, and $DyPO_4$," F. F. Y. Wang and R. S. Feigelson, in *Proc. Rare Earth Research Conf., 4th, Phoenix, Arizona,* April 1964, Leroy Eyring, ed. (Gordon and Breach, New York) p. 77.
299. "Neutron Diffraction Study on Chromium Alloy with Small Amounts of Vanadium," S. Komura and N. Kunitomi, *J. Phys. Soc. Japan* **20**, 103 (1964).
300. "Magnetic Susceptibility of Face-Centered Cubic Cobalt Just above the Ferromagnetic Curie Temperature," R. V. Colvin and S. Arajs, *J. Phys. Chem. Solids* **26**, 435 (1965).
301. "On the Magnetic and Chemical Properties of Europium Fluoride," K. Lee, H. Muir, and E. Catalano, *J. Phys. Chem. Solids* **26**, 523 (1965).
302. "The Paramagnetic Properties of the Monoborides of V, Cr, Mn, Fe, Co and Ni," N. Lundquist, H. P. Myers, and R. Westin, *Phil. Mag.* **7**, 1197 (1962).
303. "Study of the Properties of the Antiferromagnetic Compound CrSb," P. P. Kuz'menko, P. A. Suprunenko, and G. A. Galina, pp. 63–67 in *Issledovanie elektronnykh svojstr metallov i splavov* (Naukova Dumka, Kiev, 1967).
304. E. O. Wollan and W. C. Koehler, *Phys. Rev.* **100**, 545 (1955).
305. "A Study of Neodymium Substituted Yttrium Iron Garnet," T. H. Ramsey, Jr., H. Steinfink, and E. J. Weiss, *J. Phys. Chem. Solids* **23**, 1105 (1962).
306. "Magnetic Structure of Rare-Earth-Cobalt (RCo_2) Intermetallic Compounds," R. M. Moon, W. C. Koehler and John Farrell, *J. Appl. Phys.* **36**, 978 (1965).
307. "Magnetic Structure of Binary Fluorides Containing Mn^{2+}," S. J. Pickart, H. A. Alperin, and R. Nathans, BNL-7278.
308. T. Ohoyama, K. Kanematsu, and K. Yasukochi, *J. Phys. Soc. Japan* **18**, 589 (1963).
309. "Phase Transformation in Mn_5Si_3," R. P. Krentsis, I. Z. Radovskii, P. V. Gel'd, and L. P. Andreeva, *Zh. Neorgan. Khim.* **10**, No. 9, 2192-93 (1965).
310. P. Lecocq, *Ann. Chim.* **8**, 85 (1963).
311. "Susceptibility Measurements on $KCrF_3$," S. Yoneyama and K. Hirakawa, *J. Phys. Soc. Japan* **21**, 183 (1966).
312. "Magnetic Structure of $EuTiO_3$," T. R. McGuire, M. W. Shafer, and R. J. Joenk, *J. Appl. Phys.* **37**, 981 (1966).
313. "Antiferromagnetism I–Antiferromagnetism of the Spinel $MnGa_2O_4$," M. K. Wilkinson, B. Boucher, A. G. Herpin, and A. Oles, *J. Appl. Phys.* **37**, 960 (1966).
314. "Magnetic Properties of Some Rare-Earth Phosphides," G. Busch, P. Schwob, and O. Vogt, *Phys. Letters* **11**, 100 (1964).
315. "Low-Temperature Specific Heat Measurements of EuSe and EuTe," G. Busch, P. Junod, R. G. Morris, J. Muheim, and W. Stutius, *Phys. Letters* **11**, 9 (1964).
316. F. K. Lotgering, *Philips Res. Rept.* **11**, 337 (1956).
317. "Neel Temperatures of Some Antiferromagnetic Oxides with Spinel Structure," G. Blasse and J. F. Fast, *Philips Res. Rept.* **18**, 393 (1963).
318. "Magnetic Moments of the Ferromagnetic Borides Co_3B, Co_2B, and FeB," R. Fruchart, *Compt. Rend.* **256**, 15 (1963).

REFERENCES

319. "Ferromagnetic-Paramagnetic Transition in Iron," S. Arajs and R. V. Colvin, *J. Appl. Phys.* **35**, 2424 (1964).
320. "The Crystal Structure and Ferrimagnetism of Yttrium-Iron Garnet, $Y_3Fe_2(FeO_4)_3$," S. Geller and M. A. Gilleo, *J. Phys. Chem. Solids* **3**, 30 (1957).
321. "Investigations of the Structure and Magnetic Behavior of Peritectic Pr-Fe and Nd-Fe Compounds of the Type RFe_7 by Means of Neutron Diffraction," H. Weik, P. Fischer, E. Stoll, and W. Haeig, *Proc. Rare Earth Research Conf.*, Phoenix, 4th, April 1964, CONF-405-29.
322. "A Neutron-Diffraction Study of Very Pure Chromium," G. E. Bacon, *Acta Cryst.* **14**, 823 (1961).
323. "Neutron Diffraction Determination of Antiferromagnetism in Face-Centered Cubic (γ) Iron," S. C. Abrahams, L. Guttman, and J. S. Kasper, *Phys. Rev.* **127**, 2052 (1962).
324. "Propriétés Magnetiques et Structures de Manganite d'Yttrium," E. F. Bertaut, R. Pauthenet, and M. Mercier, *Phys. Letters* **7**, 110 (1963).
325. "Bismuth Substitution in Yttrium Iron Aluminum Garnets," S. Geller, H. J. Williams, R. C. Sherwood, and G. P. Espinosa, *J. Appl. Phys.* **35**, 1754 (1964).
326. "Crystal Structure, Synthesis, and Magnetic Properties of Chrysoberyl," R. E. Newnham et al., MIT Tech. Rept. 183, Contract AF 33(616)-8353, November 1963; also R. P. Santoro and R. E. Newnham, *J. Am. Ceram. Soc.* **47**, 491 (1964).
327. 'Antiferromagnetism in fcc and hcp Iron Manganese Alloys "Mossbauer Effect,"' C. Kimball, W. D. Gerber, and A. Arrott, *J. Appl. Phys.* **34**, 1046 (1963).
328. A. W. Overhauser, *J. Appl. Phys.* **34**, 1019 (1963).
329. "Specific Heat of EuS," V. L. Moruzzi and D. T. Teaney, *Solid State Commun.* **1**, 127 (1963). (26)
330. "Magnetic Properties of $GdFe_2$ and $DyFe_2$," M. Mansmann and W. E. Wallace, *J. Chem. Phys.* **40**, 1167 (1964).
331. "Sur les Propriétés Magnetiques des Differentes Phases du Systeme Fer–Etain," C. Jannin, P. Lecocq, and A. Michel, *Compt. Rend.* **257**, 1906 (1963).
332. "Mise en Envidence du Ferrimagnetisme dans l'Oxyde Mixte Mn_2VO_4," J.-C. Bernier, P. Poix, and A. Michel, *Compt. Rend.* **256**, 5583 (1963).
333. "Etude Paramagnetique de Zn_2VO_4 et Co_2VO_4," J.-C. Bernier, P. Poix, and A. Michel, *Bull. Soc. Chim. France*, 2219 (1963).
334. "The Magnetic Susceptibility of Ag-Mn and Cu-Mn Solid Solutions Between 1.2°K and 368°K," A. Van Itterbeek, W. Peelaers, and F. Steffens, *Appl. Sci. Res.*, Sec. B, **8**, 337 (1960).
335. "Modified Ferromagnetic Chromium Dioxide," B. Kubota, T. Nishikawa, A. Yanase, E. Hirota, T. Mihara, and Y. Iida, *J. Am. Ceram. Soc.* **46**, 550 (1963).
336. "Magnetic Properties of Mn-Ga Alloys with a High Coercive Force," I. Tsuboya and M. Sugihara, *J. Phys. Soc. Japan* **20**, 170 (1965).
337. L. M. Matarrese and J. W. Stout, *Phys. Rev.* **94**, 1792 (1954).
338. "$EuGd_2S_4$, a New Ferromagnetic Semiconductor," F. Hulliger and O. Vogt, *Phys. Letters* **17**, 238 (1965).
339. "Magnetic Structure of $CaMnSiO_4$," L. G. Caron, R. P. Santoro, and R. E. Newnham, *J. Phys. Chem. Solids* **26**, 927 (1965).
340. "Neutron Diffraction Study of Antiferromagnetism in UO_2," B. T. M. Willis and R. I. Taylor, *Phys. Letters* **17**, 188 (1965).
341. "Structure Magnetique et Proprietes Magnetiques de $GeNi_2O_4$," E. F. Bertaut, Vu Van Qui, R. Pauthenet, and A. Murasik, *J. Phys.* **25**, 516 (1964).
342. "Antiferromagnetic-Ferromagnetic Transition in the Compound Mn_3GaC," J.-P. Bouchaud, R. Fruchart, R. Pauthenet, M. Guillot, H. Bartholin, and F. Chaisse, *J. Appl. Phys.* **37**, 971 (1966).
343. "Magnetic Characteristics of Lanthanide-Bismuth Compounds," T. Tsuchida and W. E. Wallace, *J. Chem. Phys.* **43**, 2087 (1965).
344. "Magnetic Characteristics of Lanthanide Elements Combined with Tin, Lead, and Indium," T. Tsuchida and W. E. Wallace, *J. Chem. Phys.* **43**, 3811 (1965).
345. "Magnetic Characteristics of Compounds of Cerium and Praseodymium with Va Elements," T. Tsuchida and W. E. Wallace, *J. Chem. Phys.* **43**, 2885 (1965).

346. "Magnetic Ordering in Eu_3O_4 and $EuGd_2O_4$," L. Holmes and M. Schieber, *J. Appl. Phys.* **37**, 968 (1966).
347. "Magnetic Properties of Rare Earth Intermetallic Compounds in Gd(Ag, Cd, In) and Gd(Cu, Ag, Au) Systems," K. Sekizawa and K. Yasukochi, *J. Phys. Soc. Japan* **21**, 684 (1966).
348. "Electrical Resistivity of Antiferromagnet $MnTe_2$," A. Sawaoka and S. Miyahara, *J. Phys. Soc. Japan* **20**, 2087 (1965).
349. "Preparation and Thermomagnetic Analysis of Compounds of Ce, Pr, and Nd with Fe Having the Approximate Compositions RFe_7," A. E. Ray, K. Strnat, and D. Feldmann, CONF-20-12, *Proc. Rare Earth Conf., Clearwater, 3rd*, Apr. 21–24, 1963.
350. "Rare Earth Metals and Alloys," W. C. Koehler, *J. Appl. Phys.* **35**, 1030 (1964).
351. "Galvanomagnetic Properties of Ferromagnetic Uranium Monosulfide," M. A. Kanter, *J. Appl. Phys.* **35**, 1053 (1964).
352. "Magnetic and Electric Properties of Monosulfides and Mononitrides of Thorium and Uranium," R. Didchenko and F. P. Gortsema, *Inorg. Chem.* **2**, 1079 (1963).
353. "Effect of Fe^{4+} in the System $SrFeO_3$-$SrTiO_3$," T. R. Clevenger, Jr., *J. Am. Ceram. Soc.* **46**, 207 (1963).
354. W. Trzebiatowski, J. Niemicc, and A. Sepichowska, *Bull. Acad. Polon. Sci.* **9**, 373 (1961).
355. "Recent Magnetic Neutron Scattering Investigations at Oak Ridge National Laboratory," M. K. Wilkinson, H. R. Child, W. C. Koehler, J. W. Cable, and E. O. Wollan, *J. Phys. Soc. Japan* **17**, 27 (1962).
356. "Neutron Diffraction and Mossbauer Study of Ordered and Disordered $LiFeO_2$," D. E. Cox, G. Shirane, P. A. Flinn, S. L. Ruby, and W. J. Takei, *Phys. Rev.* **132**, 1547 (1963).
357. "Antiferromagnetic Structure of EuTe," G. Will, S. J. Pickart, H. A. Alperin, and R. Nathans, *J. Phys. Chem. Solids* **24**, 1679 (1963).
358. B. T. Matthias, R. M. Bozorth, and J. H. Van Vleck, *Phys. Rev. Letters* **7**, 160 (1961).
359. T. R. McGuire, B. E. Argyle, M. W. Shafer, and J. S. Smart, *Appl. Phys. Letters* **1**, 17 (1962).
360. "Ferromagnetism in Gadolinium Trichloride," W. P. Wolf, M. J. M. Leask, B. Mangum, and A. F. G. Wyatt, *J. Phys. Soc. Japan* **17**, 487 (1961).
361. "Spiral Structure and Moments in Gd-Dy Alloys," R. M. Bozorth and J. C. Suits, *J. Appl. Phys.* **35**, 1039 (1964).
362. "Ferromagnetism in Rare-Earth Group V_A and VI_A Compounds with Th_3P_4 Structure," F. Holtzberg, T. R. McGuire, S. Methfessel, and J. C. Suits, *J. Appl. Phys.* **35**, 1033 (1964).
363. "Coexistence of Magnetic and Electric Ordering in Crystals," G. A. Smolenskii and V. A. Bokov, *J. Appl. Phys.* **35**, 915 (1964).
364. "Exchange Inversion in Ternary Modifications of Iron Rhodium," P. H. L. Walter, *J. Appl. Phys.* **35**, 938 (1964).
365. "Silver Lanthanum Ferrite – A New Ferromagnetic Material," K. K. Laroia and A. P. B. Sinha, *Indian J. Pure Appl. Phys.* **1**, 215 (1963).
366. "Influence of Milling upon the Magnetic Properties of the Intermetallic Compound MnAlGe," W. A. J. J. Velge and K. J. de Vos, *J. Appl. Phys.* **34**, 3568 (1963).
367. "Magnetic Properties of Chromium Oxide Modified by Vanadium Oxide," B. Kubota, T. Nishikawa, and A. Yanas, *J. Phys. Soc. Japan* **16**, 2340 (1961).
368. "Magnetic and Crystal Structure of Titanium Sesquioxide," S. C. Abrahams, *Phys. Rev.* **130**, 2230 (1963).
369. "Ferromagnetic Compounds of Manganese with Perovskite Structure," G. H. Jonker and J. H. Van Santen, *Physica* **16**, 337 (1950).
370. "Neutron Diffraction Investigation of the Magnetic Order in MnI_2," J. W. Cable, M. K. Wilkinson, E. O. Wollan, and W. C. Koehler, *Phys. Rev.* **125**, 1860 (1962).
371. "Magnetic Interaction in EuS, EuSe, and EuTe," S. Van Houten, *Phys. Letters* **2**, 215 (1962).
372. "Ferromagnetism in V-Fe and Cr-Fe Alloys," M. V. Nevitt and A. T. Aldred, *J. Appl. Phys.* **34**, 463 (1963).
373. "Low-Temperature Magnetic Properties of Some Rare-Earth Garnet Compounds," M. Ball, G. Garton, M. J. M. Leask, D. Ryan, and W. P. Wolf, *J. Appl. Phys.* **32**, 267S (1961).

REFERENCES

374. "Samarium Substitutions in Yttrium Iron Garnet," J. R. Cunningham, Jr., and E. A. Anderson, *J. Appl. Phys.* **31**, 45S (1960).
375. "Neutron Diffraction Investigations of Antiferromagnetism in $CrCl_2$," J. W. Cable, M. K. Wilkinson, and E. O. Wollan, *J. Phys. Chem. Solids* **19**, 29 (1961).
376. "Magnetic Structures in the MnSb-CrSb System," W. J. Takei, D. E. Cox, and G. Shirane, *Phys. Rev.* **129**, 2008 (1963).
377. M. Foex, *Compt. Rend.* **227**, 193 (1948).
378. "The Magnetic Properties of the κ Phase in Mn-Al-Co System," I. Tsuboya and M. Sugihara, *J. Phys. Soc. Japan* **17**, 410 (1962).
379. "Curie Temperature of Some Garnets by the Differential Thermal Analysis Technique," A. Aharoni, E. H. Frei, Z. Scheidlinger, and M. Schieber, *J. Appl. Phys.* **32**, 1851 (1961).
380. "On the Ferromagnetic Phase in Manganese-Aluminum System," H. Kono, *J. Phys. Soc. Japan* **13**, 1444 (1958).
381. "New Type of Superexchange in the Spinel Structure — Some Magnetic Properties of Oxides $Me^{2+}Co_2O_4$ and $Me^{2+}Rh_2O_4$ with Spinel Structure," G. Blasse, *Philips Res. Rept.* **18**, 383 (1963).
382. High-Field Magnetization of Thulium Single Crystals, D. B. Richards and S. Legvold, *Phys. Rev.* **186**, 508-14 (1969).
383. "Magnetism and the Chemical Bond," J. B. Goodenough (Interscience, N.Y., 1963), Table 8.
384. C. Starr, F. Bitter, and A. R. Kaufmann, *Phys. Rev.* **58**, 977 (1940).
385. G. A. Smolenskii, V. M. Yudin, P. P. Syrnikov, and A. B. Sherman, *Soviet Phys.-Solid State* **8**, 2368 (1967). (The transparent, hexagonal ferrimagnet $RbNiF_3$)
386. K. Strnat, G. Hoffer, and A. E. Ray, *IEEE Trans. Magnet.* MAG2, 489 (1966). (Magnetic properties of rare earth-iron intermetallic compounds).
387. V. E. Adamyan and G. M. Loginov, *Soviet Phys.-Solid State* **8**, 2472 (1967). (Magnetic properties of cerium chalcogenides in the temperature range 4.2–77°K)
388. "High Resolution Specific Heat of Antiferromagnetic $MnCl_2 \cdot 4H_2O$ near Its Neel Point," George Sumter Dixon, Jr., Ph.D. thesis, 1967, Univ. of Ga. (available from University Microfilms, Ann Arbor, Mich., Order No. 67-16212.
389. "A New Ferrimagnetic Compound $TlNiF_3$," Kay Kohn, Reijiro Fukuda, and Shuichi Iida, *J. Phys. Soc. Japan* **22**, 333 (1967).
390. "Neutron-Diffraction Study of Dilute Chromium Alloys with Iron," A. Arrott, S. A. Werner, and H. Kendrick, *Phys. Rev.* **153**, 624 (1967).
391. "The Magnetic Structure of Fe_2As," Hisao Katsuraki and Norio Achiwa, *J. Phys. Soc. Japan* **21**, 2238 (1966).
392. "Proprietes Magnetiques des Diborures de Manganese et de Chrome: MnB_2 et CrB_2," M. C. Cadeville, *J. Phys. Chem. Solids* **27**, 667 (1966).
393. "Magneto-Crystalline Anisotropy of Co_2B," Atsushi Iga, *J. Phys. Soc. Japan* **21**, 1464 (1966).
394. "^{57}Fe hfs and Néel Temperature in $Fe_xZn_{1-x}F_2$," G. K. Wertheim, D. N. E. Buchanan, and H. J. Guggenheim, *Phys. Rev.* **152**, 527 (1966).
395. H. E. Nigh, S. Legvold, F. H. Spedding, and B. V. Beaudry, *J. Chem. Phys.* **41**, 3799 (1964).
396. "Magnetic Properties of $PbCo_{0.5}W_{0.5}O_3$ and $BaNi_{0.5}W_{0.5}O_3$," S. A. Kizhaev and V. A. Bokov, *Soviet Phys.-Solid State* **8**, 1554 (1966).
397. "Temperature Dependence of the Spontaneous Hall Effect Coefficient in Ferromagnets," K. P. Belov and E. P. Svirina, *Soviet Phys.-Solid State* **8**, 967 (1966).
398. "Paramagnetic Susceptibility of Mn-Zn Ferrites Near the Curie Temperature," K. M. Bol'shova and T. A. Elkina, *Soviet Phys.-Solid State* **8**, 2250 (1967).
399. "The Magnetic Properties of $BiFeO_3$," V. M. Yudin, *Soviet Phys.-Solid State* **8**, 217 (1966).
400. "The Preparation and Magnetic Properties of Some Complex Fluorides Having the Perovskite Structure," D. J. Machin, R. L. Martin, and R. S. Nyholm, *J. Chem. Soc. (London)*, Article 281, p. 1490 (1963).
401. "The Magnetic Structure of Uranium Monophosphite," N. A. Curry, *Proc. Phys. Soc.* **89**, 427 (1966).

402. "Determination de la Structure Magnetique du Spinelle Cubique $NiMn_2O_4$ par Diffraction de Neutrons," Bernard Boucher, Robert Buhl, and Michel Perrin, *Compt. Rend.* **263**, 344 (1966).
403. "On the Preparation of Polycrystalline $Zn_2Ba_2Fe_{12}O_{22}$ Especially with Respect to the Influence of Added Bi_2O_3," Wolfgang Tolksdorf, *Intern. Conf. Magnetics, Stuttgart*, April 20–22, 1966.
404. "Magnetische, Elektrische und Termische Eigenschaften der bcc-α-Phase im System Fe-Ga," H. Wagini, *Z. Naturforsch.* **22a**, 143 (1967).
405. "Some Physical Properties of Single Crystal Manganese Ferrites," R. Gerber, Z. Simsa, and M. Vichr, *Czech. J. Phys.* **B16**, 913 (1966).
406. "Magnetic Properties of Ce, Pr, and Nd Monochalcogenides at 4.2 to 1300°K," G. A. Smolenskii, V. P. Zhuze, V. E. Adamyan, and G. M. Loginov, *Phys. Status Solidi* **18**, 873 (1966).
407. "Elektronische Spezifische Wärme und Antiferromagnetismus in Chromlegierungen," F. Heiniger, *Phys. kondens. Materie* **5**, 285 (1966).
408. "Etudes de Diffraction Neutronique a Livermore," J. T. Sparks and T. Komoto, *J. Phys.* **25**, 567 (1964).
409. "Magnetization and Anisotropy in Gallium Iron Oxide," A. Pinto, *J. Appl. Phys.* **37**, 4372 (1966).
410. "Spontaneous Magnetization of EuO and GdN," P. Junod and F. Levy, *Phys. Letters* **23**, 624 (1966).
411. "Weak Ferromagnetism of $YCrO_3$," V. M. Judin and A. B. Sherman, *Solid State Commun.* **4**, 661 (1966).
412. "Magnetic Properties of $DyAl_2$ and $NdAl_2$," N. Nereson, C. Olsen, and G. Arnold, *J. Appl. Phys.* **37**, 4575 (1966).
413. "Etude par Effet Mossbauer et par Diffraction Neutronique du Ferrite Monocalcique $Fe_2O_3 \cdot CaO$, E. F. Bertaut, J. Chappert, A. Apostolov, and V. Semenov, *Bull Soc. Franc. Mineral. Crist.* **89**, 206 (1966).
414. "Magnetic Susceptibility and Electrical Resistivity of Au-Mn Alloys," A. Giansoldati, J. O. Linde, and G. Borelius, *J. Phys. Chem. Solids* **11**, 46 (1959).
415. "Magnetic Characteristics of Some 1:1 Compounds of the Lanthanides with Gold and Aluminum," F. Kissell and W. E. Wallace, *J. Less-Common Metals* **11**, 417 (1966).
416. "Ferromagnetic Curie Temperatures of Some Iron-Zinc Solid Solutions," H. A. Wriedt and S. Arajs, *Phys. Status Solidi* **16**, 475 (1966).
417. "Magnetic Properties of the Intermetallic Compounds of Cobalt with the Rare Earth Metals and Yttrium (Concluded)," R. Lemaire, *Cobalt* **33**, 201 (1966).
418. "Preparation, Proprietes Cristallines et Magnetiques des Spinelles de Formule $Mn^{2+}_{(1+x)} V^{3+}_{2(1-x)} V^{4+}_{x} O_4$," Gerard Villers and Andre Lecerf, *Compt. Rend.* **263**, 427 (1966).
419. "Antiferromagnetism of K_2MnF_4," D. J. Breed, *Phys. Letters* **23**, 181 (1966).
420. "The Magnetic Structures of Dilute Cr-Mn Alloys," T. J. Bastow, *Proc. Phys. Soc.* **88**, 935 (1966).
421. "Antiferroelectric and Magnetic Properties of $PbCo_{1/2}W_{1/2}O_3$," V. A. Bokov, S. A. Kizhaev, I. E. Myl'nikova, and A. G. Tutov, *Soviet Phys.-Solid State* **6**, 2419 (1965).
422. "Crystals and Magnetic Structures of CeC_2, PrC_2, and NdC_2," M. Atoji, *Phys. Letters* **22**, 21 (1966).
423. "Magnetic Structures of TbC_2 and HoC_2," M. Atoji, *Phys. Letters* **23**, 208 (1966).
424. "Curie Temperature Measurements of $Ga_xFe_{3-x}O_4$ Crystals," Helen Gamari-Seale and S. Karavelas, *Phys. Letters* **23**, 31 (1966).
425. "Magnetic Transition in EuSe," P. Schwob and O. Vogt, *Phys. Letters* **22**, 374 (1966).
426. H. Hirota and A. Yanase, *J. Phys. Soc. Japan* **20**, 1596 (1965).
427. "Antiferromagnetism in Disordered $Mn_3Pt_{1-x}Rh_x$ Alloys," E. Kren, *Phys. Letters* **21**, 383 (1966).
428. "Ferromagnetism and Anisotropy of Chromium Tribromide," J. F. Dillon, Jr., *J. Phys. Soc. Japan* **19**, 1662 (1964).
429. "Effect of Oxygen Deficiency on the Phase Transition of Copper Ferrite," Seishiro Sahara and Takashi Yamaguchi, *J. Appl. Phys.* **37**, 3324 (1966).

REFERENCES

430. "Crystalline Structure and Magnetic Properties of Ferrites Having the General Formula $5Fe_2O_3 \cdot 3M_2O_3$," F. Bertaut and R. Pauthenet, *Proc. IEE (London), Suppl. B* **104**, 261 (1957).
431. "Specific Heat of MnS Through the Néel Temperature," Donald R. Huffman and Robert L. Wild, *Phys. Rev.* **148**, 526 (1966).
432. "Magnetic Interactions of the System $Sr_3Fe_2O_{6.00-6.90}$," J. B. MacChesney, H. J. Williams, R. C. Sherwood, and J. F. Potter, *Mat. Res. Bull* **1**, 113 (1966).
433. "Magnetic Properties of $ZrZn_2$ Between 120°K and 0.1°K. Search for Superconductivity," R. L. Falge, Jr., and R. A. Hein, *Phys. Rev.* **148**, 940 (1966).
434. "Ferromagnetism in $CdCr_2Se_4$ and $CdCr_2S_4$," N. Menyuk, K. Dwight, R. J. Arnott, and A. Wold, *J. Appl. Phys.* **37**, 1387 (1966).
435. Magnetische, Elektrische und Thermische Eigenschaften von $FeGa_{1.3}$," H. Wagini, *Z. Naturforsch.* **21a**, 528 (1966).
436. "Intermetallic Compounds for Permanent Magnets," W. A. J. J. Velge and K. J. DeVos, *Z. Angew. Phys.* **21**, 115 (1966).
437. "Structural and Magnetic Transitions in $Sr_2(NiMo)O_6$," Shoichiro Nomura and Takehiko Nakagawa, *J. Phys. Soc. Japan* **21**, 1068 (1966).
438. "Vacancy Ordering in Epitaxially-Grown Single Crystals of γ-Fe_2O_3," Humihiko Takei and Shu Chiba, *J. Phys. Soc. Japan* **21**, 1255 (1966).
439. "Antiferromagnetic Structure of Uranium Diphosphide," R. Troc, J. Leciejewicz, and R. Ciszewski, *Phys. Status Solidi* **15**, 515 (1966).
440. "Oxide Crystal Growth by Flux Evaporation," W. H. Grodkiewicz and D. J. Nitti, *J. Am. Ceram. Soc.* **49**, 576 (1966). (84)
441. "Anomalous Magnetizations in the Yttrium Gallium Iron Garnet System and the Exchange Constant of Yttrium Iron Garnet," B. Luthi, *Phys. Rev.* **148**, 519 (1966).
442. "Magnetic Properties of the Intermetallic Compounds of Cobalt with the Rare Earth Metals and Yttrium," R. Lemaire, *Cobalt* **32**, 132 (1966).
443. "Magnetic and Optical Properties of Transparent $RbNiF_3$," M. W. Shafer, T. R. McGuire, B. E. Argyle, and G. J. Fan, *Appl. Phys. Letters* **10**, 202 (1967).
444. "Potential Applications of Magnetic Rare Earth Compounds," S. Methfessel, *IEEE Trans. Magnetics* **MAG 1**, 144 (1965).
445. "Antiferromagnetism in $Mn_3B_2O_6$, $Co_3B_2O_6$, and $Ni_3B_2O_6$," R. E. Newnham, R. P. Santoro, P. F. Seal, and G. R. Stallings, *Phys. Status Solidi* **16**, K17 (1966).
446. "Magnetocrystalline Anisotropy of Single Crystal CrO_2," D. S. Rodbell, *J. Phys. Soc. Japan* **21**, 1224 (1966).
447. "Magnetic Properties and Phase Transitions of Fe_3Ge," Kazuo Kanematsu, Ko Yasukochi, and Tetuo Ohoyama, *J. Phys. Soc. Japan* **18**, 920 (1963).
448. "Magnetic Properties of Intermetallic Compounds in Iron-Germanium System: $Fe_{1.67}Ge$ and $FeGe_2$," Ko Yasukochi, Kazuo Kanematsu, and Tetuo Ohoyama, *J. Phys. Soc. Japan* **16**, 429 (1961).
449. "Magnetic Susceptibilities of Gadolinium Orthophosphate ($GdPO_4$)," F. F. Y. Wang, *Phys. Status Solidi* **14**, 193 (1966).
450. "Magnetic and Crystallographic Studies on Rare Earth Germanides," Kazuko Sekizawa, *J. Phys. Soc. Japan* **21**, 1137 (1966).
451. J. P. Turbil, F. A. Fatsea, and Y. Lecocq, *Compt. Rend.* **263c**, 807 (1966).
452. "Magnetic Properties of Eu_3O_4," F. F. Y. Wang, *Phys. Status Solidi* **14**, 189 (1966).
453. "Ferromagnetic Resonance in Calcium Ferrite Containing Gd^{3+}," Noboru Ichinose, Shu Chiba, and Kazunobu Kurihara, *Japan. J. Appl. Phys.* **3**, 335 (1964).
454. "Electrical Resistivity, Thermal Conductivity and Magnetic Susceptibility of Polycrystalline Samarium at Low Temperatures," Sigurds Arajs and G. R. Dunmyre: *Z. Naturforsch.* **21**, 1856 (1966).
455. "Properties of Rare Earth and Transition Group Ions," Professor B. Bleaney, Clarendon Laboratory, Oxford University, Oxford, England, July 10, 1965, final report on Contract No. AF 61(052)-677.

REFERENCES

456. "Magnetic and Electrical Properties of Copper Containing Sulphides and Şelenides with Spinel Structure," F. K. Lotgering and R. P. van Stapele, *Solid State Commun.* **5**, 143 (1967).
457. "Temperature Dependence of the Magnetization of an Amorphous Ferromagnet," C. C. Tsuei, G. Longworth, and S. C. H. Lin, *Phys. Rev.* **170**, 603-6 (1968).
458. "Magnetic and Crystallographic Properties of Substituted Yttrium-Iron Garnet, $3Y_2O_3 \cdot xM_2O_3 \cdot (5-x) Fe_2O_3$," M. A. Gilleo and S. Geller, *Phys. Rev.* **110**, 73 (1958).
459. "The Transformation in Manganese Arsenide at About 40°C," Z. S. Basinski, R. O. Kornelsen and W. B. Pearson, *Trans. Indian Inst. Metals* **13**, 141 (1960).
460. V. V. Klyushin, V. V. Kelarev, I. Y. Getman, V. E. Arkhipov, and S. K. Sidorov, *Izv. Akad. Nauk SSSR, Ser. Fiz.* **30**, 968 (1966).
461. "Neutron Diffraction Studies of Single Crystal Terbium-Holmium Alloys," F. H. Spedding, Y. Ito, and R. G. Jordan, in *Ann. Summary Rept., Ames Lab.*, IS-1900, July 1968.
462. E. Kren and J. Solyom, *Phys. Letters* **22**, 273 (1966).
463. N. N. Sirota and G. I. Makovetskii, *Dokl. Akad. Nauk SSSR* **170**, 1300 (1966).
464. V. I. Nikolaev, S. S. Yakimov, I. A. Dubovtsev and Z. G. Gavrilova, *Soviet Phys.-JETP Letters* **2**, 235 (1965).
465. B. Andron, G. Berodias, M. Chevreton and P. Mollard, *Compt. Rend.* **B263**, 621 (1966).
466. R. H. Busey and E. Sonder, *J. Appl. Phys.* **36**, 93 (1962).
467. "Properties of Ferrimagnetic Materials for Microwave Applications," A. S. Boxer, J. F. Ollom, and R. F. Rauchmiller, attachment to Technical Doc. Rept. No. ML-TDR-64-224, July 1964, Contract No. AF 33(657)-8439, Project No. 7371, Task No. 737103.
468. "Structural and Magnetic Properties of Copper-Substituted Manganese-Aluminum Alloy," Makoto Sugihara and Ichiro Tsuboya, *Japan. J. Appl. Phys.* **2**, 373 (1963).
469. "Eigenschaften der Materie in Ihren Aggregatzuständen. 9. Teil, Magnetische Eigenschaften I," ed. by Karl-Heinz Hellwege and Anne Marie Hellwege, Springer-Verlag, Berlin, 1962.
470. "Electric and Magnetic Properties of 'VO'," Shinji Kawano, Koji Kosuge, and Sukeji Kachi, *J. Phys. Soc. Japan* **21**, 2744 (1966).
471. "Influence du Manganese sur les Propriétés Magnétiques et Structurales du Germaniure de fer Fe_3Ge," Yvonne Lecocq, Pierre Lecocq, and André Michel, *Compt. Rend.* **256**, 4913 (1963). (94)
472. "Mössbauer Study of Magnetic Properties in Ferrous Compounds," Kazuo Ono and Atsuko Ito, *J. Phys. Soc. Japan* **19**, 899 (1964).
473. "Un Nouveau Corps Magneto-Electrique: $LiMnPO_4$," M. Mercier and J. Gareyte, *Solid State Commun.* **5**, 139 (1967).
474. "Propriétés des Ferrites de Lithium Partiellement Substitués par l'Aluminum Destines aux Dispositifs Micro-Ondes," A. Vassiliev and A. Lagrange, *IEEE Trans. Magnetics* MAG **2**, 707 (1966).
475. A Neutron Diffraction Study of NpO_2, D. E. Cox and B. C. Frazer, BNL-11266; *J. Phys. Chem. Solids* **28**, 1649-50 (1967).
476. "Etude Magnetique et Structurale de Fe_5Si_3," Yvonne Lecocq, Pierre Lecocq, and Andre Michel, *Compt. Rend.* **258**, 5655 (1964).
477. "Etude de la Phase Fe_5Si_3 Dans Son Etat Paramagnetique," Erich Ubelacker and Pierre Lecocq, *Compt. Rend.* **262**, 793 (1966).
478. "Etude Structurale et Magnetique des Arseniures de Chrome," Lazlo Hollan, Pierre Lecocq, and Andre Michel, *Compt. Rend.* **260**, 2233 (1965).
479. "Etude Magnetique et Structurale de Fe_3Si," Pierre Lecocq and Andre Michel, *Compt. Rend.* **258**, 1817 (1964).
480. "Magnetic Structure Transformation in MnPt," E. Kren, M. Cselik, G. Kadar, and L. Pal, *Phys. Letters* **A24**, 198 (1967); Hungarian Academy of Sciences, Central Research Institute for Physics, Budapest KFKI 3/1967.
481. "Low Temperature Thermoelectric Power and Magnetic Susceptibility of Rare-Earth Metals in Gold and Silver," D. Gainon, P. Donze, and J. Sierro, *Solid State Commun.* **5**, 151 (1967).
482. "Etude de la Structure Magnetique des Chromites d'Erbium et de Neodyme par Diffraction Neutronique," E. F. Bertaut and J. Mareschal, *Solid State Commun.* **5**, 93 (1967).

REFERENCES

483. "The Ordering Temperature of Cu_2MnAl," T. Ohoyama, P. J. Webster, and R. S. Tebble, *Brit. J. Appl. Phys. (J. Phys. D), Ser.* 2, 1, 951-52 (1968).
484. J. M. Mays, *Phys. Rev.* 131, 38 (1963).
485. M. Yuzuri, *J. Phys. Soc. Japan* 15, 2007 (1960).
486. "Magnetic-Susceptibility Studies on the Magneli Phases of the Titanium-Oxygen System," L. K. Keys and L. N. Mulay, *J. Appl. Phys.* 38, 1466 (1967).
487. G. Blasse, *Proc. Intern. Conf. Magnetism, Nottingham*, September 1964 (Phys. Soc. London, 1965), p. 350.
488. "Magnetic Behavior of the System Mn_2O_3-Fe_2O_3," S. Geller, R. W. Grant, J. A. Cape, and G. P. Espinosa, *J. Appl. Phys.* 38, 1457 (1967).
489. "Pressure Dependence of the Magnetic Transitions in Fe-Rh Alloys," R. C. Wayne, *Phys. Rev.* 170, 523-27 (1968).
490. E. Banks, E. Kostiner, and G. K. Wertheim, *J. Chem. Phys.* 45, 1189 (1966).
491. "Magnetic Structure of $Fe_{0.9}Mn_{0.9}Ge$," Toshiro Suzuoka, E. Adelson, and A. E. Austin, *Acta Cryst.* A24, 513-17 (1968).
492. "Magnetic Properties of Substituted $Ca_2Fe_2O_5$," R. W. Grant, H. Wiedersich, S. Geller, U. Gonser, and G. P. Espinosa, *J. Appl. Phys.* 38, 1455 (1967).
493. "The Magnetic Susceptibility of Neptunium Oxide and Carbide Between 4.2° and 350°K," J. W. Ross and D. J. Lam, *J. Appl. Phys.* 38, 1451 (1967).
494. "Magnetic Properties of $PrAl_2$," C. E. Olsen, G. Arnold, and N. Nereson, *J. Appl. Phys.* 38, 1395 (1967).
495. "Magnetic Structure of Er_2O_3 and Yb_2O_3," R. M. Moon, H. R. Child, W. C. Koehler, and L. J. Raubenheimer, *J. Appl. Phys.* 38, 1383 (1967).
496. "Magnetic Properties of Ferrite Single Crystals with the Y Structure," J. Verweel, *J. Appl. Phys.* 38, 1111 (1967).
497. "Magnetic Anisotropy of Some Nickel Zinc Ferrite Crystals," A. Broese van Groenou, J. A. Schulkes, and D. A. Annis, *J. Appl. Phys.* 38, 1133 (1967).
498. "NMR Studies of Magnetic Properties of Light Rare-Earth Hydrides," J. P. Kopp and D. S. Schreiber, *J. Appl. Phys.* 38, 1373 (1967).
499. "Ferromagnetism and Superconductivity in $TiFe_xCo_{1-x}$," B. F. DeSavage and J. F. Goff, *J. Appl. Phys.* 38, 1337 (1967).
500. "Antiferromagnetic Resonance in Cubic $TlMnF_3$," D. E. Eastman and M. W. Shafer, *J. Appl. Phys.* 38, 1274 (1967).
501. "Investigation of the First-Order Magnetic Transformation in Mn_3Pt," E. Kren, G. Kadar, L. Pal, and P. Szabo, *J. Appl. Phys.* 38, 1265 (1967).
502. "Effects of Mechanical and Thermal Treatment on the Structure and Magnetic Transitions in FeRh," J. M. Lommel and J. S. Kouvel, *J. Appl. Phys.* 38, 1263 (1967).
503. "Magnetic Properties of the Iron-Group Metal Phosphides," R. J. Gambino, T. R. McGuire, and Y. Nakamura, *J. Appl. Phys.* 38, 1253 (1967).
504. "A Family of New Cobalt-Base Permanent Magnet Materials," K. Strnat, G. Hoffer, J. Olson, W. Ostertag, and J. J. Becker, *J. Appl. Phys.* 38, 1001 (1967).
505. "Temperature Dependence of the ^{57}Fe hfs in FeF_2 Below the Neel temperature," G. K. Wertheim, *J. Appl. Phys.* 38, 971 (1967).
506. "Propriétés Magnetiques des Composes Intermetalliques Entre le Nickel et les Terres Rares de Formule T_3Ni," Jacques-Louis Feron, Remy Lemaire, Dominique Paccard, and Rene Pauthenet, *Compt. Rend.* 267, 371-74 (1968).
507. "Magnetic Properties of the Systems $HgCr_2S_4$-$CdCr_2S_4$ and $ZnCr_2Se_4$-$CdCr_2Se_4$," P. K. Baltzer, M. Robbins, and P. J. Wojtowicz, *J. Appl. Phys.* 38, 953 (1967).
508. "Crystal and Magnetic Structure of $PbCrO_3$," W. L. Roth and R. C. DeVries, *J. Appl. Phys.* 38, 951 (1967).
509. "Some Magnetic Properties of Fe_2SiO_4 from 4°K to 300°K," Walter Kundig, J. A. Cape, R. H. Lindquist, and G. Constabaris, *J. Appl. Phys.* 38, 947 (1967).

510. P. K. Baltzer, P. J. Wojtowicz, M. Robbins, and E. Lopatin, *Phys. Rev.* **151**, 367 (1966).
511. "Ferromagnetism in Compounds with Pyrochlore Structure," P. F. Bongers and E. R. Van Meurs, *J. Appl. Phys.* **38**, 944 (1967).
512. "Magnetic Properties of Rare-Earth Compounds," G. Busch, *J. Appl. Phys.* **38**, 1386 (1967).
513. S. A. Kizhaev, A. G. Tutov, V. A. Bokov, *Fiz. Tverd. Tela* **7**, 2868 (1965).
514. K. Aoyagi and N. Uchida, *J. Phys. Soc. Japan* **20**, 617 (1963).
515. "Magnetic and Structural Characteristics of Some Equiatomic Rare-Earth Germanides," K. H. J. Buschow and J. F. Fast, *Phys. Status Solidi* **16**, 467 (1966).
516. "Factors Controlling the Occurrence of Laves Phases and AB_5 Compounds Among Transition Elements," A. E. Dwight, *Trans. ASM* **53**, 479 (1961).
517. "YCo_5 — A Promising New Magnet Material," Karl J. Strnat and Gary I. Hoffer, AFML-TR-65-446, May 1966.
518. "Magnetic Investigations of New Ternary Rare-Earth Chalcogenides," F. Hulliger and O. Vogt, *Phys. Letters* **21**, 138 (1966).
519. "Etude des Propriétés Magnetostatiques et des Structures Magnetiques des Chromites des Terres Rares et d'Yttrium," E. F. Bertaut, J. Mareschal, G. DeVries, R. Aleonard, R. Pauthenet, J. P. Rebouillat, and V. Zarubicka, *IEEE Trans. Magnetics MAG* **2**, 453 (1966).
520. "Research on High-Temperature Dielectric Materials," F. Chernow, W. B. Westphal, and R. E. Newnham, AFML-TR-67-27, November 1966, AF-33(615)-2199.
521. "Research to Investigate the Microstructure of the Internal Magnetic Field in Selected Magnetic Materials," Joseph J. Becker, Charles D. Graham, Jr., and John S. Kasper, AFML-TR-67-28, March 1967, AF-33(615)-1490.
522. "Some Electric and Magnetic Properties of Rare Earth Monosulfides and Nitrides," R. Didchenko and F. P. Gortsema, *J. Phys. Chem. Solids* **24**, 863 (1963).
523. "Electrical Resistivity of Some Chromium-Silicon Alloys," Sigurds Arajs and William E. Katzenmeyer, *J. Phys. Soc. Japan* **23**, 932-36 (1968).
524. "Ferrites," J. Smit and H. P. J. Wijn, Philips' Technical Library, Eindhoven-Netherlands, 1959.
525. "Crystallographic and Magnetic of Several Spinels Containing Trivalent JA-1044 Manganese," D. G. Wickham and W. J. Croft, *J. Phys. Chem. Solids* **7**, 351 (1958).
526. R. V. Pisarev, A. I. Belyaeva, and P. P. Syrnikov, *Soviet Phys.-Solid State* **8**, 627 (1966).
527. "Neutron Diffraction Study of Antiferromagnetism in UO_2," B. T. M. Willis and R. I. Taylor, AERE-R-4999, July 1965.
528. "Temperature Dependence of the Magnetic Ordering in $Gd-GdFe_2$ Phase Mixtures," S. J. Sivonen and E. J. Suoninen, *Phys. Stat. Sol.* **29**, K171-73 (1968).
529. "The Magnetic Property of $Ni(OH)_2$," Toshio Takada, Yoshichika Bando, Masao Kiyama, Kiroki Miyamoto, and Tsuyoshi Sato, *J. Phys. Soc. Japan* **21**, 2745 (1966).
530. "Crystal Structures and Magnetic Properties of the Intermetallic Compound Fe_2Zr," Kenzo Kai, Takuro Nakamichi, and Mikio Yamamoto, *J. Phys. Soc. Japan* **25**, 1192 (1968).
531. "Effect of Hydrostatic Pressure on the Magnetic Transition Temperature of Dy," Tetsuhiko Okamoto, Nobuo Iwata, Shoji Ishida, and Eiji Tatsumoto, *J. Phys. Soc. Japan* **21**, 2727 (1966).
532. "Optical Diffraction by Magnetic Domains in Europium Chalcogenides," J. C. Suits, *J. Appl. Phys.* **38**, 1498 (1967).
533. "Magnetic Properties of Fe_xO as Related to the Defect Structure," F. B. Koch and M. E. Fine, *J. Appl. Phys.* **38**, 1470 (1967).
534. "Magnetic Equation of State of YIG near the Curie Point," Kohji Ohbayashi and Shuichi Iida, *J. Phys. Soc. Japan* **25**, 1187 (1968).
535. "Magnetic and Crystal Structures of CeC_2, PrC_2, NdC_2, TbC_2, and HoC_2 at Low Temperatures," M. Atoji, *J. Chem. Phys.* **46**, 1891 (1967).
536. "Thermal Conductivity, Electrical Resistivity, and Lorenz Function of Polycrystalline Erbium Between 5 and 300°K," S. Arajs and G. R. Dunmyre, *Physica* **31**, 1466 (1965).
537. "Crystallographic and Magnetic Investigation of the Rare Earth-Cobalt Compounds R_2Co_{17}," Werner Ostertag, Karl Strnat, and Gary I. Hoffer, AFML-TR-66-420, February 1967.

REFERENCES

538. "Proprietes Cristallographiques et Magnetiques des Alliages de Formule TNi_3, Dans Laquelle T Designe un Metal de Terre Rare ou L'yttrium," Dominique Paccard and Rene Pauthenet, *Compt. Rend.* **264**, 1056 (1967).
539. "Uber die Vanadintelluride und Ihre Magnetischen Eigenschaften," Erling Rost, Liv Gjersten, and Haakon Haraldsen, *Z. Anorg. Allgem. Chem.* **333**, 301 (1964).
540. "Electrical and Magnetic Properties of Holmium Single Crystals," D. L. Strandburg, S. Legvold, and F. H. Spedding, *Phys. Rev.* **127**, 2046 (1962).
541. "Some Magnetic Properties of Mixed Metal Oxides with Ordered Perovskite Structure," G. Blasse, *Philips Res. Rept.* **20**, 327 (1965).
542. "Mossbauer Effect in Some Iron-Rare Earth Intermetallic Compounds," G. K. Wertheim and J. H. Wernick, *Phys. Rev.* **125**, 1937 (1962).
543. "Ferromagnetism in EuH_2," R. L. Zanowick and W. E. Wallace, *Phys. Rev.* **126**, 537 (1962).
544. "Magnetic Properties of Ordering Heavy Rare-Earth Arsenides," G. Busch, O. Vogt, and F. Hulliger, *Phys. Letters* **15**, 301 (1965).
545. "Structures Magnetiques de $TbFeO_3$," E. F. Bertaut, J. Chappert, J. Mareschal, J. P. Rebouillat, and J. Sivardiere, *Solid State Commun.* **5**, 293 (1967).
546. "Ferromagnetism in the Intermetallic Phase Ni_3Al," F. R. De Boer, J. Biesterbos, and C. J. Schinkel, *Phys. Letters* **24A**, 355 (1967).
547. "Magnetization and Electrical Resistivity of Gadolinium Single Crystals," H. E. Nigh, S. Legvold, and F. H. Spedding, IS-661, May 1963.
548. "Magnetic Properties of $KMnF_3$. II. Weak Ferromagnetism," A. J. Heeger, Olof Beckman, and A. M. Portis, *Phys. Rev.* **123**, 1652 (1961).
549. H. Katsuraki and K. Suzuki, *J. Appl. Phys.* **36**, 1094 (1965).
550. N. A. Curry, *Proc. Phys. Soc. (London)* **86**, 1193 (1965).
551. G. Drabkin, E. I. Mal'tsev, and V. P. Plakhtii, *Fiz. Tverd. Tela* **7**, 1241 (1965).
552. R. Kohlhaas, W. D. Weiss, and K. Saupe, *Phys. Status Solidi* **9**, K189 (1965).
553. H. Hirota and H. Yanase, *J. Phys. Soc. Japan* **20**, 1960 (1965).
554. M. Atoji, *J. Chem. Phys.* **43**, 222 (1965).
555. C. D. Graham, *J. Appl. Phys.* **36**, 1135 (1965).
556. K. P. Belov, Yu. V. Ergin, L. I. Koroleva, R. Z. Levitin, and A. V. Ped'ko, *Phys. Status Solidi* **12**, 219 (1965).
557. "Ferro- and Antiferromagnetism of the Laves Phase Compound in Fe-Ti Alloy System," Takuro Nakamichi, *J. Phys. Soc. Japan* **25**, 1189 (1968).
558. K. P. Belov and L. I. Koroleva, *Phys. Metals Metallog.* **19**, 126 (1965).
559. A. Ito and K. Ono, *J. Phys. Soc. Japan* **20**, 784 (1965).
560. H. Bizette, C. Terrier, and B. Tsai, *Compt. Rend.* **261**, 653 (1965).
561. "Nouvelle Preparation et Etude de l'Oxyde ϵ-Fe_2O_3," Jean-Marie Trautmann and Hubert Forestier, *Compt. Rend.* **261**, 4423 (1965).
562. "Ferromagnetic Interactions in Non-Metallic Perovskites," G. Blasse, *J. Phys. Chem. Solids* **26**, 1969 (1965).
563. "Insulating Ferromagnetic Spinels," P. K. Baltzer, H. W. Lehmann, and M. Robbins, *Phys. Rev. Letters* **15**, 493 (1965).
564. E. Legrand and M. Verschueren, *J. Phys. (Paris)* **25**, 578 (1964).
565. J. P. Bouchaud, R. Fruchart, M. Guillot, H. Bartholin, and F. Chaisse, *Compt. Rend.* **261**, 655 (1965).
566. J. Loriers and G. Villers, *Compt. Rend.* **252**, 1590 (1961).
567. "Mise en Evidence d'une Transition Magnetique du Premier Ordre dans la Phase Mn_3GaC," Jean-Pierre Bouchard and Robert Fruchart, *Compt. Rend.* **261**, 458 (1965).
568. "Mise en Evidence d'une Transition du Premier Ordre dans le Nitrure Mn_3GaN," Jean-Pierre Bouchard, Eliane Fruchart, Gerard Lorthioir, and Robert Fruchart, *Compt. Rend.* **262**, 640 (1966).
569. "Ferromagnetism in Au_4V," L. Creveling, H. L. Luo, and G. S. Knapp, *Phys. Rev. Letters* **18**, 851 (1967).

570. "Transverse Magnetoresistance of High Purity Chromium Foils," T. J. Bastow and R. Street, *Phil. Mag.* **10**, 269 (1964).
571. "Rare-Earth and Yttrium-Free Ferrimagnetic Garnet with 493°K Curie Temperature," S. Geller, G. P. Espinosa, H. J. Williams, R. C. Sherwood, and E. A. Nesbitt, *Appl. Phys. Letters* **3**, 60 (1963).
572. "Magnetic Properties of the Planar Antiferromagnet Rb_2FeF_4," G. K. Wertheim, H. J. Guggenheim, H. J. Levenstein, D. N. E. Buchanan, and R. C. Sherwood, *Phys. Rev.* **173**, 614-16 (1968). (136)
573. A. Giansoldati, *J. Phys. Radium* **16**, 342 (1955).
574. G. E. Bacon and R. Street, *Proc. Phys. Soc. London* **72**, 490 (1958).
575. "An Antiferromagnetic Heusler Alloy, Cu_2MnSb," D. P. Oxley, C. T. Slack, R. S. Tebble, and K. C. Williams, *Nature* **195**, 465 (1962).
576. "Antiferromagnetic Resonance in $MnTiO_3$," J. J. Stickler and G. S. Heller, *J. Appl. Phys.* **33**, 1302 (1962).
577. K. Endo, *Sci. Rept. Tohoku Univ.* **25**, 879 (1937).
578. "Variation Avec la Pression du Point de Neel de l'Oxyde de Manganese," Henri Bartholin, Daniel Bloch, and Roland Georges, *Compt. Rend.* **264**, 360 (1967).
579. "Valency Changes in Cerium Compounds," G. Busch and O. Vogt, *Phys. Letters* **20**, 152 (1966).
580. "Antiferromagnetism of Disilicides of Heavy Rare Earth Metals," K. Sekizawa and K. Yasukochi, *J. Phys. Soc. Japan* **21**, 274 (1966).
581. "Propriétés Magnetiques des Composés Intermetalliques de Formule T_3Al_2 Entre l'Aluminum et les Terres Rares," Bernard Barbara, Christian Becle, Remy Lemaire, and Rene Pauthenet, *Compt. Rend.* **267**, 309-12 (1968).
582. "Studies of the Normal Antiferromagnetic Spinel $MnGa_2O_4$ by Neutron Diffraction and Magnetic Measurements," B. Boucher and A. Oles, *J. Phys.* **27**, 51 (1966).
583. "Ferrimagnetic Structure of Magnetoelectric $Ga_{2-x}Fe_xO_3$," R. B. Frankel, N. A. Blum, S. Foner, A. J. Freeman, and M. Schieber, *Phys. Rev. Letters* **15**, 958 (1965).
584. "Structure and Magnetic Properties of $Dy_2Mn_4O_9$," I. Nowik, H. J. Williams, L. G. van Uitert, and H. J. Levinstein, *J. Appl. Phys.* **37**, 970 (1966).
585. "Thermal Conductivity of Ho and Er at 2 to 100°K," N. G. Aliev and N. V. Volkenshtejn, *Fiz. Tverd. Tela, SSSR* **7**, 2560 (1965).
586. "Magnetic Properties of Eu_3O_4," A. A. Samokhvalov, V. G. Babburov, N. V. Volkenshtein, T. D. Zotov, A. A. Ivakin, Yu. N. Morozov, and M. I. Simonova, *Fiz. Metal. i Metalloved.* **20**, 308 (1965).
587. "Susceptibility Hysteresis and Hyperfine Interactions in MnSe," E. D. Jones, *Phys. Letters* **18**, 98–9 (1965).
588. "Applications of Miniature Coil, Pulsed-Field Techniques to Studies of Magnetic Phase Transition," R. B. Flippen, *J. Appl. Phys.* **34**, 2026 (1963).
589. "The Magnetic Susceptibility of Tantalum Diselenide," Rod K. Quinn, Robert Simmons, and John J. Banewicz. *J. Phys. Chem.* **70**, 230 (1966).
590. "Magnetic Anisotropy of the Alloy MnHg," A. Oles, *Phys. Status Solidi* **14**, K39 (1966).
591. "The Magnetic Susceptibility of $GdCl_{1.6}$ and the Influence of Minor Amounts of Magnetic Impurities," J. D. Greiner, J. F. Smith, J. D. Corbett, and F. J. Jelinek, *J. Inorg. Nucl. Chem.* **28**, 971 (1966).
592. "Uranium Monosulfide. The Ferromagnetic Transition. The Heat Capacity and Thermodynamic Properties from 1.5°K to 350°K," Edgar F. Westrum, Jr., Robert P. Walters, Howard E. Flotow, and Darrell W. Osborne, *J. Chem. Phys.* **48**, 155-61 (1968).
593. "Magnetic Studies of Gadolinium Compounds with the CsCl Structure," G. T. Alfieri, E. Banks, and K. Kanematsu, *J. Appl. Phys.* **37**, 1254 (1966).
594. "Electrical Transport Properties of the Insulating Ferromagnetic Spinels $CdCr_2S_4$ and $CdCr_2Se_4$," H. W. Lehmann and M. Robbins, *J. Appl. Phys.* **37**, 1389 (1966).
595. "Magnetic Structure of MnP," G. P. Felcher, *J. Appl. Phys.* **37**, 1056 (1966).
596. "Magnetic Susceptibility of Cobalt and Iron-Nickel Alloys in the Vicinity of the Curie Point," Gerhard Develey, *Compt. Rend.* **262**, 103 (1966).

REFERENCES

597. "Paramagnetic Susceptibility of $ZrZn_2$," E. P. Wohlfarth, *Phys. Letters* **20**, 253 (1966).
598. "Magnetic Structure of Iron Chromite $FeCr_2O_4$ at Low Temperatures by the Moessbauer Effect," Pierre Imbert and Eliane Martel, *Compt. Rend.* **261**, 5404 (1965).
599. "Structure and Magnetic Properties of Antiferromagnetic $CaFe_2O_4$," Y. Allain, B. Boucher, P. Imbert, and M. Perrin, *Compt. Rend.* **263**, B9 (1966).
600. "The Magnetic Susceptibilities of the Cobalt-Sulfur System," Robert F. Heidelberg, Alan H. Luxem, Sami Talhouk, and John J. Banewicz, *Inorg. Chem.* **5**, 194 (1966).
601. "Structure and Magnetic Properties of $BiMnO_3$," V. A. Bokov, I. E. Myl'nikova, S. A. Kizhaev, M. F. Bryzhina, and N. A. Grigoryan, *Fiz. Tverd. Tela* **7**, 3695 (1965).
602. "Effect of Pressure on the Magnetic Transition Point of Manganese Telluride," K. Ozawa, S. Anzai, and Y. Hamaguchi, *Phys. Letters* **20**, 132 (1966).
603. "Pressure Effect on Magnetic Transitions in MnAs and Cr-Modified Mn_2Sb," Takajiko Kamigaichi, Kanichi Masumoto, and Tadamiki Hihara, *J. Sci. Hiroshima Univ., Ser. A-II*, **28**, 53 (1965).
604. "Magnetic Susceptibilities of Some 3d Transition Metal Boracites," Hans Schmid, Harry Rieder, and Edgar Ascher, *Solid State Commun.* **3**, 327 (1965).
605. "Some Properties of Ferromagnetoelectric Nickel-Iodine Boracite, $Ni_3B_7O_{13}I$," E. Ascher, H. Rieder, H. Schmid, and H. Stossel, *J. Appl. Phys.* **37**, 1404 (1966).
606. F. K. Lotgering, *Philips Res. Rept.* **11**, 190, 337 (1956).
607. "Determination of Neel Temperatures in fcc Iron," U. Gonser, C. J. Meechan, A. H. Muir, and H. Wiedersich, *J. Appl. Phys.* **34**, 2373 (1963).
608. H. Bizette and B. Tsai, *Compt. Rend.* **240**, 2213 (1955).
609. N. Perakis and A. Serres, *J. Phys. Radium* **18**, 47 (1957).
610. H. Bizette, *Ann. Phys. (Paris)* **1**, 316 (1946).
611. R. Benoit, *J. Chim. Phys.* **52**, 119 (1955).
612. F. Bertaut et al., *Compt. Rend.* **251**, 1733 (1960).
613. R. Aleonard and R. Pauthenet, *Compt. Rend.* **251**, 1730 (1960).
614. K. Motizuki, *J. Phys. Soc. Japan* **14**, 759 (1959).
615. "Etude du Systeme $Li_{0.5}Fe_{2.5}O_4$-Fe_2CuO_4," Michel Lenglet, *Compt. Rend.* **267**, 367-70 (1968). (136)
616. L. Corliss et al., *Phys. Rev.* **117**, 929 (1960).
617. E. Hirota and B. Kubota, *J. Phys. Soc. Japan* **15**, 1715 (1960).
618. F. Lotgering and E. Gorter, *J. Phys. Chem. Solids* **3**, 238 (1957).
619. L. Corliss et al., Sheffield Magnetism Conference, September 1959.
620. B. Frazer and P. Brown, Reported in Antiferromagnetic Compounds, H. A. Alperin and S. J. Pickart, NOLTR 61-18, AD-259738, April 1961.
621. R. Bozorth and D. Davis, Reported in Antiferromagnetic Compounds, H. A. Alperin and S. J. Pickart, NOLTR-61-18, AD-259738, April 1961.
622. V. Scatturin et al., Reported in Antiferromagnetic Compounds, H. A. Alperin and S. J. Pickart, NOLTR 61-18, AD 259738, April 1961.
623. R. Martin et al., *Chem. Ind.* **3**, 38 (1956).
624. S. Pickart, Reported in Antiferromagnetic Compounds, H. A. Alperin and S. J. Pickart, NOLTR-61-18, AD-259738, April 1961.
625. B. Figgis and R. Martin, *J. Chem. Soc.* 3837 (1956).
626. N. Perakis et al., *J. Phys. Radium* **17**, 134 (1956).
627. A. Gilmour and R. Pink, *J. Chem. Soc.* 2198 (1953).
628. "Ferromagnetic Material," B. T. Matthias, Bell Telephone Labs, U.S. Patent 2,989,480, June 20, 1961.
629. J. Cable et al., *Bull. Am. Phys. Soc.* **5**, 458 (1960).
630. "Magnetic Transition Temperatures of Intermetallic Compounds of Rare Earths with Cobalt and Iron," Lee R. Salmans, Jr., Karl Strnat, and Gary I. Hoffer, *AFML-TR-68-159*, Sept. 1968. (This document is subject to special export controls and each transmittal to foreign governments or foreign nationals may be made only with prior approval of the AF Materials Lab (MAYE), WPAFB, Ohio 45433.)

631. R. Alikhanov, *J. Exptl. Theoret. Phys. USSR* **36**, 1690 (1959); *Soviet Phys. JETP* **9**, 1204 (1959).
632. S. Pickart, *Bull. Am. Phys. Soc.* **5**, 59 (1960).
633. C. Guillaud, *J. Phys. Radium* **12**, 489 (1951).
634. C. Shull et al., *Phys. Rev.* **83**, 333 (1951).
635. L. Corliss et al., *Phys. Rev.* **93**, 893 (1954).
636. I. Tsubokawa and S. Chiba, *J. Phys. Soc. Japan* **14**, 1120 (1959).
637. J. Llewellyn and T. Smith, *Proc. Phys. Soc.* **74**, 65 (1959).
638. D. Finlayson et al., *Proc. Phys. Soc.* **74**, 75 (1959).
639. "Magnetic Properties of Silicides of Iron Group Transitional Elements," Daizaburo Shinoda and Sizuo Asanabe, *J. Phys. Soc. Japan* **21**, 555 (1966).
640. R. Pauthenet and P. Blum, *Compt. Rend.* **239**, 33 (1954).
641. G. Wilkinson et al., *J. Inorg. Nucl. Chem.* **2**, 95 (1956).
642. A. Borovik-Romanov, *J. Exptl. Theoret. Phys. USSR* **31**, 579 (1956); *Soviet Phys. JETP* **4**, 531 (1957).
643. "Magnetic and Magneto-Optic Properties of EuO Films Doped with Trivalent Rare-Earth Oxide," K. Y. Ahn and T. R. McGuire, *J. Appl. Phys.* **39**, 5061-64 (1968).
644. S. Pickart, *Bull. Am. Phys. Soc.* **5**, 357 (1960).
645. R. Erickson, *Phys. Rev.* **90**, 779 (1953).
646. E. Wollan et al., *Phys. Rev.* **112**, 1132 (1958).
647. K. Yasukochi et al., *J. Phys. Soc. Japan* **14**, 1820 (1959).
648. G. Bacon et al., *Proc. Roy. Soc.* **217A**, 252 (1953).
649. A. Yoshimori, *J. Phys. Soc. Japan* **14**, 807 (1959).
650. "Propriétés Magnetiques et Structurales des Carbonitrures de Manganese et des Perowskites Mn_3GaC et Mn_3GaN," J.-P. Bouchaud, *Ann. Chim., Fr.* **3**, No. 1, 81-105 (1968).
651. J. Banewicz et al., *Phys. Rev.* **117**, 736 (1959).
652. L. Corliss et al., *Phys. Rev.* **104**, 924 (1956).
653. "Magnetic Properties of $ZrFe_2$ and $TiFe_2$ Within Their Homogeneity Range," W. Bruckner, R. Perthel, K. Kleinstuck, and G. E. R. Schulze, *Phys. Stat. Sol.* **29**, 211 (1968).
654. T. Hirone et al., *J. Phys. Soc. Japan* **11**, 1083 (1956).
655. T. Swoboda et al., *Phys. Rev. Letters* **4**, 509 (1960).
656. A. Serres, *J. Phys. Radium* **8**, 146 (1947).
657. R. Lindsey, *Phys. Rev.* **84**, 569 (1951).
658. R. Heikes et al., Reported in Antiferromagnetic Compounds, H. A. Alperin and S. J. Pickart, NOLTR-61-18, AD-259738, April 1961.
659. "Propriétés Electroniques du Monocarbure et du Mononitrure de Neptunium," C. H. de Novion and R. Lorenzelli, *J. Phys. Chem. Solids* **29**, 1901-05 (1968).
660. H. Kammerlingh Onnes and E. Costerhuis, *Commun. Kamerlingh Onnes Lab. Univ. Leiden*, 132e (1913).
661. N. Grazhdankina and D. Gurfel', *J. Exptl. Theoret. Phys. USSR* **35**, 907 (1958); *Soviet Phys. JETP* **8**, 631 (1959).
662. L. Corliss et al., Reported in Antiferromagnetic Compounds, H. A. Alperin and S. J. Pickart, NOLTR 61-18, AD-259738, April 1961.
663. "Magnetic Properties of Indium-Substituted Cobalt Ferroxplana Ferrite," N. N. Efimova, Yu. A. Mamalui, and A. A. Murakhovskii, *Soviet Phys.-Solid State* **10**, 1271-74 (1968).
664. "Antiferromagnetic-Piezoelectric Crystals: $BaMF_4$ (M = Mn, Fe, Co and Ni)," M. Eibschutz and H. J. Guggenheim, *Solid State Commun.* **6**, 737-39 (1968).
665. "Some Antiferromagnetic Resonance Measurements in α-Fe_2O_3," P. R. Elliston and G. J. Troup, *J. Phys. C (Proc. Phys. Soc.)*, Ser. 2, **1**, 169-78 (1968).
666. S. Ogawa, *J. Phys. Soc. Japan* **14**, 1114 (1959).
667. "Investigation of Magnetic Interactions in $TbAlO_3$ by Optical Spectroscopy," S. Hufner, L. Holmes, F. Varsanyi, and L. G. van Uitert, *Phys. Rev.* **171**, 507-13 (1968).
668. R. Flippen and S. Friedberg, *J. Appl. Phys.* **31**, 338 S (1960).
669. T. Haseda and M. Date, *J. Phys. Soc. Japan* **14**, 175 (1958).

REFERENCES

670. T. Sugawara, *J. Phys. Soc. Japan* **14**, 1248 (1959).
671. R. Alikhanov, *J. Exptl. Theoret. Phys. USSR* **37**, 1145 (1959); *Soviet Phys. JETP* **10**, 814 (1960).
672. W. Roth, *Phys. Rev.* **111**, 772 (1958).
673. I. Tsubokawa, *J. Phys. Soc. Japan* **13**, 1432 (1958).
674. J. Singer, *Phys. Rev.* **104**, 929 (1956).
675. "Antiferromagnetism in a 20% Ho-80% Tb Alloy Single Crystal," B. Lebech, *Solid State Commun.* **6**, 761 (1968).
676. "Antiferromagnetic Transitions in MnS_2 and $MnTe_2$," M. S. Lin and H. Hacker, Jr., *Solid State Commun.* **6**, 687-89 (1968).
677. P. K. Iyengar, B. A. Dasannacharya, P. R. Vijayaraghavan, and A. P. Roy, *J. Phys. Soc. Japan* **17**, B-III, 41 (1962).
678. "Effect of Non-Magnetic Scandium Octahedral Substitution on the Magnetization and Initial Permeability of Yttrium Iron Garnet," J. Richard Cunningham, Jr., NOLTR 63-206, December 1963.
679. "Magnetic Properties of $CaCoSiO_4$ and $CaFeSiO_4$," R. E. Newnham, L. G. Caron, and R. P. Santoro, *J. Am. Ceram. Soc.* **49**, 284 (1966).
680. I. Tsubokawa, *J. Phys. Soc. Japan* **11**, 662 (1956).
681. C. Guillaud and S. Barbezat, *Compt. Rend.* **222**, 386 (1946).
682. "Temperature Dependence of the Spontaneous Radiation of Gadolinium Ferrite," N. I. Lyashenko, and L. L. Stakhurskii, *Soviet Phys.-Solid State* **10**, 1498 (1968).
683. "Formation of Ultra Fine Spinel Ferrite," Toshihiko Sato, Makato Sugihara, and Minoru Saito, *Rev. Elec. Commun. Lab.* **11**, 26 (1963).
684. "Etude de $LaErO_3$ par Diffraction Neutronique," J. M. Moreau, J. Mareschal, and E. F. Bertaut, *Solid State Commun.* **6**, 751-56 (1968).
685. J. W. Cable, M. K. Wilkinson, E. O. Wollan, and W. C. Koehler, *Phys. Rev.* **127**, 714 (1962).
686. R. A. Alikhanov, *J. Phys. Soc. Japan* **17**, B-III, 58 (1962).
687. "Pressure Effect on Magnetic Transitions in $CrS_{1.17}$," Takahiko Kamigaichi, Tetsuhiko Okamoto, Nobuo Iwata, and Eiji Tatsumoto, *J. Phys. Soc. Japan* **21**, 2730 (1966).
688. "Equiatomic Transition Metal Alloys of Manganese. V. On the Magnetic Properties of the PtMn Phase," A. F. Andresen, A. Kjekshus, R. Mollerud, and W. B. Pearson, *Acta Chem. Scand.* **20**, 2529 (1966).
689. "Antiferromagnetism in $GdAlO_3$," K. W. Blazey and H. Rohrer, *Helv. Phys. Acta* **40**, 370 (1967).
690. N. J. Poulis and G. E. G. Hardeman, *Physica* **18**, 201 (1952).
691. I. Tsubokawa, *J. Phys. Soc. Japan* **15**, 1664 (1960).
692. R. Naya, M. Murakami, and E. Hirahara, *J. Phys. Soc. Japan* **15**, 360 (1960).
693. K. Hirakawa, *J. Phys. Soc. Japan* **12**, 929 (1957).
694. "Effect of Deuteration on the Neel Temperature of $CoCl_2 \cdot 6H_2O$," D. S. Sahri, *Phys. Letters* **19**, 625 (1966).
695. "Magnetic Properties of Rare Earth Aluminum Compounds with $MgCu_2$ Structure," H. J. Williams, J. H. Wernick, E. A. Nesbitt, and R. C. Sherwood, *J. Phys. Soc. Japan* **17**, Suppl. B-1, 92 (1962).
696. "Ultrasonic Relaxation at the Neel Temperature and Nuclear Acoustic Resonance in MnTe," K. Walther, *Solid State Commun.* **5**, 399 (1967).
697. "Synthesis of $PbCrO_3$," R. C. DeVries and W. L. Roth, General Electric Research and Development Rept. 67-C-187, Schenectady, N.Y., May 1967.
698. E. A. Skrabek and W. E. Wallace, *J. Appl. Phys.* **34**, 1356 (1963).
699. "The Effect of Pressure on the Curie Temperature and Resistivity of Some Rare-Earth Metals and Heusler Alloys," I. G. Austin and P. K. Mishra, *Phil. Mag.* **15**, 529 (1967).
700. "Ordered Magnetism in Solids," Shmuel Shtrikman and David Treves, AFML-TR-66-279, AD-810646, December 1966.
701. "The Magnetic Structure and Hyperfine Field of $FeGe_2$," J. B. Forsyth, C. E. Johnson, and P. J. Brown, *Phil. Mag.* **10**, 713 (1964).

702. "Chemical and Magnetic Order in Platinum-Rich Pt-Fe Alloys," G. E. Bacon and J. Crangle, *Proc. Roy. Soc.* **272A**, 387 (1963).
703. "Development of Microwave Ferrites," J. K. Sinha, Pran Kishan, G. P. Sharma, and R. Kaur, *J. Inst. of Telecommun. Engrs.* **11**, 396 (1965).
704. O. Beckman, *Phys. Rev.* **121**, 376 (1961).
705. A. Okazaki and Y. Suemune, *J. Phys. Soc. Japan* **16**, 4 (1961).
706. "Mossbauer Studies of Fe^2 in Paramagnetic Fayalite (Fe_2SiO_4)," M. Eibschutz and U. Ganiel, *Solid State Commun.* **5**, 267 (1967).
707. "Magnetic Properties of $NaNiF_3$," V. M. Judin and A. B. Sherman, *Phys. Status Solidi* **20**, 759 (1967).
708. "X-Ray Diffraction and Magnetic Studies of Solid Solutions $Bi_{1-x}Ca_xMnO_3$," V. A. Bokov, N. A. Grigoryan, and M. F. Bryzhina, *Phys. Status Solidi* **20**, 745 (1967).
709. "Magnetic and Crystallographic Studies of Substituted Gadolinium Iron Garnets," S. Geller, H. J. Williams, R. C. Sherwood, and G. P. Espinosa, *J. Appl. Phys.* **36**, 88 (1965).
710. "Antiferromagnetism in $LiFePO_4$," R. P. Santoro and R. E. Newnham, *Acta Cryst.* **22**, 344 (1967).
711. "Magnetic Properties of Some Alloys of Thorium with Holmium and Erbium," W. C. Koehler, H. R. Child, J. W. Cable, and R. M. Moon, *J. Appl. Phys.* **38**, 1384 (1967).
712. "Propriétés Magnetiques des Alliages de Terres Rares Avec le Nickel, de Formule T_2Ni_{17}," Jean Laforest, Remy Lemaire, Dominique Paccard, and Rene Pauthenet, *Compt. Rend.* **264**, 676 (1967).
713. "Ordering and Magnetic Transformations in an Iron-Aluminum Alloy," M. V. Dekhtyar, *Fiz. Metal. i Metalloved.* **23** (1), 37–42 (1967).
714. "Magnetic Properties of $ZrZn_2$ – Itinerant Electron Ferromagnet," Shinji Ogawa and Nobuhiko Sakamoto, *J. Phys. Soc. Japan* **22**, 1214 (1967).
715. A. Aharoni, E. H. Frei, and M. Scheiber, *Phys. Rev.* **127**, 439 (1962).
716. "Etude par Diffraction Neutronique de $FeCr_2Se_4$ Antiferromagnetique," M. Maurice Chevreton and Bernadette Andron, *Compt. Rend.* **264**, 316 (1967).
717. "Influence of Crystal Fields on the Magnetic Properties of the Rare-Earth Nitrides," G. Busch, P. Junod, F. Levy, A. Menth, and O. Vogt, *Phys. Letters* **14**, 264 (1965).
718. McNiff, Sadagopan, and S. Foner, Massachusetts Institute of Technology National Magnet Laboratory QTSR-28, June 1967.
719. "Structure Cristalline du Terbium a 120–300°K," V. A. Finkel, Ju. N. Smirnov, and V. V. Vorobev, *Zh. Eksperim. i Teor. Fiz., SSSR* **51**, 32–7 (1966).
720. "The Neel Temperature of MnF_2 in the Two-Spin Cluster-Variation Limit," L. F. Libelo and T. Tanaka, NOLTR-66-193, AD-644-897, Oct. 3, 1966.
721. P. Heller, *Phys. Rev.* **146**, 403 (1966).
722. "The Ferromagnetic Domain Structures in Haematite," B. Gustard, *Proc. Roy. Soc. (GB)A* **297**, 269 (1967).
723. "Magnetic and Paramagnetic Resonance Properties of $EuAl_4$," J. H. Wernick, H. J. Williams, and A. C. Gossard, *J. Phys. Chem. Solids* **28**, 271–3 (1967).
724. "The Magnetic Properties of $Ni(OH)_2$ and β-$Co(OH)_2$," H. Miyamoto, *Bull. Inst. Chem. Res. Kyoto Univ. (Japan)* **44**, 420–9 (1966).
725. "Diffusion Magnetique des Neutrons par MnF_2 Pres du Point de Neel," M. Antonini, *J. Phys. Chem. Solids* **28**, 11–16 (1967).
726. "Ferrimagnetic and Antiferromagnetic Structures of Cr_5S_6," B. van Laar, *Phys. Rev.* **156**, 654 (1967).
727. "Structure Magnetique des Solutions Solides $Mn_xCr_{1-x}S$," Paul Burlet and Erwin Felix Bertaut, *Compt. Rend.* **264**, 323 (1967).
728. "Magnetic Anisotropy Measurement of MnO Single Crystal," Enji Uchida, Hisamoto Kondoh, Yoshihide Nakazumi, and Takeo Nagamiya, *J. Phys. Soc. Japan* **15**, 466 (1960).
729. "Study of the (^{57}Fe) Mossbauer Effect of the System $(Fe_xMn_{1-x})TO_3$ (T = Rare Earth or Y), ($0 \leq x \leq 1$)," J. Chappert, *J. Phys. (France)* **28**, 81–8 (1967).

REFERENCES

730. "Magnetic Transition in $TiCl_3$," Shinji Ogawa, *J. Phys. Soc. Japan* **15**, 1901 (1960).
731. W. P. Wolf, M. Ball, M. T. Hutchings, M. J. M. Leask, and A. F. G. Wyatt, *J. Phys. Soc. Japan* **17**, Suppl. B-1, 443 (1962).
732. "Magnetic Ordering in Palladium-Iron Alloys," J. A. Mydosh, J. I. Budnick, M. P. Kawatra, and S. Skalski, *Phys. Rev. Letters* **21**, 1346-49 (1968).
733. A. Herpin and P. Meriel, *Compt. Rend.* **259**, 2416, 30 (1964).
734. "Magnetic Properties of $Mn_{3-x}Ni_xB_4$ and $Mn_{3-x}Co_xB_4$," Hozumi Hirota and Akira Yanase, *J. Phys. Soc. Japan* **21**, 433 (1966).
735. "Ferromagnetism in Solid Solutions of Scandium and Indium," B. T. Matthias, A. M. Clogston, H. J. Williams, E. Corenzwit, and R. C. Sherwood, *Phys. Rev. Letters* **7**, 7 (1961).
736. "Magnetic Properties of Intermetallic Compounds Between the Lanthanides and Nickel or Cobalt," John Farrell and W. E. Wallace, *Inorg. Chem.* **5**, 105 (1966).
737. R. J. Weiss and A. S. Marotta, *J. Phys. Chem. Solids* **9** (1959).
738. "Low-Temperature Heat Capacity and Thermodynamic Functions of Ferromagnetic Uranium Phosphide U_3P_4," B. Stalinski, Z. Bieganski, and R. Troc, *Phys. Status Solidi* **17**, 837 (1966).
739. "Experimental Study of the Body-Centered-Cubic Heisenberg Ferromagnet," A. R. Miedema, R. F. Wielinga, and W. J. Huiskamp, *Physica* **31**, 1585 (1965).
740. "Magnetic Properties of Nickel- and Cobalt-Iodate," H. C. Meijer and J. van den Handel, *Physica* **30**, 1633 (1964).
741. "The Rare Earth Garnets," L. Neel, R. Pauthenet, and B. Dreyfus, Chap. 7 in *Progress in Low Temperature Physics*, Vol. 4, C. J. Gorter, ed., North Holland, Amsterdam, 1964, p. 344.
742. "Thermal, Structural and Magnetic Studies of Metals and Intermetallic Compounds," W. E. Wallace and R. S. Craig, NYO-3454-13, May 15, 1967, Contract AT-(30-1)-3454.
743. "High-Pressure Study of the First-Order Phase Transition in MnAs," J. B. Goodenough and J. A. Kafalas, *Phys. Rev.* **157**, 389 (1967).
744. "Solid State Research Quarterly Technical Summary-Feb. 1, 1967 Through April 30, 1967," Alan L. McWhorter, Lincoln Laboratory, M.I.T., ESD-TR-67-266, Contract AF-19(628)-5167, Project 649L, May 15, 1967.
745. "Contribution to the Temperature Dependence of the Magnetic Properties of Permanent Magnetic Materials," Hermann Dietrich, *Cobalt* **35**, 78 (1967).
746. "Sublattice Magnetization in FeF_3 Near the Neel Point," G. K. Wertheim, H. J. Guggenheim, and D. N. E. Buchanan, *Solid State Commun.* **5**, 537 (1967).
747. "Thermal Conductivity and Specific Heat of Yttrium-Calcium Garnets," E. D. Devyatkova and V. V. Tikhonov, *Soviet Phys.-Solid State* **9**, 604 (1967).
748. G. A. Smolenskii, V. P. Polyakov, and V. M. Yudin, *Izv. Akad. Nauk USSR, Ser. Fiz.* **11**, 1396 (1961).
749. "The System Neodymium-Manganese (Structures, Magnetic Properties, Phase Diagram). The Phase Diagrams Yb-Hg and Tb-Hg," Prof. Dr. Franz Lihl, Scientific Report No. 2 from the Institut fuer Angewandte Physik, Technische Hochschule, Vienna, Austria, U.S. Contract 61(052)-942, June 20, 1967.
750. "Mossbauer Effect Studies on the $x(Rh_2O_3)_{1-x}(Fe_2O_3)$ System," I. Dezsi, Gy. Erlaki, and L. Keszthelyi, *Phys. Status Solidi* **21**, K121 (1967).
751. "Magnetic Properties of UTe_2 and UTe_3 Single Crystals," A. V. Pechennikov, V. I. Chechernikov, M. E. Barykin, G. V. Ellert, V. K. Slovyanskikh, and E. I. Yarembash, *Izv. Akad. Nauk SSSR, Neorg. Mater.* **4**, No. 8, 1342-3 (1968).
752. "The Preparation, Crystal Structure and Magnetic Properties of $Na_3Fe_5O_9$," C. Romers, C. J. M. Rooymans, and R. A. G. deGraaf, *Acta Cryst.* **22**, 766 (1967).
753. "Ueber die Anfangssuszeptibilität Oberhalb und die Spontane Magnetisierung Unterhalb der Curie-Temperatur von Eisen, Kobalt und Nickel," Kark Kryno Geissler, Rudolf Kohlhaas, and Heinrich Lange, *Z. Naturforsch.* **22a**, 830 (1967).
754. W. Rocker and R. Kohlhaas, *Z. Naturforsch.* **22a**, 291 (1967).
755. "Metamorphism of Titaniferous Iron Ores of the Ufalei Group of Deposits," G. N. Vertushkov, Yu. A. Sokolov, and V. I. Yakshin, *Zap. Vses. Mineralog. Obshchestva* **95**, 10–17 (1966).

756. "Critical Points and Moments of Some Binary Rare Earth Alloys," R. M. Bozorth and R. J. Gambino (IBM, Yorktown Heights, N.Y.), *Intern. Aerospace Abstr.* **5**, 2851 (1965) (IAA Accession No. A65-29951, 3 p., 1964).

757. "Perovskites with Ferromagnetic Properties," L. I. Shvorneva and Yu. N. Venevtsev, *Zh. Eksperim. i Teor. Fiz.* **49**, 1038-41 (1965).

758. "Neutron Diffraction Studies on the (1-x) Fe_2O_3-xRh_2O_3 System," E. Kren, P. Szabo, and G. Konczos, *Phys. Letters* **19**, 103 (1965).

759. "Ferrimagnetism of the Rare-Earth-Cobalt Intermetallic Compounds R_2Co_{17}," K. Strnat, G. Hoffer, W. Ostertag, and J. C. Olson, *J. Appl. Phys.* **37**, 1252 (1966).

760. "Heat Conductivity of Pure Iron Between -180 and $1000°$ with Special Reference to Phase Transformations," Friedhelm Richter and Rudolf Kohlhaas, *Arch. Eisenhuettenw.* **36**, 827 (1965).

761. "Existence of a Magnetic Transition in U Sesquicarbide," C. de Novion, P. Costa, and G. Dean, *Phys. Letters* **19**, 455 (1965).

762. "Magnetic Properties of Yttrium Manganite," F. Bertaut, R. Pauthenet, and M. Mercier, *Phys. Letters* **18**, 13 (1965).

763. D. O. Smith, *Phys. Rev.* **102**, 959 (1956).

764. "Antiferromagnetism of Uranium and Thorium Phosphides," V. I. Chechernikov, T. M. Shavishvili, V. A. Pletyushkin, and V. K. Slovyanskikh, *Zh. Eksper. Teor. Fiz. (USSR)* **55**, No. 1, 151-6 (Jan. 1968).

765. "Magnetic Properties and Paramagnetic Relaxation of Dysprosium in $DyFeO_3$ Below the Curie Point," I. Nowik and H. J. Williams, *Phys. Letters* **20**, 154 (1966).

766. "Electrical Resistivity of Equiatomic Rare-Earth-Noble-Metal Compounds," Chang-Chih Chao, *J. Appl. Phys.* **37**, 2081-4 (1966).

767. "The Magnetic Properties of Cadmium Manganite," S. K. Dey and J. C. Anderson, *Phil. Mag.* **12**, 975 (1965).

768. "Magnetic and Electrical Properties of $NdNbO_4$ and $GdNbO_4$," F. F. Y. Wang and R. L. Gravel, *Phys. Status Solidi* **12**, 609 (1965).

769. "Magnetic Properties of Cr Rich Fe–Cr Alloys at Low Temperatures," Y. Ishikawa, R. Tournier, and J. Filippi, *J. Phys. Chem. Solids* **26**, 1727 (1965).

770. "Magnetic Properties of Some Oxides with Spinel Structure," G. Blasse, *Philips Res. Rept.* **20**, 528 (1965).

771. H. Watanabe and M. Fukase, *J. Phys. Soc. Japan* **16**, 1181 (1961).

772. "On the Spontaneous Magnetization of $MnFe_2O_4$," F. K. Lotgering, *Philips Res. Rept.* **20**, 320 (1965).

773. Preparation, Etudes Cristallines et Magnetiques du Manganite de Nickel $NiMn_2O_4$, Gerard Villers and Robert Buhl, *Compt. Rend.* **260**, 3406 (1965); Structures Magnetiques et Etude des Propriétés Magnetiques des Spinelles Cubiques $NiMn_2O_4$, B. Boucher, R. Buhl, and M. Perrin, *J. Phys. Chem. Solids* **31**, 363-83 (1970).

774. "The Antiferromagnetic Curie Point in α-Fe_2O_3," S. Freier, M. Greenshpan, P. Hillman, and H. Shechter, *Phys. Letters* **2**, 191 (1962).

775. "Mossbauer Effect in Ferroelectric-Antiferromagnetic $BiFeO_3$," V. G. Bhide and M. S. Multani, *Solid State Commun.* **3**, 271 (1965).

776. "Magnetic Properties of $M_{1-x}Li_xFe_2O_{4-x}F_x$," C. Okazaki, E. Hirota, Y. Neichi, H. Okazaki, and S. Nakajima, *J. Phys. Soc. Japan* **21**, 199–200 (1966).

777. Magnetic Properties of Uranium Tellurides, V. I. Chechernikov, A. V. Pechennikov, M. E. Barykin, V. K. Slovyanskikh, E. I. Yarembash, and G. Ellert, *Zh. Eksperim. i Teor. Fiz.* **52**, 854 (1967); English transl.: *Soviet Phys. JETP* **25**, 560-61 (1967).

778. "Crystal Growth of CrO_2," B. L. Chamberland, *Mater. Res. Bull.* **2**, 827–35 (1967).

779. "Magnetic Properties of $Eu_{1-x}Ca_xO$ Solid Solutions," A. A. Samokhvalov, T. D. Zotov, N. V. Volkenshtein, V. G. Bamburov, A. A. Ivakin, Yu. N. Morozov, and M. I. Simonova, *Soviet Phys.-Solid State* **9**, 555 (1967).

780. "Specific Heat Singularities of the Ising Antiferromagnets $CoCs_3Cl_5$ and $CoCs_3Br_5$," R. F. Wielinga, H. W. J. Blote, J. A. Roest, and W. J. Huiskamp, *Physica* **34**, 223–40 (1967).

781. "Magnetic Characteristics of Some Lanthanide Nitrides," David P. Schmacher and W. E. Wallace, *Inorg. Chem.* **5**, 1563 (1966).

REFERENCES

782. "Recent Developments in the Chemical Thermodynamics of the Uranium Chalcogenides," E. F. Westrum, Jr., *Thermodynamics, Proc. Symp., Vienna, 1965* **2**, 497–510, discussion 571 (Pub. 1966).
783. "The Thermoelectric Power in Chromium and Vanadium," A. R. Mackintosh and L. Sill, *J. Phys. Chem. Solids* **24**, 501 (1963).
784. "Magnetic Properties of Several Metasilicates and Metagermanates with Pyroxene Structure," Akira Sawaoka, Syohei Miyahara, and Syun-iti Akimoto, *J. Phys. Soc. Japan* **25**, 1253-57 (1968).
785. "Magnetic Properties of Some Intra-Rare Earth Alloys," Paul Edward Roughan, thesis, Iowa State Univ., Ames, Iowa, 1962, 111 pages.
786. "Long-Range Antiferromagnetism in Disordered Fe-Ni-Mn Alloys," J. S. Kouvel and J. S. Kasper, *J. Phys. Chem. Solids* **24**, 529 (1963).
787. "Permanent Magnetic Properties of Iron-Cobalt-Phosphides," K. J. de Vos, W. A. J. J. Velge, M. G. van der Steeg, and H. Zijlstra, *J. Appl. Phys.* **33** Suppl., 1320 (1962).
788. "Antiferromagnetic-Ferromagnetic Transition in Gold-Manganese-Indium Alloys," D. P. Morris, G. W. Davies, and C. D. Price, *J. Phys. Chem. Solids* **23**, 109 (1962).
789. "Neutron Diffraction Studies of Various Transition Elements," C. G. Shull and M. K. Wilkinson, *Rev. Mod. Phys.* **25**, 100 (1953).
790. "Antiphase Antiferromagnetic Structure of Chromium," R. J. Weiss, L. M. Corliss, and J. M. Hastings, *Phys. Rev. Letters* **3**, 211 (1959).
791. "Magnetic Properties of Chromium," Bykov et al., *Soviet Phys. Doklady English Transl.* **4**, 1070 (1960).
792. "Influence d'Oxygen sur les Propriétés Magnetiques du Composé de la Phase Laves de Fer et Titane," Mototaka Okazaki, Kenzo Kai, Takuro Nakamichi, and Mikio Yamamoto, *J. Phys. Soc. Japan* **25**, 1509-10 (1968).
793. "Saturation Magnetization and Crystal Chemistry of Ferrimagnetic Oxides," E. W. Gorter, *Philips Res. Rept.* **9**, 295–320 (1954).
794. "Research on Spontaneous Magnetization in Solid Bodies," I. S. Jacobs, D. S. Rodbell, and W. L. Roth, ASD Tech. Rept. 61-630, February 1962, Contract AF-33(616)-7396, Project 7371.
795. "Polarized Neutron Study of Antiferromagnetic Domains in MnF_2," H. A. Alperin, P. J. Brown, R. Nathans, and S. J. Pickart, *Phys. Rev. Letters* **8**, 237 (1962).
796. "Crystal Structure of Paramagnetic $DyMn_2O_5$ at 298°K," S. C. Abrahams and J. L. Bernstein, *J. Chem. Phys.* **46**, 3776 (1967).
797. "Heat Capacity of Palladium and Dilute Palladium: Iron Alloys from 1.4 to 100°K," Boyd W. Veal and John A. Rayne, *Phys. Rev.* **135**, A442 (1964).
798. "The Magnetization of Some Nickel Alloys in Magnetic Fields up to 15 kOe Between 0°K and 300°K," H. C. van Elst, B. Lubach, and G. J. van den Berg, *Physica* **28**, 1297 (1962).
799. "Magnetic Moments and Unpaired Spin Densities in the Fe-Rh Alloys," G. Shirane, R. Nathans, and C. W. Chen, *Phys. Rev.* **134**, A1547 (1964).
800. "Oxides. Synthesis and Properties of Ferromagnetic Chromium Oxide," T. J. Swoboda, Paul Arthur, Jr., N. L. Cox, J. N. Ingraham, A. L. Oppegard, and M. S. Sadler, *J. Appl. Phys.* **32**, Suppl., 374S (1961).
801. "Electron Ordering Transitions in Several Chromium Spinel Systems," Ronald J. Arnott, Aaron Wold, and Donald B. Rogers, *J. Phys. Chem. Solids* **25**, 161 (1964).
802. "The Crystal Structures and Magnetic Properties of the Rare-Earth Nickel (RNi) Compounds," S. C. Abrahams, J. L. Bernstein, R. C. Sherwood, J. H. Wernick, and H. J. Williams, *J. Phys. Chem. Solids* **25**, 1069 (1964).
803. "Neutron Diffraction Studies at the Puerto Rico Nuclear Center," I. Almodovar, H. J. Bielen, B. C. Frazer, and M. I. Kay, *J. Phys. (France)* **25**, 442 (1964).
804. "Concerning the Magnetic Susceptibility of Praseodymium Oxides," J. B. MacChesney, H. J. Williams, R. C. Sherwood, and J. F. Potter, *J. Chem. Phys.* **41**, 3177 (1964).
805. "Note on the Preparation of Th and U Oxychalcogenides," H. U. Boelsterli and F. Hulliger, *J. Mater. Sci.* **3**, 664-65 (1968).

806. "Etude du Systeme $CuFe_2O_4$-$CuGa_2O_4$," Michel Lenglet and Jean-Claude Tellier, *Compt. Rend.* **267**, 525-27 (1968).
807. "Neutron Diffraction Study of the Intermetallic Compound FeSi," Hiroshi Watanabe, Hisao Yamamoto, and Ken-ichi Ito, *J. Phys. Soc. Japan* **18**, 995 (1963).
808. "Parametres des Manganites de Terres Rares Perovskites et Structure Magnetique du Manganese dans $MnPrO_3$ et $MnNdO_3$ par Diffraction Neutronique," S. Quezel-Ambrunaz, *Bull. Soc. fr. Mineral. Crist.* **91**, 339-43 (1968).
809. "Antiferromagnetic Structures of Manganese and Cobalt Fluosilicates Below 1°K," Akio Ohtsubo, *J. Phys. Soc. Japan* **20**, 82 (1965).
810. "Structural and Magnetic Properties of a Bi-MnBi Composite," J. M. Noothoven van Goor and H. Zijlistra, *J. Appl. Phys.* **39**, 5471-74 (1968).
811. R. G. Shulman and B. J. Wyluda, *J. Chem. Phys.* **35**, 1498 (1961).
812. "Magnetic Characteristics of Gadolinium, Praseodymium, and Thulium Nitrides," David P. Schumacher and W. E. Wallace, *J. Appl. Phys.* **36**, 984 (1965).
813. "Antiferromagnetism and the Magnetic Phase Diagram of $GdAlO_3$," K. W. Blazey and H. Rohrer, *Phys. Rev.* **173**, 574-80 (1968).
814. K. G. Srivastava, *Phys. Letters* **4**, 55 (1963).
815. "High Field Magnetic Moment and Antiferromagnetic Resonance Measurements in Alpha-Fe_2O_3, CoF_2, FeF_2 and $(MnF_2)_{1-x}(ZnF_2)_x$," S. Foner, *Proc. Intern. Conf. Magnetism, Nottingham*, Sept. 7-11, 1964, p. 438.
816. "A Resistance-Measurement Study of Ordering in Iron-Silicon Alloys, FeSi- and Fe_3Si-Type Superstructures," Roger N. Dokken, *Trans. AIME* **233**, 1187 (1965).
817. "Magnetic Properties of $MnCr_2S_4$," N. Menyuk, K. Dwight, and A. Wold, *J. Appl. Phys.* **36**, 1088 (1965).
818. "The Temperature Dependence of Optical Absorption in the Antiferromagnetic $KNiF_3$," R. V. Pisarev and S. D. Prokhorova, *Soviet Phys.-Solid State* **10**, 1668-72 (1969).
819. "Distribution of Magnetic Moments in Pd-3d and Ni-3d Alloys," J. W. Cable, E. O. Wollan, and W. C. Koehler, *Phys. Rev.* **138**, A755 (1965).
820. "First-Order Magnetic Phase Change in Chromium at 38.5°C," A. Arrott, S. A. Werner, and H. Kendrick, *Phys. Rev. Letters* **14**, 1022 (1965).
821. "Magnetization Distribution in a Palladium-Rich FePd Alloy," Walter C. Phillips, *Phys. Rev.* **138**, A1649 (1965).
822. "Initial Permeability and Derived Anisotropy of Scandium-Substituted Yttrium Iron Garnets," J. Richard Cunningham, Jr., *J. Appl. Phys.* **36**, 2491 (1965).
823. "Calorimetric Study of Several Rare-Earth Gallium Garnets," D. G. Onn, Horst Meyer, and J. P. Remeika, *Phys. Rev.* **156**, 663 (1967).
824. "Preparation and Magnetic Properties of Cobalt Disulfide," B. Morris, V. Johnson, and A. Wold, *J. Phys. Chem. Solids* **28**, 1565 (1967).
825. "Antiferromagnetism in Dilute Iron Chromium Alloys," Yoshikazu Ishikawa, Sadao Hoshino, and Yasuo Endoh, *J. Phys. Soc. Japan* **22**, 1221 (1967).
826. "A Neutron Diffraction Study of the Magnetic Behavior of Gadolinium," J. W. Cable and E. O. Wollan, *Phys. Rev.* **165**, 733-34 (1968).
827. "Effect of Pressure on the Curie Temperature of CrTe and MnSb Compounds of the Nickel Arsenide Type," Hideaki Ido, Takejiro Kaneko, and Kazuo Kamigaki, *J. Phys. Soc. Japan* **22**, 1418 (1967).
828. "Mossbauer Studies of Fe^{57} in Orthoferrites," M. Eibschutz, S. Shtrikman, and D. Treves, *Phys. Rev.* **156**, 562 (1967).
829. "Antiferromagnetism of Alpha-FeOOH Investigated with the Mossbauer Effect," A. Z. Hrynkiewicz, D. S. Kulgawczuk, and K. Tomala, *Phys. Letters* **17**, 93 (1965).
830. "Magnetic Properties of Uranium Compounds with Elements of the V A and VI A Groups. I. Compounds of UX Type," J. Grunzweig-Genossar, M. Kuznietz, and F. Friedman, *Phys. Rev.* **173**, 562-73 (1969).

REFERENCES

831. "Oxygen Stoichiometry in the Barium Ferrates; Its Effect on Magnetization and Resistivity," J. B. MacChesney, J. F. Potter, R. C. Sherwood, and H. J. Williams, *J. Chem. Phys.* **43**, 3317 (1965).
832. "Thermodynamic Properties of Uranium Compounds. III. Low-Temperature Heat Capacity and Entropy of UP and U_3P_4," J. F. Counsell, R. M. Dell, A. R. Junkison, and J. F. Martin, *Trans. Faraday Soc.* **63**, 72 (1967).
833. "Specific Heat of $Fe_3(PO_4)_2 \cdot 8H_2O$," H. Forstat, N. D. Love, and J. McElearney, *Phys. Rev.* **139**, A1246 (1965).
834. J. T. Sparks and T. Komoto, *J. Appl. Phys.* **37**, 1040 (1966).
835. J. M. Leger, C. Susse, R. Epain, and B. Vodar, *Solid State Commun.* **4**, 197 (1966).
836. J. R. Shane, R. W. Kedzie, and M. Kestigian, *J. Appl. Phys.* **37**, 1134 (1966).
837. I. N. Kalinkina and V. N. Kostryukov, *Soviet Phys.-Solid State* **8**, 137 (1966).
838. N. Ichinose and T. Yoshioka, *J. Phys. Soc. Japan* **21**, 1471 (1966).
839. H. Umebayashi and Y. Ishikawa, *J. Phys. Soc. Japan* **21**, 1281 (1966).
840. J. van den Handel and H. C. Meijer, *Proc. Intern. Conf. Low Temperature Physics, 9th, Columbus, 1964*, Plenum, New York, 1965, p. 873.
841. R. Z. Levitin and B. K. Ponomarev, *Soviet Phys.-JETP* **23**, 984 (1966).
842. M. C. Montmory, E. F. Bertaut, and P. Mollard, *Solid State Commun.* **4**, 749 (1966).
843. J. Skalyo, A. F. Cohen, and S. A. Friedberg, *Proc. Intern. Conf. Low Temperature Physics, 9th, Columbus, 1964*, Plenum, New York, 1965, Pt. B, p. 884.
844. "Magnetic Properties of Ordering Rare-Earth Antimonides," G. Busch, O. Marincek, A. Menth, and O. Vogt, *Phys. Letters* **14**, 262 (1965).
845. G. Fabri, E. Germagnoli, M. Musci, and V. Svelto, *Phys. Rev.* **138**, A178 (1965).
846. "Magnetic Properties of $LiCoPO_4$ and $LiNiPO_4$," R. P. Santoro, D. J. Segal, and R. E. Newnham, *J. Phys. Chem. Solids* **27**, 1192 (1966).
847. I. Tsubokawa, *J. Phys. Soc. Japan* **15**, 2243 (1960).
848. E. F. Bertaut, A. Delapalme, F. Forrat, G. Roult, F. de Bergevin, and R. Pauthenet, *J. Appl. Phys. Suppl.* **33**, 1123 (1962).
849. G. Rassman and H. Wich, *Arch. Eisenhuettenw.* **33**, 115 (1963).
850. "Untersuchungen zur Kristallstruktur und Magnetischen Struktur des Ferberits $FeWO_4$," Dincer Ulku, *Z. Krist.* **124**, 192 (1967).
851. "The Magnetic Properties of $EuLu_2O_4$ at Low Temperatures," A. A. Samokhvalov, Yu. N. Morozov, V. G. Bamburov, N. V. Volkenshtein, and T. D. Zotov, *Soviet Phys-Solid State* **10**, 1729-30 (1969).
852. A. Giansoldati, J. O. Linde, and G. Borelius, *J. Phys. Chem. Solids* **9**, 183 (1959).
853. I. S. Jacobs, J. S. Kouvel, and P. E. Lawrence, *J. Phys. Soc. Japan* **17**, Suppl. B-I, 157 (1962).
854. T. Ohoyama, *J. Phys. Soc. Japan* **16**, 1995 (1961).
855. R. Wendling, *Compt. Rend.* **252**, 3207 (1961).
856. K. Yasukochi, K. Kanematsu, and T. Ohoyama, *J. Phys. Soc. Japan* **16**, 1123 (1961).
857. S. T. Lin and A. R. Kaufmann, *Phys. Rev.* **108**, 1171 (1957).
858. "Suszeptibilitätsmessungen an $DyAlO_3$ und $HoFeO_3$," I. Grambow, P. Kronauer, J. Schneider, and H. Schuchert, *Z. Naturforsch.* **22a**, 828 (1967).
859. S. D. Margolin and I. G. Fakidov, *Phys. Metals Metallog. (USSR) (English Transl.)* **9**, 22 (1960).
860. A. J. P. Meyer and R. Asfeld, *Compt. Rend.* **254**, 4266 (1962).
861. M. Asanuma, *J. Phys. Soc. Japan* **15**, 1343 (1960).
862. E. P. Warekois, *J. Appl. Phys.* **31**, 346 S (1960). (Suppl.).
863. R. J. Chandross and D. P. Shoemaker, *J. Phys. Soc. Japan* **17**, Suppl. B-III, 16 (1962).
864. E. Piegger and R. S. Craig, *J. Chem. Phys.* **39**, 137 (1963).
865. C. P. Bean and R. S. Rodbell, *Phys. Rev.* **126**, 104 (1962).
866. A. J. P. Meyer, *Compt. Rend.* **244**, 2028 (1957); *J. Phys. Radium* **20**, 430 (1959).
867. I. Tsuboya and M. Sugihara, *J. Phys. Soc. Japan* **18**, 143 (1963).
868. R. Fontaine and R. Pauthenet, *Compt. Rend.* **254**, 650 (1962).

869. "Etude Radiocristallographie à Basse Temperature de la Transition Magnetique de l'Oxyde d'Uranium UO_2," Rostislav de Kouchkovsky and Marcel Lecomte, *Compt. Rend.* **267**, 620-22 (1968).
870. K. Aoyagi and M. Sugihara, *J. Phys. Soc. Japan* **17**, 1072 (1962).
871. T. Taoka and T. Ohtsuka, *J. Phys. Soc. Japan* **9**, 723 (1954).
872. J. S. Kouvel, C. D. Graham, Jr., and J. J. Becker, *J. Appl. Phys.* **29**, 518 (1958).
873. C. Guillaud, *Compt. Rend.* **235**, 468 (1952).
874. M. Asanuma, *J. Phys. Soc. Japan* **17**, 300 (1962).
875. L. V. Cherry and W. E. Wallace, *J. Appl. Phys. Suppl.* **32**, 340 (1961).
876. S. Tezuka, S. Sakai, and Y. Nakagawa, *J. Phys. Soc. Japan* **15**, 931 (1960).
877. Y. Nakagawa, S. Sakai, and T. Hori, *J. Phys. Soc. Japan* **17**, Suppl. B-I, 168 (1962).
878. C. E. Olsen, *J. Appl. Phys. Suppl.* **31**, 340 (1960).
879. S. Komura, N. Kunitomi, Poh-Kun Tseng, N. Shikazono, and H. Takekochi, *J. Phys. Soc. Japan* **16**, 1479 (1961).
880. W. M. Hubbard, E. Adams, and J. V. Gilfrich, *J. Appl. Phys. Suppl.* **31**, 368 (1960).
881. "The Magnetic and Chemical Ordering of the Heusler Alloys Pd_2MnSn and Pd_2MnSb," P. J. Webster and R. S. Tebble, *Phil. Mag.* **16**, 347 (1967).
882. "Magnetic and Chemical Order in Pd_2MnAl in Relation to Order in the Heusler Alloys Pd_2MnSn, Pd_2MnIn, and Pd_2MnSb," P. J. Webster and R. S. Tebble, presented at the International Congress on Magnetism, Boston-Cambridge, Mass., Sept. 11–15, 1967.
883. V. B. Compton and B. T. Matthias, *Acta Cryst.* **12**, 651 (1959).
884. D. P. Morris, R. R. Preston, and I. Williams, *Proc. Phys. Soc. (London)* **73**, 520 (1959).
885. "Magnetic Properties of the Spinel System $Fe_{1-x}Cu_xCr_2S_4$," G. Haacke and L. C. Beegle, *J. Phys. Chem. Solids* **28**, 1699 (1967).
886. J. B. Goodenough, *J. Appl. Phys.* **31**, 359 S (1960) (Suppl.).
887. "Ferrimagnetism in the $RbMg_{1-x}Co_xFe_3$ System," M. W. Shafer and T. R. McGuire, *Phys. Letters* **27A**, 676-77 (1968).
888. "Magnetic Structure of DyAg," G. Arnold, N. Nereson, and C. Olsen, *J. Chem. Phys.* **46**, 4041 (1967).
889. "Magnetism of the Titanium-Oxygen System," L. K. Keys and L. N. Mulay, *Japan. J. Appl. Phys.* **6**, 122 (1967).
890. "Neutron Diffraction Study of FeSn," Kazuyuki Yamaguchi and Kiroshi Watanabe, *J. Phys. Soc. Japan* **22**, 1210 (1967).
891. "Superparamagnetic Magnesioferrite Precipitates from Dilute Solutions of Iron in MgO," G. P. Wirtz and M. E. Fine, *J. Appl. Phys.* **38**, 3729 (1967).
892. "On the Magnetocrystalline Anisotropy of Iron Selenide Fe_7Se_8," Takashi Kamimura, Kazuo Kamigaki, Tokutaro Hirone, and Kiyoo Sato, *J. Phys. Soc. Japan* **22**, 1235 (1967).
893. R. F. Wielinga, Kamerlingh Onnes Laboratory, Leiden, Netherlands, private communication, September 1967.
894. S. J. Pickart and H. A. Alperin, in Landolt-Boernstein, 6th ed., Magnetic Properties I, Springer, Berlin, 1962, Vol. II/9, pp. 3-143 to 3-154.
895. "The Magnetic Structure of K_2IrCl_6," M. T. Hutchings and C. G. Windsor, *Proc. Phys. Soc.* **91**, 928 (1967).
896. "Study of the System Magnesium Ferrate-Magnesium Gallate," Jean-Claude Tellier and Maurice Lensen, *Compt. Rend.* **255**, 125 (1962).
897. "Preparation and Crystal Chemistry of Divalent Europium Compounds," M. W. Shafer, *J. Appl. Phys.* **36**, 1145 (1965).
898. "Magnetic and Structural Characteristics of Pseudo-Binary Systems Based on Laves Phases," W. Loser and R. S. Craig, in *Thermal, Structural and Magnetic Studies of Metals and Intermetallic Compounds*, Annual Report to USAEC, May 15, 1965, to May 15, 1966, W. E. Wallace and R. S. Craig, NYO-3454-6, Contract No. AT(30-1) 3454.
899. "Magnetic Susceptibilities of Sodium Neodymium Tungstate and Sodium Terbium Tungstate," F. F. Y. Wang and M. Kestigian, *Phys. Status Solidi* **23**, 289 (1967).

REFERENCES

900. "Sodium Iron Fluoride, a Transparent Ferrimagnet," E. G. Spencer, S. B. Berger, R. C. Linares, and P. V. Lenzo, *Phys. Rev. Letters* **10**, 236 (1963).
901. "Magnetic Properties of Some Refractory Uranium Compounds," M. Allbutt, A. R. Junkison, and R. M. Dell, Met. Soc. AIME, Petroleum Eng., Inst. Metals Div. Special Rept., Ser. 10, No. 13, pp. 65–81 (1964).
902. Private communication, Dr. Thomas B. Reed, Lincoln Laboratory, Massachusetts Institute of Technology, Lexington, Massachusetts 02173.
903. "Magnetic Properties of a Single Crystal of Fe_2TiO_4," Y. Ishikawa, *Phys. Letters* **24A**, 725 (1967).
904. "Neutron Diffraction Study of Ce_2C_3 at Low Temperatures," Masao Atoji, *J. Chem. Phys.* **46**, 4148 (1967).
905. "Susceptibility of MnO Measured by the NMR Method," K. Tompa, F. Toth, and G. Gruner, *Phys. Status Solidi* **22**, K11 (1967).
906. "Crystal Structure and Magnetic Properties of Some Rare Earth Germanides," K. H. J. Buschow and J. F. Fast, *Phys. Status Solidi* **21**, 593 (1967).
907. Y. Lecocq, M. Laridjani, and P. Lecocq, *Compt. Rend.* **258**, 1344 (1964).
908. "The Constitution of $CuFe_5O_8$," C. F. Jefferson, *J. Appl. Phys.* **36**, 1165 (1965).
909. "Magnetic Properties and Optical and Paramagnetic Spectra of Divalent Nickel in $Sr_2(NiW)O_6$," Shoichiro Nomura and Takehiko Nakagawa, *J. Phys. Soc. Japan* **21**, 1679 (1966). (87)
910. "Magnetic Study of the Manganous Sulfates $MnSO_4$ and $MnSO_4 \cdot H_2O$," Y. Allain, J. P. Krebs, and J. de Gunzbourg, *Intern. Congr. Magnetism, Boston*, Sept. 10–16, 1967.
911. "Ferrimagnetism of Mn_4N and of Some Mixed Crystals of Mn_4N," Robert Juza and Heinrich Puff, ANL-TRANS-230 translated from *Z. Elektrochem.* **61**, 810–19 (1957).
912. M. V. Dekhtjar, *Fiz. Tverd. Tela* **5**, 3138 (1963).
913. M. Schieber, R. G. Gordon, and S. L. Hou, in *Proc. Intern. Conf. Magnetism, Nottingham, 1964*, The Institute of Physics and the Physical Society, London, 1965, p. 499.
914. S. Miyahara and S. Horiuti, in *Proc. Intern. Conf. Magnetism, Nottingham, 1964*, The Institute of Physics and the Physical Society, London, 1965, p. 550.
915. F. R. de Boer, J. Biesterbos, and C. J. Schinkel, private communication, November 1967.
916. "Magnetic Properties of Borides with a $Cr_{23}C_6$ Structure," Hozumi Hirota, *J. Phys. Soc. Japan* **23**, 512 (1967).
917. W. Trzebiatowski and A. Zygmunt, *Bull. Acad. Polon. Sci., Ser. sci. chim.* **14**, 495 (1966).
918. W. Trzebiatowski, A. Sepichowska, and A. Zygmunt, *Bull. Acad. Polon. Sci., Ser. sci. chim.* **10**, 687 (1964).
919. F. F. Y. Wang and M. Kestigian, *Bull. Am. Phys. Soc.* **10**, 352 (1965).
920. "Magnetic Characteristics of Gadolinium, Terbium, and Ytterbium Hydrides in Relation to the Electronic Nature of the Lanthanide Hydrides," W. E. Wallace, Yoshio Kubota, and R. L. Zanowick, p. 122 in *Nonstoichiometric Compounds, Symp. Sponsored by Division of Inorganic Chemistry at 141st Meeting of Am. Chem. Soc., Washington, D.C., March 21–23, 1962*, Roland Ward, chairman, Advances in Chemistry Series 39 (American Chemical Society, Washington, D.C., 1963).
921. "Chemical and Magnetic Order in PtNi Alloy," Masao Watanabe and Syohei Miyahara, *J. Phys. Soc. Japan* **23**, 451 (1967).
922. "Magnetic Effects in $NiI_2 \cdot [(NH_2)_2CS]_6$," H. Forstat, N. D. Love, and J. N. McElearney, *J. Phys. Soc. Japan* **23**, 229 (1967).
923. "On the Existence of an Antiferromagnetic Phase in RFe_7 Compounds," H. Weik, E. Susedik, and M. Turner, in *Oak Ridge National Laboratory 6th Rare Earth Research Conf.*, 1967, pp. 556–67.
924. Y. Kubota and W. E. Wallace, *J. Appl. Phys. Suppl.* **33**, 1348 (1962).
925. "Magnetic Characteristics of Praseodymium, Neodymium, and Samarium Hydrides," Yoshio Kubota and W. E. Wallace, *J. Appl. Phys.* **34**, 1348 (1963).

926. "Structural and Magnetic Characteristics of Dysprosium-Yttrium Solid Solutions," S. Weinstein, R. S. Craig, and W. E. Wallace, *J. Appl. Phys.* **34**, 1354 (1963).

927. "Preparation and Magnetic Properties of Nickel-Cobalt-Oxygen Spinels," M. W. Shafer, pp. 568–71 in *Proc. Intern. Conf. on Magnetism* (London: The Institute of Physics and the Physical Society, 1965).

928. "Antiferromagnetism in $Rb_2MnCl_4 \cdot 2H_2O$," H. Forstat, N. D. Love, and J. N. McElearney, *Phys. Letters* **25A**, 253-4 (1967).

929. "Susceptibility Measurements of the Antiferromagnetic α-Goethite," A. Szytuka, A. Burewicz, K. Dyrek, A. Hrynkiewicz, D. Kulgawczuk, Z. Obuszko, H. Rzany, and A. Wanic, *Phys. Stat. Sol.* **17**, K 195-7 (1966).

930. "Magnetic Properties of Cr_5S_6 in Chromium Sulfides," K. Dwight, R. W. Germann, N. Menyuk, and A. Wold, *J. Appl. Phys.* **33**, Suppl., 1341 (1962).

931. "Anomalies in the Physical Properties of Vanadium. The Role of Hydrogen." D. G. Westlake, *Phil. Mag.* **16**, 905 (1967).

932. "Etude Structurale et Magnetique d'Arseniures Mixtes M_2As," Laszlo Hollan, *Ann. Chim.* **1**, 437–48 (1966).

933. "Magnetostriction in Ferromagnetic $CdCr_2Se_4$," D. E. Eastman and M. W. Shafer, *J. Appl. Phys.* **38**, 4761 (1967).

934. "Antiferromagnetism in a Cr-0.35% Mn Alloy," J. A. Oberteuffer and J. A. Marcus, *Bull. Am. Phys. Soc.* **11**, 473 (1966).

935. "Transformation du Second Ordre dans Fe_2O_3 a 685°C," M. Schneider and C. E. Beaulieu, *Canad. Metallurg. Quart.* **6**, No. 1, 1-7 (1967).

936. "Electrical Resistance of Ferromagnetic Chromium-Germanium Alloys at 20-300°K," N. I. Davidenko and I. G. Fakidov, *Fiz. Met. i Metallov.* **24**, No. 1, 190-92 (1967).

937. "Magnetic Properties of CsCl Type Compounds of Rare Earth Metals with Cadmium," G. T. Alfieri, E. Banks, K. Kanematsu, and T. Ohoyama, *J. Phys. Soc. Japan* **23**, 507 (1967).

938. "Magnetic Phase Equilibrium in Chromium-Substituted Calcium Ferrite," L. M. Corliss, J. M. Hastings, and W. Kunnmann, *Phys. Rev.* **160**, 408-13 (1967).

939. "Structure Magnetique de $CoCr_2S_4$," C. Colominas-Broquetas, Vu Van Qui, and E.-F. Bertaut, *Bull. Soc. Fr. Mineral. Crist.* **90**, 109-110 (1967).

940. "Effect of Alloying of Ti, Y, Nb and Hf on the Band Ferromagnetism in $ZrZn_2$," S. Ogawa, *Phys. Letters* **25A**, 516 (1967).

941. "Calorimetric Evidence for the Absence of a Magnetic Phase Transition in Gallium near 1.7°K," J. E. Neighbor and C. A. Shiffman, *Phys. Rev. Letters* **19**, 640-1 (1967).

942. "Mossbauer Effect in FeF_3," U. Bertelsen, J. M. Knudsen, and H. Krogh, *Phys. Stat. Sol.* **22**, 59-64 (1967).

943. "A Low Temperature Magnetic Transition in Plutonium Monocarbide," J. L. Green, G. P. Arnold, J. A. Leary and N. G. Nereson, *J. Nucl. Materials* **23**, 231-2 (1967).

944. "Magnetic Study of the Manganate Phases: $CaMnO_3$, $Ca_4Mn_3O_{10}$, $Ca_3Mn_2O_7$, Ca_2MnO_4," J. B. MacChesney, H. J. Williams, J. F. Potter, and R. C. Sherwood, *Phys. Rev.* **164**, 779-785 (1967).

945. "Magnetic Torsion Measurements on Hexagonal FeGe," K. A. Blom, O. Beckman, and M. Richardson, *Solid State Commun.* **5**, 977-79 (1967).

946. "New Transparent Ferrimagnets and their Magnetic Properties near Curie Point," L. P. Boky, P. P. Syrnikov, V. M. Yudin, and G. A. Smolensky, *Solid State Commun.* **5**, 927-31 (1967).

947. "Antiferromagnetisme de la Solution Solide de l'Aluminium dans le Chrome," A. Kallel and F. de Bergevin, *Solid State Commun.* **5**, 955-58 (1967).

948. "The Magnetic Structure of $TbAu_2$," M. Atoji, *Phys. Letters* **25A**, 528 (1967).

949. "New Ferromagnetic 5:4 Compounds in the Rare Earth Silicon and Germanium Systems," F. Holtzberg, R. J. Gambino, and T. R. McGuire, *J. Phys. Chem. Solids* **28**, 2283-89 (1967).

950. "Oxygen Content and Thermomagnetic Properties in $Cu_{1-x}Mg_xFe_2O_4$," Kohji Ohbayashi and Shuichi Iida, *J. Phys. Soc. Japan* **23**, 776 (1967).

REFERENCES

951. "Pressure Dependence of Magnetic Transitions in Spinels," J. A. Kafalas, N. Menyuk, K. Dwight, Jr., and J. B. Goodenough, *ESD-TR-67-562*, Lincoln Lab., Massachusetts Institute of Technology, p. 19.

952. Evolution de la Transition Magnetique du Premier Ordre dans la Solution Solide $Mn_3Cu_{1-x}Zn_xN$, Roland Madar, Michel Barberon, Gerard Lorthioir, Eliane Fruchart, and Robert Fruchart, *Compt. Rend., Ser. C*, **267**, 1404-06 (1968).

953. W. Rudorff, J. Kandler, and D. Babel, *Z. Anorg. Allg. Chem.* **317**, 261 (1962).

954. "Proprietes Magnetiques des Alliages de Formule T_2Ni_7, dans Laquelle T Designe un Metal de Terre Rare ou l'Yttrium," Remy Lemaire, Dominque Paccard, and Rene Pauthenet, *Compt. Rend.* **265**, 1280-82 (1967).

955. "First Order Magnetic Transitions in the Fe-Mn-As System," R. M. Rosenberg, W. H. Cloud, F. J. Darnell, and R. B. Flippen, *Phys. Letters* **25A**, 723 (1967).

956. "Magnetic Properties of Molybdenum- and Wolfram-Modified Mn_3B_4," Atsushi Iga and Yoshio Tawara, *J. Phys. Soc. Japan* **24**, 28 (1968).

957. "Magnetic Dilution of Rare-Earth Aluminum Cubic Laves Phases RAl_2," K. H. J. Buschow, J. F. Fast, A. M. van Diepen, and H. W. de Wijn, *Phys. Stat. Sol.* **24**, 715 (1967).

958. "Magnetic Properties of Al-Substituted Yttrium Orthoferrite," F. F. Y. Wang and M. Kestigian, *Phys. Stat. Sol.* **25**, 119 (1968).

959. "Magnetic Transitions in Alloys of Gadolinium and Dysprosium," Frederick Milstein, and Lawrence Baylor Robinson, *Phys. Rev.* **159**, 466-72 (1967).

960. "Magnetic Transition Metal-Rare Earth Alloys, Final Report," Franz Lihl (Institut fur Angewandte Physik, Technische Hochschule, Vienna, Austria), October 27, 1967, Contract 61(052)-942 (Air Force Materials Lab., Wright Patterson AFB, Ohio).

961. "Proton Magnetic Resonance Study of Magnetic Ordering in Two Cupric Salts," S. Wittekoek and N. J. Poulis, *J. Appl. Phys.* **39**, 1017 (1968).

962. "Observation of Ferri- and Antiferromagnetic Resonance in Insulating Magnetic Spiral Structures," J. J. Stickler and H. J. Zeiger, *J. Appl. Phys.* **39**, 1021 (1968).

963. "Magnetic Properties of FeS_2 and CoS_2," Syohei Miyahara and Teruo Teranishi, *J. Appl. Phys.* **39**, 896 (1968).

964. "Magnetic and Crystallographic Properties of the System $MnCr_2S_4$-$MnInCrS_4$," L. Darcy, P. K. Baltzer, and E. Lopatin, *J. Appl. Phys.* **39**, 898 (1968).

965. "Magnetic Properties of Uranium Compounds with NaCl Structure," J. Grunzweig and M. Kuznietz, *J. Appl. Phys.* **39**, 905 (1968).

966. "Magnetic Properties of Some Compounds with Pyrite Structure," K. Adachi, K. Sato, and M. Takeda, *J. Appl. Phys.* **39**, 900 (1968).

967. "Interpretation of the Magnetic Properties of the Rare Earth Chromites and the Rare Earth Manganites," R. Aleonard, R. Pauthenet, J. P. Rebouillat, and C. Veyret, *J. Appl. Phys.* **39**, 379 (1968).

968. "NMR in Antiferromagnetic Complex Hydrated Manganese Chlorides," R. D. Spence, J. A. Casey, and V. Nagarajan, *J. Appl. Phys.* **39**, 1011 (1968).

969. "Mossbauer Effect in $(Zr_xNb_{1-x})Fe_2$," Midori Tanaka, Naoko Iio, T. Tokoro, and K. Kanematsu, *J. Phys. Soc. Japan* **25**, 1541 (1968).

970. "Magnetic and Crystallographic Studies of Compound $MnAs_{0.9}P_{0.1}$," Hideaki Ido, *J. Phys. Soc. Japan* **25**, 1543 (1968).

971. "Magnetic Moments and Hyperfine Interactions in Carbon-Stabilized Fe_5Si_3," C. E. Johnson, J. B. Forsyth, G. H. Lander, and P. J. Brown, *J. Appl. Phys.* **39**, 465 (1968).

972. "Magnetism and Crystal Structure of Zirconium Compound with Laves Structure," K. Kanematsu, *J. Appl. Phys.* **39**, 465 (1968).

973. "Magnetic Structures and Phase Transformations in Mn-Based CuAu-I Type Alloys," L. Pal, E. Kren, P. Szabo, and T. Tarnoczi, *J. Appl. Phys.* **39**, 538 (1968).

974. "Structures and Phase Transformations in the Mn-Ni System near Equiatomic Concentration," E. Kren, E. Nagy, I. Nagy, L. Pal, and P. Szabo, *J. Phys. Chem. Solids* **29**, 101-8 (1968).

975. "Mossbauer and Magnetic Properties of Several Europium Intermetallic Compounds," H. H. Wickman, J. H. Wernick, R. C. Sherwood, and C. F. Wagner, *J. Phys. Chem. Solids* **29**, 181-2 (1968).

976. "Growth and Properties of Single Crystals of the Ferromagnetic Semiconductor $CdCr_2Se_4$," Henning von Philipsborn, *Helv. Phys. Acta* **40**, 810 (1967).
977. "Low-Temperature Magnetization in Very Dilute Palladium-Iron Alloys," M. McDougald and A. J. Manuel, *J. Appl. Phys.* **39**, 961 (1968).
978. J. H. Colwell and B. W. Mangum, *J. Appl. Phys.* **38**, 1468 (1967).
979. "Magnetic Properties and Electrical Conduction of Copper-Containing Sulfo- and Selenospinels," F. K. Lotgering and R. P. van Stapele, *J. Appl. Phys.* **39**, 417 (1968).
980. "Antiferromagnetic Properties of Light Rare Earth Monochalcogenides," G. A. Smolensky, V. E. Adamjan, and G. M. Loginov, *J. Appl. Phys.* **39**, 786 (1968).
981. "Weakly Coupled Moments in the Strong Antiferromagnet, $MnSn_2$," J. S. Kouvel and I. S. Jacobs, *J. Appl. Phys.* **39**, 467 (1968).
982. "Nuclear Spin-Lattice Relaxation of Several Nuclei in $Rb_2MnCl_4 \cdot 2H_2O$," C. E. Taylor and J. A. Cowen, *J. Appl. Phys.* **39**, 498 (1968).
983. "Specific Heat of $DyFeO_3$ from 1.2°-80°K," A. Berton and B. Sharon, *J. Appl. Phys.* **39**, 1367 (1968).
984. "Ferromagnetic Halo-Chalcogenide Spinels ($CuCr_2X_3Y$) and Some Properties of the Systems $CuCr_2Se_3Br-CuCr_2Se_4$ and $CuCr_2Te_3I-CuCr_2Te_4$," M. Robbins, P. K. Baltzer, and E. Lopatin, *J. Appl. Phys.* **39**, 662 (1968).
985. "Neutron Diffraction Study of the Ferromagnetic Mexed-Anion Spinel $CuCr_2Se_3Br$," J. G. White and M. Robbins, *J. Appl. Phys.* **39**, 664 (1968).
986. "Neutron-Diffraction Data on Ti_2O_3 and V_2O_3," H. Kendrick, A. Arrott, and S. A. Werner, *J. Appl. Phys.* **39**, 585 (1968).
987. "Magnetic Susceptibility of $(U_{1-x}Th_x)O_2$," J. B. Comly, *J. Appl. Phys.* **39**, 716 (1968).
988. "Magnetic Structure of $MnSeO_4$," H. Fuess and G. Will, *J. Appl. Phys.* **39**, 628 (1968).
989. "Preparation and Properties of the $SrTi_{1-x}Fe_xO_{3-x}\phi_{x/2}$ System," L. H. Brixner, *Mat. Res. Bull.* **3**, 299-308 (1968).
990. "Comportement Magnetique de Composés $La_{1-x}T.R._xRu_2$ (T.R. Terre Rare)," P. Donze, and M. Peter, *Helv. Phys. Acta* **40**, 357–9 (1967).
991. "Etude des Spectres d'Emission de RX de Ti dans Ti_2O_3 et Ti_3O_5 en Relation avec les Anomalies de Leurs Propriétés Electriques et Magnetiques," V. I. Chirkov and E. E. Vajnshtejn, *Izv. Akad. Nauk SSSR, Neorg. Mater.* **3**, No. 6, 1017-21 (1967).
992. "Attempts to Prepare $LiF \cdot Fe_2O_3$," F. W. Harrison and G. K. Lang, *J. Phys. Soc. Japan* **25**, 1609 (1968).
993. "The Preparation and Properties of Some Vanadium Spinels," D. B. Rogers, R. J. Arnott, A. Wold, and J. B. Goodenough, *J. Phys. Chem. Solids* **24**, 347-60 (1963).
994. "Low Temperature Specific Heat of EuSe," H. W. White, D. C. McCollum, and J. Callaway, *Phys. Letters* **25A**, 388 (1967).
995. "On the Néel Temperature of $Ca_2Fe_2O_5$," S. Geller, R. W. Grant, U. Gonser, H. Wiedersich, and G. P. Espinosa, *Phys. Letters* **25A**, 722 (1967).
996. "Magnetic Properties of Some Rare-Earth-Aluminum Alloys," B. Barbara, C. Becle, R. Lemaire, and R. Pauthenet, *J. Appl. Phys.* **39**, 1084 (1968).
997. "Magnetic Properties of $SrRuO_3$ and $CaRuO_3$," J. M. Longo, P. M. Raccah, and J. B. Goodenough, *J. Appl. Phys.* **39**, 1327 (1968).
998. "Magnetic Properties and Nuclear Magnetic Resonance of Diluted $(Gd,Zr)Zn_2$," M. Asanuma and T. Yamadaya, *J. Appl. Phys.* **39**, 1244 (1968).
999. "Magnetic Properties of the Intermetallic Compounds $Dy_xY_{1-x}Fe_2$," A. R. Piercy and K. N. R. Taylor, *J. Appl. Phys.* **39**, 1096 (1968).
1000. "Magnetic Properties of Rare Earth Hydroxides," W. P. Wolf, H. Meissner, and C. A. Catanese, *J. Appl. Phys.* **39**, 1134 (1968).
1001. "Magnetic Properties of the Compound Series $Y(Mn_xFe_{1-x})_2$ and $Y_6(Mn_xFe_{1-x})_{23}$," Hans R. Kirchmayr, *J. Appl. Phys.* **39**, 1088 (1968).
1002. "Magnetic Properties of $Ba_4Zn_2Fe_{36}O_{60}$ Single Crystals," A. J. Kerecman, A. Tauber, T. R. AuCoin, and R. O Savage, *J. Appl. Phys.* **39**, 726 (1968).

REFERENCES

1003. "Magnetic Properties of Antiferromagnetic $GdAlO_3$," J. D. Cashion, A. H. Cooke, J. F. B. Hawkes, M. J. M. Leask, T. L. Thorp, and M. R. Wells, *J. Appl. Phys.* **39**, 1360 (1968).
1004. "Magnetic Ordering of Terbium in Some Perovskite Compounds," J. Mareschal, J. Sivardiere, G. F. de Vries, and E. F. Bertaut, *J. Appl. Phys.* **39**, 1364 (1968).
1005. "Magnetic Properties of Hexagonal $RbNiF_3$ with Substituted Ions," T. R. McGuire and M. W. Shafer, *J. Appl. Phys.* **39**, 1130 (1968).
1006. "Neutron Scattering from Paramagnetic Chromium Halides," L. Madhav Rao, N. S. Satya Murthy, G. Venkataraman, and P. K. Iyengar, *J. Appl. Phys.* **39**, 1113 (1968).
1007. "Magnetic Properties of an Antiferromagnetic Orthoferrite," G. Gorodetsky, B. Sharon, and S. Shtrikman, *J. Appl. Phys.* **39**, 1371 (1968).
1008. "Magnetic Properties of Rare Earth-Lanthanum Alloys," W. C. Koehler, H. R. Child, E. O. Wollan, and J. W. Cable, *J. Appl. Phys.* **39**, 1331 (1968).
1009. H. J. Williams, J. H. Wernick, R. C. Sherwood, and G. K. Wertheim, *J. Appl. Phys.* **37**, 1256 (1966).
1010. "Magnetic Properties of fcc Rare Earth-Thorium Alloys," H. R. Child, W. C. Koehler, and A. H. Millhouse, *J. Appl. Phys.* **39**, 1329 (1968).
1011. "Low-Temperature Magnetic Properties of Dysprosium Single Crystals," R. G. Jordan and E. W. Lee, *Proc. Phys. Soc.* **92**, 1074 (1967).
1012. "Spin Orientation and Magnetic Properties of Ca_2FeAlO_5," R. W. Grant, S. Geller, H. Wiedersich, U. Gonser, and L. D. Fullmer, *J. Appl. Phys.* **39**, 1122 (1968).
1013. "Metamagnetism in $TbAlO_3$," L. Holmes, R. Sherwood, and L. G. van Uitert, *J. Appl. Phys.* **39**, 1373 (1968).
1014. "Magnetic Properties of Er-Dy and Ho-Dy Alloys," A. H. Millhouse, *Bull. Am. Phys. Soc.* **13**, 440 (1968).
1015. "Equiatomic Transition Metal Alloys of Manganese. VI. Structural and Magnetic Properties of Pd-Mn Phases," A. Kjekshus, R. Mollerud, A. P. Andresen, and W. B. Pearson, *Phil. Mag.* **16**, 1063-83 (1967).
1016. "Optical Properties of Ni-Cu Alloys at 300° and 77°K from 0.2 to 11 eV," W. J. Scouler, J. Feinleib, and J. Hanus, *Bull. Am. Phys. Soc.* **13**, 387 (1968).
1017. "Propriétés Magnetiques et Transferts d'Etats Electroniques dans les Systemes Fe-Be, Co-Be, Ni-Be," Antoine Herr and Andre-J.-P. Meyer, *Compt. Rend.* **265**, 1165-68 (1967).
1018. "Heat Capacity of $Mn(NH_4)_2(SO_4)_2 \cdot 6H_2O$ near Its Critical Point," Martin Rayl, O. E. Vilches, and J. C. Wheatley, *Phys. Rev.* **165**, 692-97 (1968).
1019. "Heat Capacity of $K_3Fe(CN)_6$ and $CuK_2(SO_4)_2 \cdot 6H_2O$ Below 1°K," Martin Rayl, O. E. Vilches, and J. C. Wheatley, *Phys. Rev.* **165**, 698-702 (1968).
1020. "The Néel Temperatures and Stability of Magnetic Phases in the Disturbed Body-Centered Tetragonal Lattice. Uranium Compounds," J. Przystawa, *Phys. Stat. Sol.* **24**, 313 (1967).
1021. "Etude Cristallographique et Magnetique d'un Oxyde Trirutile Antiferromagnetique," Jean-Claude Bernier and Paul Poix, *Compt. Rend.* **265**, 1247-49 (1967).
1022. "Magnetic and Crystallographic Transitions in the System $(Mn_{1-x}Fe_x)_2O_3$," R. W. Grant, J. A. Cape, S. Geller, and G. P. Espinosa, *Bull. Am. Phys. Soc.* **13**, 462 (1968).
1023. "Magnetic Properties of $PrAl_2$ and $ErAl_2$," Clayton E. Olsen, Norris Nereson, and George Arnold, *Bull. Am. Phys. Soc.* **13**, 460 (1968).
1024. "Magnetic Properties of Intermetallic Compounds Between the Lanthanides and Platinum," W. E. Wallace and Y. G. Vlasov, *Inorg. Chem.* **6**, 2216-19 (1967).
1025. "Magnetic Susceptibility of $CsMnCl_3 \cdot 2H_2O$ at Low Temperatures," T. Smith and S. A. Friedberg, *Bull. Am. Phys. Soc.* **13**, 461 (1968).
1026. "The Magnetic Structures of the Alloys $Au_2(Mn,Al)_2$," G. E. Bacon and E. W. Mason, *Proc. Phys. Soc.* **92**, 713-25 (1967).
1027. "Magnetic Properties of Monoclinic $Na_2Mn_2Si_2O_7$ Single Crystal," O. V. Kachalov and N. M. Kreines, *Zh. Eksper. Teor. Fiz. (USSR)* **53**, 858-65 (1967).
1028. "Pade Approximant Estimates of the Magnetic Properties of EuO and EuS," J. D. Patterson, Peter C. Y. Chen, and Alfred L. Broz, *J. Appl. Phys.* **39**, 1629 (1968).

1029. "Magnetic Properties of $La_{0.5}Sr_{0.5}CoO_3$ near Its Curie Temperature," N. Menyuk, P. M. Raccah, and K. Dwight, *Phys. Rev.* **166**, 510-13 (1968).

1030. "Magnetic and Crystallographic Properties of some Rare Earth Cobalt Compounds with $CaZn_5$ Structure," W. A. J. J. Velge and K. H. J. Buschow, *J. Appl. Phys.* **39**, 1717 (1968).

1031. "Elastic Moduli and Ultrasonic Attenuation of Polycrystalline Alpha-Mn from 4.2-300°K," M. Rosen, *Phys. Rev.* **165**, 357 (1968).

1032. "Structure and Properties of the Chromium Sulfides," C. F. van Bruggen and F. Jellinek, p. 31 in *Propriétés Thermodynamiques Physiques et Structurales des Derivés Semimetalliques*, No. 157, *Orsay, Sept. 28–Oct. 1, 1965, Colloq. Intern. Du Centre National de la Recherche Scientifique* (Editions du Centre National de la Recherche Scientifique, Paris, 1967).

1033. "Analogie des Systemes Fe-C et Mn-C. Isomorphisme des Phases Fe_3C et Mn_3C-Fe_5C_2 et Mn_5C_2-Fe_7C_3 et Mn_7C_3. Relation Electronique dans les Carbures, Borures et Phosphures T_3X et les Phases Derivées," M. R. Fruchart, A. M. Blanc, E. Fruchart, MM. Bouchard, and J. P. Senateur, p. 95 in *Propriétés Thermodynamiques Physiques et Structurales des Derivés Semi-metalliques* (see ref 1032).

1034. "Etude Experimentale de la Solution Solide de Silicium dans le fer," Y. Lecocq and P. Lecocq, p. 118 in *Propriétés Thermodynamiques Physiques et Structurales des Derivés Semi-metalliques* (see ref 1032).

1035. "Etude Magnetique et Structurale des Solutions Solides de Silicium, de Germanium et d'Etain dans le Nickel," C. Djega, Y. Lecocq, and Pierre Lecocq, p. 159 in *Propriétés Thermodynamiques Physiques et Structurales des Derivés Semi-metalliques* (see ref 1032).

1036. "Etude des Carbosiliciures de Manganese," J. P. Senateur, P. Spinat, and R. Fruchart, p. 127 in *Propriétés Thermodynamiques Physiques et Structurales des Derivés Semi-metalliques* (see ref 1032).

1037. "Etude Magnetique et Structurale des Composés Ternaires à Base de Chrome, de Manganese, de Fer et d'Arsenic," Lazlo Hollan and Pierre Lecocq, p. 237 in *Propriétés Thermodynamiques Physiques et Structurales des Derivés Semi-metalliques* (see ref 1032).

1038. "Magnetisme des Carbures de Terres Rares a Basse Temperature," R. Lallement, p. 343 in *Propriétés Thermodynamiques Physiques et Structurales des Derivés Semi-metalliques* (see ref 1032).

1039. "Le Ferro et Antiferromagnetisme des Composés d'Uranium avec les Elements du Veme Groupe," W. Trzebiatowski, T. Palewski, A. Sepichowska, R. Troc, A. Misiuk, W. Wojciechowski, and A. Zygmunt, p. 499 in *Propriétés Thermodynamiques Physiques et Structurales des Derivés Semi-metalliques* (see ref 1032).

1040. "Magnetic Structure of $Ca_2Fe_2O_5$," Takayoshi Takeda, Yasuo Yamaguchi, Shoichi Tomiyoshi, Masahiro Fukase, Mitsuo Sugimoto, and Hiroshi Watanabe, *J. Phys. Soc. Japan* **24**, 446 (1968).

1041. "Etude de la Structure Magnetique du Composé TbAl par Diffraction Neutronique," C. Becle, R. Lemaire, and E. Parthe, *Solid State Commun.* **6**, 115-23 (1968).

1042. "Magnetic Structure of Cr_2O_3-Fe_2O_3 System," J. K. Srivastava, G. K. Shenoy, and R. P. Sharma, *Solid State Commun.* **6**, 73-76 (1968).

1043. L. M. Corliss, J. M. Hastings, W. Kunnmann and E. Banks, *Acta Cryst.* **21**, Suppl. A, 95 (1966).

1044. "Characteristic Features of the Ordering of Rare-Earth Ions in Dysprosium and Terbium Orthoferrites," K. P. Belov, A. M. Kadomtseva, T. M. Ledneva, T. L. Ovchinnikova, Ya. G. Ponomarev, and V. A. Timofeeva, *Soviet Phys.-Solid State* **9**, 2193 (1968).

1045. "Spontaneous Magnetostriction of the Compound $MnAu_2$," V. A. Gordienko and V. I. Nikolaev, *Soviet Phys.-Solid State* **9**, 2216 (1968).

1046. "Antiferromagnetism of Ordered Au_5Mn_2," Kazuko Inoue, Yoji Nakamura, Katsumi Yamamoto, and Saiyu Maruyama, *J. Phys. Soc. Japan* **24**, 646 (1968).

1047. "Ferromagnetism in $CrBe_{12}$," Norman M. Wolcott and R. L. Falge, Jr., *Bull. Am. Phys. Soc.* **13**, 572 (1968).

1048. "The Magnetic Susceptibilities of Some Rare-Earth Garnets," A. H. Cooke, T. L. Thorp, and M. R. Wells, *Proc. Phys. Soc.* **92**, 400-7 (1967).

REFERENCES

1049. "Etude Magnetique et Cristallographique des Solutions Solides $(Fe_{1-x}Cr_x)_3P$ et de la Phase Ferromagnetique Fe_5B_2P," Anne-Marie Blanc, Eliane Fruchart, and Robert Fruchart, *Ann. Chim.* **2**, 251-54 (1967).

1050. "Etude par Diffraction Neutronique et Mesures Magnetiques de $UFeO_4$ Ferromagnetique," Madeleine Bacmann, Erwin Felix Bertaut, Alain Blaise, *Compt. Rend.* **266**, 45-48 (1968).

1051. "Chemische und Magnetische Eigenschaften der Barium-fluorometalle (II) Ba_2MF_6 mit M Zn, Cu, Ni, Co, Fe," H. G. von Schnering, *Z. anorg. allg. Chem., Dtsch.* **353**, No. 1-2, 1-12 (1967). (120)

1052. "Heat Capacity Singularities in Two Gadolinium Salts Below 1°K," R. F. Wielinga, J. Lubbers, and W. J. Huiskamp, *Physica* **37**, 375-92 (1967).

1053. "Electrical Conductivity of Vanadium Oxide at Low Temperatures," S. M. Ariia, B. la. Brach, and V. A. Vladimirova, *N68-17385* (Lockheed Missiles and Space Co., Sunnyvale, Calif.) translated from *Vestn. Leningr. Univ., Ser. Fiz. i Khim. (Leningrad)*, No. **22**, 157-59 (1967).

1054. "Mixed Crystals Between Binary Sulphides or Selenides with Spinel Structure," F. K. Lotgering, *J. Phys. Chem. Solids* **29**, 699-709 (1968).

1055. "Etude de $FeLiO_2$ (Phase Beta) par Effet Mossbauer," Georges A. Fatseas and Simone Lefebvre, *Compt. Rend.* **266**, 374-76 (1968).

1056. "Néel Transformation in near-Stoichiometric Fe_xO," M. E. Fine and F. B. Koch, *J. Appl. Phys.* **39**, 2478 (1968).

1057. "$CaCrO_3$ – A New Antiferromagnetic Perovskite," J. M. Longo, J. B. Goodenough, J. A. Kafalas, and D. A. Batson, p. 28-29 in *ESD-TR-68-17* (Lincoln Laboratory, Mass. Institute of Technology), April 1968.

1058. "Ferrimagnetism of Hexagonal Ferrites," T. M. Perekalina and V. P. Cheparin, *Soviet Phys.-Solid State* **9**, 2524-26 (1968).

1059. "Pressure Effect on the Magnetic Properties of Au_4Mn," Minoru Matsumoto, Takejiro Kaneko, and Kazuo Kamigaki, *J. Phys. Soc. Japan* **24**, 953 (1968).

1060. "Magnetic and Electrical Properties of Ordered Perovskite $Sr_2(FeMo)O_6$ and Its Related Compounds," Takehiko Nakagawa, *J. Phys. Soc. Japan* **24**, 806-11 (1968).

1061. "Heat Capacity of Potassium Cobalt Trifluoride," C. Deenadas, H. V. Keer, R. V. Gopalarad, and A. B. Biswas, *Brit. J. Appl. Phys.* **18**, 1833-4 (1967).

1062. "The Antiferromagnetic Structure of ζPd_3Mn_2," J. A. Gonzalo and M. I. Kay, *Acta Cryst.* **21**, Pt. 7, Suppl. A 93 (Dec. 30, 1966) (Seventh Intern. Congr. and Symp. Intern. Union of Crystallography, Moscow, 1966).

1063. "Magnetic Properties of Rare Earth and Yttrium Chromites," J. Mareschal, G. F. de Vries, R. Aloenard, R. Pauthenet, J. P. Rebouillat, and V. Zarubicka, *Acta Cryst.* **21**, Pt. 7, A 98 (Dec. 30, 1966).

1064. "Magnetic and Crystallographic Characteristics of Praseodymium Hydrides," W. E. Wallace and K. H. Mader, *J. Chem. Phys.* **48**, 84-9 (1968).

1065. "Magnetic Structure of $DyCrO_3$," E. F. Bertaut, J. Mareschal, *J. Phys.* **29**, 67-73 (1968).

1066. "Magnetic Transitions in Alpha Hematite," J. Lielmezs and A. C. D. Chaklader, *Science* **60**, 1137 (1968).

1067. "Magnetic Properties of $CuCr_2Se_xS_{4-x}$," Kohji Ohbayashi, Yasunori Tominaga, and Shuichi Iida, *J. Phys. Soc. Japan* **24**, 1173-74 (1968).

1068. "Magnetization Measurements and Pressure Dependence of the Curie Point of the Phase Sc_3In," W. E. Gardner, T. F. Smith, B. W. Howlett, C. W. Chu, and A. Sweedler, *Phys. Rev.* **166**, 577-88 (1968).

1069. "Permanent Magnetic Properties of Rare Earth Cobalt Compounds (RCo_5)," W. A. J. J. Velge and K. H. J. Buschow, *Conf. on Magnetic Materials and Their Applications, Proc. IEE Conf.*, London, Sept. 26-28th, 1967, pp. 45–50.

1070. "The Magnetic Structure and Hyperfine Field of Goethite (α-FeOOH)," J. B. Forsyth, I. G. Hedley, and C. E. Johnson, *J. Phys. C. (Proc. Phys. Soc.)* Ser. 2, Vol. 1, 179-88 (1968).

1071. "Classification Structurale des Divers Composes Fluores du Fer et de Cations Monovalents. Application a leurs Propriétés Magnetiques," J. Portier, A. Tressaud, R. de Pape, and P. Hagenmuller, *Mat. Res. Bull.* **3**, 433-36 (1968).

1072. "Magnetic Properties and Crystal Distortions of $BiMnO_3$ and $BiCrO_3$," Fuyuhiko Sugawara, Shuichi Iiida, Yasuhiko Syono, and Syun-iti Akimoto, *J. Phys. Soc. Japan* **25**, 1553 (1968).

1073. "Etude Cristallographique et Magnetique des Oxydes Doubles de Terre Rare de Type $LaTO_3$; T = Ho, Y, Er, Tm, Yb, Lu," Jean Michel Moreau, *Mat. Res. Bull.* **3**, 427-32 (1968).

1074. "Magnetic Structure Properties of Alloys of Tb, Ho, and Er with Sc," H. R. Child and W. C. Koehler, preprint submitted to the *Phys. Rev.*, May 1968.

1075. "On the Magnetostriction of Au_2Mn at High Magnetic Fields," Moriaki Kazama, Tokutaro Hirone, Kazuo Kamigaki, and Takejiro Kaneko, *J. Phys. Soc. Japan* **24**, 980-83 (1968).

1076. "Magnetization and Magnetic Structure of Mn-Zn and Mn-Zn-Ga Alloys of CsCl-Type Structure," Tomiei Hori, Yasuaki Nakagawa, and Junji Sakurai, *J. Phys. Soc. Japan* **24**, 971-76 (1968).

1077. A. Aharoni and M. Schieber, *Phys. Rev.* **123**, 807 (1961).

1078. R. Au, J. A. Cowen, and H. van Till, *Proc. Intern. Conf. Low Temp. Phys., Columbus, Ohio, 1964* (Plenum Press, New York, 1965), Part B, p. 877.

1079. R. Ballestracci, E. F. Bertaut, J. Coing-Boyat, A. Delapalme, W. James, R. Lemaire, R. Pauthenet, and G. Roult, *J. Appl. Phys.* **34**, 1333 (1963).

1080. E. Banks, M. Robbins, and A. Tauber, *J. Phys. Soc. Japan* **17**, Suppl. B-1, 196 (1962).

1081. E. F. Bertaut, G. Buisson, A. Durif, J. Mareschal, M. C. Monmory, S. Quezel-Ambrunaz, *Bull. Soc. Chim. France*, p. 1132 (1965).

1082. "Neutron Diffraction Investigation of the Spiral Magnetic Structure in Cr_2BeO_4," D. E. Cox, B. C. Frazer, R. E. Newnham, and R. P. Santoro, *Brookhaven National Laboratory BNL 12945*, 1968.

1083. "Structural, Electrical, and Magnetic Properties of High-Pressure $CdCr_2Se_4$," M. D. Banus and M. C. Lavine, pp. 30–33 in *ESD-TR-68-353, Solid State Research, 1968*, Lincoln Laboratory, Massachusetts Institute of Technology.

1084. F. T. J. Smith, pp. 23–25 in *ESD-TR-68-353* (see ref 1083).

1085. G. Blasse and E. W. Gorter, *J. Phys. Soc. Japan* **17**, Suppl. B-1, 176 (1962).

1086. "High-Pressure Forms of $CsNiF_3$," J. M. Longo and others, pp. 27–29 in ESD-TR-68-353 (see ref 1083).

1087. T. Okamoto, H. Fujii, Y. Hidaka, and E. Tatsumoto, *J. Phys. Soc. Japan* **24**, 951 (1968).

1088. W. L. Roth, *Acta Cryst.* **13**, 140 (1960).

1089. H. Forstat, N. D. Love, and J. McElearney, *Bull. Am. Phys. Soc.* **10**, 32 (1965); *J. Chem. Phys.* **43**, 1626 (1965).

1090. S. Geller, H. J. Williams, R. C. Sherwood, and G. P. Espinosa, *J. Appl. Phys.* **33**, 1195 (1962).

1091. S. Geller, H. J. Williams, R. C. Sherwood, and G. P. Espinosa, *J. Phys. Chem. Solids* **23**, 1525 (1962).

1092. S. Geller, H. J. Williams, R. C. Sherwood, G. P. Espinosa, and E. A. Nesbitt, *Appl. Phys. Letters* **3**, 60 (1963).

1093. S. Geller, H. J. Williams, R. C. Sherwood, and G. P. Espinosa, *Bell System Tech. J.* **43**, 565 (1964).

1094. S. Geller, H. J. Williams, R. C. Sherwood, and G. P. Espinosa, *J. Appl. Phys.* **35**, 542 (1964).

1095. S. Geller, H. J. Williams, R. C. Sherwood, G. P. Espinosa, and E. A. Nesbitt, *J. Appl. Phys.* **35**, 520 (1964).

1096. K. K. Loroia and K. P. Sinha, *Indian J. Pure Appl. Phys.* **1**, 215 (1963).

1097. E. W. Gorter, *J. Appl. Phys.* **34**, 1253 (1963).

1098. S. L. Hou, Thesis, Harvard University, 1964.

1099. A. Z. Hrynkiewicz and D. S. Kulgawchuk, *Acta Phys. Polon.* **24**, 680 (1963).

1100. N. Ichinose and K. Kurihara, *J. Phys. Soc. Japan* **18**, 1700 (1963).

1101. H. Dachs, *J. Phys. Radium* **25**, 563 (1964).

1102. K. Kosuge, T. Takada, and S. Kachi, *J. Phys. Soc. Japan* **18**, 318 (1963).

1103. *Landolt-Bornstein, Tabellen Magnetischer und Elektrischer Eigenschaften 9-II* (Springer-Verlag, Berlin, 1963).

1104. J. Longo and R. Ward, *J. Am. Chem. Soc.* **83**, 2816 (1961).

REFERENCES

1105. F. K. Lotgering, *Proc. Intern. Conf. Magnetism, Nottingham, 1964* (Phys. Soc., London, 1965), p. 533.
1106. F. K. Lotgering, *Solid State Commun.* **3**, 347 (1965).
1107. J. B. MacChesney, H. J. Williams, R. C. Sherwood, and J. F. Potter, *J. Chem. Phys.* **44**, 596 (1966).
1108. E. H. Frei and others, *Phys. Rev.* **118**, 657 (1960).
1109. N. Menyuk, A. Wold, and D. Rogers, and K. Dwight, *J. Appl. Phys.* **33**, Suppl., 1144 (1962).
1110. C. Nowlin, Thesis, Harvard University, 1963.
1111. M. B. Parma-Vittorelli, M. U. Palma, and F. Persico, *J. Phys. Soc. Japan* **17**, Suppl. B-1, 475 (1962).
1112. M. S. Robbins, S. Lerner, and E. Banks, *J. Phys. Chem. Solids* **24**, 759 (1963).
1113. P. K. Baltzer and E. Lopatin, *Proc. Intern. Conf. Magnetism, 1964*, p. 564 (Phys. Soc. London, 1965).
1114. "Magnetic Properties of Normal Spinels with Only A-A Interactions," W. L. Roth, *General Electric Res. Laboratories Rept. 63 RL*, 1963, p. 3438.
1115. J. B. Goodenough and others, *J. Appl. Phys.* **29**, 382 (1958).
1116. W. L. Roth, *J. Phys. Radium* **25**, 507 (1964).
1117. W. Rudorff, G. Lincke, and D. Babel, *Z. Anorg. Allgem. Chem.* **320**, 150 (1963).
1118. A. Sawaoka, S. Miyahara, S. Akimoto, and H. Fujisawa, *J. Phys. Soc. Japan* **19**, 1750 (1964).
1119. M. Schieber, *J. Appl. Phys.* **35**, 1074 (1964).
1120. "Experimental Magnetochemistry, Nonmetallic Magnetic Materials," Michael M. Schieber, (North Holland Publishing Company, Amsterdam, 1967), p. 265.
1121. M. Schieber, R. B. Frankel, N. Blum, and S. Foner, *Acta Cryst.* **20**, Suppl. (1966).
1122. K. Siratori and S. Iida, *J. Phys. Soc. Japan* **15**, 210 (1960).
1123. K. G. Srivastava and E. W. Gorter, *J. Appl. Phys.* **34**, 1256 (1963).
1124. J. W. Stout and W. O. J. Boo, *J. Appl. Phys.* **37**, 966 (1966).
1125. T. Takada, M. Kiyama, Y. Bando, T. Nakamura, M. Shiga, T. Shinjo, N. Yamamoto, Y. Endoh, and H. Takaki, *J. Phys. Soc. Japan* **19**, 1747 (1964).
1126. A. Tauber, J. A. Kohn, and R. O. Savage, *J. Appl. Phys.* **34**, 1265 (1963).
1127. J. Thery and R. Collongues, *Compt. Rend.* **250**, 1070 (1960); **254**, 685 (1962).
1128. P. van Dalen, H. M. Gijsman, N. Love, and H. Forstat, *Proc. Ninth Intern. Conf. Low Temp. Phys., Columbus, Ohio, 1964* (Plenum Press, New York, 1965), Part B, p. 888.
1129. G. Villers, R. Pauthenet, and J. Loriers, *J. Phys. Radium* **20**, 382 (1959).
1130. G. Villers, P. Lombard, and J. Loriers, *Compt. Rend.* **257**, 2419 (1963).
1131. T. Watanabe, *J. Phys. Soc. Japan* **17**, 1856 (1962).
1132. D. E. Cox, W. J. Takei, R. C. Miller, and G. Shirane, *J. Phys. Chem. Solids* **23**, 863 (1962).
1133. "Mossbauer Effect Studies on Binary Iron-Tin Alloys," E. Both, G. Trumpy, J. Traff, and P. Ostergaard, in *Hyperfine Structure and Nuclear Radiations* (Proc. Intern. Conf. on Hyperfine Interactions Detected by Nuclear Radiation, Asilomar Conference Grounds, California, 1967) (North Holland Publishing Company, Amsterdam, 1968).
1134. G. Villers, J. Loriers, and F. Clerc, *Compt. Rend.* **255**, 1196 (1962).
1135. J. Thery, Thesis, University of Paris, 1962.
1136. S. C. Abrahams and H. J. Williams, *J. Chem. Phys.* **39**, 7923 (1963).
1137. H. Forstat and D. R. McNealey, *J. Chem. Phys.* **35**, 594 (1961).
1138. "Ferromagnetism in the System $(Zr_{1-x}Hf_x)Zn_2$," H. J. Blythe, *Phys. Letters* **27A**, 42-43 (1968).
1139. "Thermal Anomalies and Antiferromagnetic Ordering $KCoF_3$, $KNiF_3$, and $KCuF_3$," C. Deenadas, H. V. Keer, R. V. G. Rao, and A. B. Biswas, *Indian J. Pure Appl. Phys.* **5**, 147-48 (1967).
1140. "Neutron Diffraction and Electrical Transport Properties of $CuCr_2Se_4$," M. Robbins, H. W. Lehmann, and J. G. White, *J. Phys. Chem. Solids* **28**, 897-902 (1967).

1141. "Application of Intense Continuous Magnetic Fields to the Study of Antiferro-Ferromagnetic Transitions and of Their Ferromagnetic Interactions in Dilute Alloys of Iron in Gold," W. E. Henry, in *Les Champs Magnetiques Intenses, leur Production et leur Applications* (Paris, Centre National de la Recherche Scientifique, 1967), pp. 151–67.

1142. "Magnetic Structure of EuSe," S. J. Pickart and H. A. Alperin, *J. Phys. Chem. Solids* **29**, 414-16 (1968).

1143. "Ferromagnetic Materials," Thomas R. McGuire and Merrill W. Shafer, International Business Machines Corp., U.S. Patent 3, 371, 042, Feb. 27, 1968.

1144. "The Temperature Dependence of the Paramagnetic Susceptibility of Ni–I Boracite," B. Heinrich, J. Zitkova, and J. Kaczer, *Phys. Stat. Sol.* **26**, 443 (1968).

1145. "Magnetic Structures of $TbAu_2$," Masao Atoji, *J. Chem. Phys.* **48**, 560-64 (1968).

1146. "Magnetic Properties of Iron Ions in CoO(I) and CoO(II)," Hang Nam Ok and James G. Mullen, *Phys. Rev.* **168**, 563-74 (1968).

1147. "Evidence of Two Forms of Cobaltous Oxide," Hang Nam Ok and James G. Mullen, *Phys. Rev.* **168**, 550-62 (1968).

1148. "Magnetic Susceptibility and EPR Study of Rare-Earths Diluted in $LaRu_2$," H. Cottet, P. Donze, J. Dupraz, B. Giovannini, and M. Peter, *Z. angew. Phys.* **24**, 249-54 (1968).

1149. "Super-Transferred Hyperfine Interactions at Diamagnetic Ions in Ferrimagnetic Insulators," S. L. Ruby, B. J. Evans, and S. S. Hafner, *Solid State Commun.* **6**, 277-80 (1968).

1150. "Studies on the Heusler Alloys – III. The Antiferro-Magnetic Phase in the Cu-Mn-Sb System," R. H. Forster and G. B. Johnston, and D. A. Wheeler, *J. Phys. Chem. Solids* **29**, 855-861 (1968).

1151. "Magnetic Properties of R_4Co_3-Type Intermetallic Compounds of Cobalt with Rare-Earth Metals and Yttrium," C. Berthet-Colominas, J. Laforest, R. Lemaire, R. Pauthenet, and J. Schweizer, *Cobalt* **39**, 97-101 (1968).

1152. "The Weak Ferromagnetic Behaviour of Ludlamite at Low Field Strengths," H. C. Meijer, T. W. Adair, III, and J. van den Handel, *Physica* **38**, 233-40 (1968).

1153. "Magnetic Properties of UAs-USe Solid Solutions," W. Trzebiatowski, A. Misiuk, and T. Palewski, *Bull. Acad. Polon. Sci.* **15**, No. 11, 543-47 (1967).

1154. "Etude par Diffraction Neutronique de la Forme Ordonne de l'Orthotitanate de Manganese-Structure Cristalline et Structure Cristalline et Structurale Magnetique," E. F. Bertaut and H. Vincent, *Solid State Commun.* **6**, 269-75 (1968).

1155. "Study of the Low-Temperature Specific Heat of Cobalt Chromite," A. S. Shibanov and N. A. Smol'kov, *Soviet Phys.-Solid State* **9**, 2910-11 (1968).

1156. "Magnetic Properties of Uranium Germanides," Wlodzimierz Trzebiatowski and Andrzej Misiuk, Roczniki Chemii, *Ann. Soc. Chim. Polonorum* **42**, 161-62 (1968).

1157. J. W. Stout and R. C. Chisholm, *J. Chem. Phys.* **36**, 979 (1962).

1158. "Resonance Properties of $Ni_{1-x}Ge_xFe_{2-2x}O_4$ Single Crystals," R. Krishnan, *J. Appl. Phys.* **39**, 1340-42 (1968).

1159. "The Magnetic Structure of Mn_2P," M. Yessik, *Phil. Mag.* **17**, 623-32 (1968).

1160. "Preparation et Propriétés des Monocristaux de Ferrites Type Spinelle et Grenat," Ramanathan Krishnan, Thesis, Faculte of Science of University of Paris; also in *Metaux, Corrosion, Industrie*, No. **484**, Dec. 1965; No. **485**, Jan. 1966; No. **486**, Feb. 1966.

1161. "Chaleurs Specifiques entre 1.2° et 5°K de Quelques Perovskites de Terres Rares," A. De Combarieu, J. Mareschal, J. C. Michel, and J. Sivardiere, *Solid State Commun.* **6**, 257-59 (1968).

1162. "Etude par Diffraction Neutronique et Effet Mossbauer du Tellurate de Fer Fe_2TeO_6," M. C. Montmory, M. Belakhovsky, R. Chevalier, and R. Newnham, *Solid State Commun.* **6**, 317-21 (1968).

1163. "Structures Magnetiques des Composés Trirutiles Cr_2TeO_6 et Cr_2WO_6," M. C. Montmory and R. Newnham, *Solid State Commun.* **6**, 323-26 (1968).

REFERENCES

1164. "Magnetic Ordering in the Rare-Earth Hexaborides," T. H. Geballe, B. T. Matthias, K. Andres, J. P. Maita, A. S. Cooper, and E. Corenzwit, Science 160, 1443-4 (1968).

1165. Private communication from Hagai Shaked, Nuclear Research Centre-Negev, Atomic Energy Commission, Beer Sheva, Israel, July 3, 1968.

1166. "Preparation and Properties of Some Copper Chromium Halochalcogenides with the Spinel Structure," A. W. Sleight and H. S. Jarrett, J. Phys. Chem. Solids 29, 868-70 (1968).

1167. "Thermal Research into the Magnetostriction of an FeCo Alloy," K. K. Bogma and V. V. Zubov, Izv. Vysshikh Uchebn. Zaveden., Fiz. 6, 145-7 (1966).

1168. "Propriétés Magnetiques et Structure Magnetique du Composé HoAl," Christian Becle, Remy Lemaire and Rene Pauthenet, Compt. Rend. 266, 994-97 (1968).

1169. "Dielectric and Magnetic Properties of $Pr(Fe_{1/2}Te_{1/2})O_3$," S. Nomura, H. Takabayashi and T. Nakagawa, Japan. J. Appl. Phys. 7, 600-04 (1968).

1170. "Susceptibility Derived Sublattice Magnetization in Cr_2O_3," H. Shaked and S. Shtrikman, Solid State Commun. 6, 425-6 (1968).

1171. "Magnetic Susceptibility of Neodymium Hexaboride," H. Hacker, Jr., and M. S. Lin, Solid State Commun. 6, 379-81 (1968).

1172. "Electrical Resistivity of Some Dilute Gold-Rare-Earth Alloys," L. R. Edwards and Sam Legvold, J. Appl. Phys. 39, 3250-52 (1968).

1173. "Magnetic and Optical Properties of Cobalt-Substituted $RbNiF_3$," J. C. Suits, T. R. McGuire, and M. W. Shafer, Appl. Phys. Letters 12, 406-8 (1968).

1174. M. J. Besnus and A. J. P. Meyer, Proc. Intern. Conf. Magnetism 1964, p. 507 (Proc. Phys. Soc. London, 1965).

1175. "Mossbauer Study of Metamagnetic Eu_3O_4 and Ferromagnetic EuO," H. H. Wickman and E. J. Catalano, J. Appl. Phys. 39, 1248-9 (1968).

1176. "Structures and Magnetic Properties of Some Metal (I) Chromium (III) Sulfides and Selenides," P. F. Bongers, C. F. van Bruggen, J. Koopstra, W. P. F. A. M. Omloo, G. A. Wiegers, and F. Jellinek, J. Phys. Chem. Solids 29, 977-84 (1968).

1177. "Neutron Diffraction Study of Magnetic Ordering in K_2IrCl_6, K_2ReBr_6, and K_2ReCl_6," V. J. Minkiewicz, G. Shirane, B. C. Frazer, R. G. Wheeler, and P. B. Dorian, J. Phys. Chem. Solids 29, 881-84 (1968).

1178. "Stability, Infrared Spectrum and Magnetic Properties of $FeBO_3$," J. C. Joubert, T. Shirk, W. B. White, and Rustum Roy, Mat. Res. Bull. 3, 671-76 (1968).

1179. "Proprietes Magnetiques des Boracites des Metaux de Transition (3d)," G. Quezel and H. Schmid, Solid State Commun. 6, 447-51 (1968).

1180. "Magnetic and Crystallographic Characteristics of $(Pr,La)Al_3$, $(Pr,Y)Al_3$, $(Pr,La)Al_2$ and $(Pr,Y)Al_2$," K. H. Mader, E. Segal, and W. E. Wallace, NYO-3454-22, May 1968 (Dept. of Chem., University of Pittsburgh).

1181. "Magnetic Characteristics of 2-17 Lanthanide-Nickel Compounds," Paul D. Carfagna and W. E. Wallace, J. Appl. Phys. 39, 5259-62 (1968).

1182. "Magnetic Characteristics of Lanthanide-Aluminum Compounds," K. H. Mader and W. E. Wallace, J. Less-Common Metals 16, 85-90 (1968).

1183. "Structural and Magnetic Characteristics of $TiCr_2$, $ZrCr_2$, $HfCr_2$ and the $TiCo_2$-$ZrCo_2$ and YFe_2-YCo_2 Alloy Systems," Alan W. Abel and R. S. Craig (Dept. of Chem., Univ. of Pittsburgh), NYO-3454-22, May 1968.

1184. "Magnetic Characteristics of $CeAl_3$ and $CeAl_4$," K. H. Mader and W. M. Swift, J. Phys. Chem. Solids. 29, 1759-64 (1968).

1185. "Magnetic Structure of DyC_2," Masao Atoji, J. Chem. Phys. 48, 3384-8 (1968). (128-A)

1186. "Magnetic Properties of $Zr_{1-x}Ti_xZn_2$, $Zr_{1-x}Y_xZn_2$, $Zr_{1-x}Nb_xZn_2$ and $Zr_{1-x}Hf_xZn_2$," Shinji Ogawa, J. Phys. Soc. Japan 25, 109-119 (1968).

1187. "Structure Cristalline et Propriétés Magnetiques du Composé GdPd," Jacques Pierre and Etiennette Siaud, Compt. Rend. 266, 1483-85 (1968).

1188. "Magnetic, Mossbauer, and Structural Studies on Three Modifications of $FeMoO_4$," A. W. Sleight, B. L. Chamerland, and J. F. Weiher, *Inorg. Chem.* **7**, 1093 (1968).

1189. "On the Effect of Isothermal Magnetic Annealing on the Magnetization and Magnetostriction in a Fe_3Cr Superlattice Alloy at High and Room Temperatures," Hakaru Masumoto, Hideo Saito, and Minoru Takahashi, *Sci. Rep. Research Inst., Tohoku Univ.* **A-Vol. 19**, 283-93 (1968).

1190. "Synthesis of Magnetic Fe_3S_4," Masayuki Uda, *Sci. Papers Inst. Phys. Chem. Res.* **62**, 14-23 (1968).

1191. "Crystal Growth and Magnetic Susceptibility of Some Rare-Earth Compounds. Part 2. Magnetic Susceptibility Measurements on a Number of Rare-Earth Compounds," J. D. Cashion, A. H. Cooke, M. J. M. Leask, T. L. Thorp, and M. R. Wells, *J. Materials Sci.* **3**, 402-7 (1968).

1192. "Etude par Diffraction Neutronique et Mesures Magnetiques des Oxysulfures de Terres Rares T_2O_2S," R. Ballestracci, E. F. Bertaut, and G. Quezel, *J. Phys. Chem. Solids* **29**, 1001-14 (1968).

1193. "Some Magnetic Properties of Ferromagnetoelectric Nickel Iodine Boracite," J. Kaczer, T. Shalnikova, and Z. Hauptman, *Czech. J. Phys.* **B18**, 734-37 (1968).

1194. "Magnetic Characteristics of Laves Phase Compounds Containing Two Lanthanides with Aluminum," W. M. Swift and W. E. Wallace, *J. Phys. Chem. Solids* **29**, 2053-61 (1968).

1195. Magnetic Characteristics of Europium-Aluminum Intermetallics, K. H. Mader and W. E. Wallace, *J. Chem. Phys.* **49**, 1521-25 (1968).

1196. "Giant Magnetic Moments in Dilute Alloys of Fe in Ni_3Ga," C. J. Schinkel, F. R. de Boer and J. Biesterbos, *Phys. Letters* **26A**, 501-02 (1968).

1197. "Neutron Diffraction Study of Antiferromagnetic Mn_2N," Mamoru Mekata, Junsuke Haruna, and Hideo Takaki, *J. Phys. Soc. Japan* **25**, 234-38 (1968).

1198. Y. Ishikawa, *Phys. Letters* **24A**, 725 (1967).

1199. "Magnetic Properties of the Stoichiometric Laves Phase Compound in Cobalt-Titanium System," Takuro Nakamicmi, Yoshihira Aoki, and Mikio Yamamoto, *J. Phys. Soc. Japan* **25**, 77-81 (1968).

1200. "Magnetic Properties of the hcp Iron-Ruthenium Alloys," Hideo Ohno, Mamoru Mekata, and Hideo Takaki, *J. Phys. Soc. Japan* **25**, 283 (1968).

1201. "Magnetic Properties of Mixed Chalcogenoselenides of Chromium," V. N. Ikorskii, L. M. Doronina, and S. S. Batsanov, *Zh. Strukt. Khim.* **9**, No. 1, 143-4 (1968).

1202. "Electronic Structure of Iron Nitrides Studied by Electron Diffraction. I. $\gamma'-Fe_4N$," Sigemaro Nagakura, *J. Phys. Soc. Japan* **25**, 488-98 (1968).

1203. "Magnetic Properties of Single Crystal Chalcogenide Spinels; $CuCr_2X_3Y$ (X = S, Se, and Te, Y = Cl, Br and I) System," Kazuo Miyatani, Yasuo Wada, and Fumio Okamoto, *J. Phys. Soc. Japan* **25**, 369-72 (1968).

1204. "On the Pressure Effect of the Neel Temperature in the Compound AuMn," Minoru Matsumoto, Takejiro Kaneko, and Kazuo Kamigaki, *J. Phys. Soc. Japan* **25**, 631 (1968).

1205. "Magnetic Anisotropy of $Mn_{53}Sb_{47}$," Hideaki Ido, *J. Phys. Soc. Japan* **25**, 625 (1968).

1206. "Lead Chromium Sulfide, $PbCr_2S_4$, and Some Isotypic Compounds," W. P. F. A. M. Omloo and F. Jellinek, *Recueil des Traveaux Chimiques des Pays-Bas* **87**, No. 6, 545-48 (1968).

1207. "Electrical Conductivity and Thermoelectric Power of Uranium Monosulfide Near Curie Temperature," Hirotaka Furuya, *Japan. J. Appl. Phys.* **7**, 779 (1968).

1208. "Magnetic Properties of $NaCrS_2$," K. W. Blazey and H. Rohrer, *Helv. Phys. Acta* **41**, Fasciculus quartus (1968).

1209. "Neutron Diffraction Studies of α-FeOOH," A. Szytula, A. Burewicz, Z. Dimitrijevic, S. Krasnicki, H. Rzany, J. Todorovic, A. Wanic, and W. Wolski, *Phys. Stat. Sol.* **26**, 429-34 (1968).

1210. "Ferromagnetism of $CrBe_{12}$," N. M. Wolcott and R. L. Fagle, Jr., *Phys. Rev.* **171**, 591-5 (1968).

1211. "Magnetic Properties and Crystal Structure of the Solid Solution System $Mg_2SnO_4-MgFe_2O_4$," V. A. Bokov, G. V. Novikov, O. B. Proskuryakov, Yu. G. Saksonov, V. A. Trukhtanov, and S. I. Yushchuk, *Soviet Phys.-Solid State* **10**, 855-58 (1968).

1212. D. G. Wickham, N. Menuyk, and K. Dwight, *J. Phys. Chem. Solids* **20**, 316 (1961).

1213. "Effect of Lattice Distortions on the Magnetic Behaviour of Perovskite-Type Manganites," V. A. Bokov, N. A. Grigoryan, M. F. Bryzhina, and V. V. Tikhonov, *Phys. Stat. Sol.* **28**, 835 (1968).

1214. "Propriétés Metamagnetiques des Composés Tb_3Al_2 et Dy_3Al_2," Bernard Barbara, Christian Becle, Jacques-Louis Feron, Remy Lemaire, and Rene Pauthenet, *Compt. Rend.* **267**, 244-47 (1968).

1215. "Magnetic Behaviour of FeF_3 Close to the Curie Temperature," L. M. Levinson, *J. Phys. Chem. Solids* **29**, 1331-36 (1968).

1216. "On the Magnetic Properties of Mn_5Ge_3," K. Sato, K. Adachi, and Y. Tawara, *Mem. Fac. Engineering., Nagoya Univ. (Japan)* **19**, No. 1, 167-70 (May 1967).

1217. "Magnetic Investigation in the System Tin-Manganese in the Liquid and Solid State," E. Wachtel and R. Ulrich, *Z. Metallkde.* **59**, No. 3, 227-35 (March 1968).

1218. "Cation Distributions in Octahedral and Tetrahedral Sites of the Ferrimagnetic Spinel $CoFe_2O_4$," G. A. Sawatzky, F. Van Der Woude, and A. H. Morrish, *J. Appl. Phys.* **39**, No. 2, Pt. 2, 1204-6 (Feb. 1968) (Proc. Intern. Congr. Magnetism, Boston and Cambridge, 1967).

1219. "Thermomagnetic Properties of Ferrites $Cu_{0.5}Fe_{2.5}O_4$," Z. Simsa and V. Houdek, preprint No. 37 (Institute of Solid State Physics, Czechslovak Academy of Sciences, Prague), October 1968, to be published in *J. Phys. Chem. Solids*.

1220. "Magnetic Property of the Compound DyAu," Takejiro Kaneko, *J. Phys. Soc. Japan* **25**, 905 (1968).

1221. "Magnetic Structure of TbAg," Masao Atoji, *J. Chem. Phys.* **48**, No. 8, 3380-83 (1968).

1222. "Antiferromagnetism of CuMnSb," Keizo Endo, Tetuo Ohoyama, and Ren'iti Kimura, *J. Phys. Soc. Japan* **25**, 907-8 (1968).

1223. "On the Magnetic Anisotropy and Susceptibility of $CoF_2 5HF,6H_2O$ and $NiF_2 5HF,6H_2O$," S. K. Dutta Roy and B. Ghosh, *J. Phys. Chem. Solids* **29**, 1511-18 (1968).

1224. "Mechanisms for Metal-Nonmetal Transitions in Transition-Metal Oxides and Sulfides, David Adler, *Rev. Mod. Phys.* **40**, 714-36 (1968).

1225. Magnetically Induced Quadrupole Interactions in $FeCr_2S_4$, G. R. Hoy and K. P. Singh, *Phys. Rev.* **172**, 514-19 (1968).

1226. Spin-Flopping in $CoBr_2 \cdot 6H_2O$, H. Forstat, J. N. McElearney, and P. T. Bailey, *Phys. Letters* **27A**, 70-1 (1968).

1227. Magnetic Properties of Uranium Oxyselenide, A. Murasik, W. Suski, R. Troc, and J. Leciejewicz, *Phys. Stat. Sol.* **30**, 61-6 (1968).

1228. Magnetic Properties of Manganese Phosphide, T. Komatsubara, *Sci. Rep. Tohoku Univ., First Ser. (Japan)* **1**, No. 2, 69-91 (1967).

1229. Recherche d'une Structure Antiferromagnetique dans U_3O_8, A. Oles, *Nukleonika, Polska* **13**, No. 1, 85-7 (1968).

1230. Propriétés Magnetiques et Structure Magnetique du Composé ErAl, C. Becle and R. Lemaire, *Phys. Letters* **27A**, 541-42 (1968).

1231. Effect of Pressure on Magnetic Phase Transitions of Europium Chalcogenides: EuO, EuS, and EuSe, V. C. Srivastava and R. Stevenson, *Can. J. Phys.* **46**, 2703-13 (1968).

1232. Susceptibility and Magnetization of β- and δ-FeOOH, D. S. Kulgawczuk, Z. Obuszko, and A. Szytula, *Phys. Stat. Sol.* **26**, K83-85 (1968).

1233. Magnetic Interactions and Crystal Structure in the $(La,Gd)_2CoMnO_6$ System, A. Marsh and C. C. Clark, *Proc. Brit. Ceram. Soc.* No. 10, 285-97 (1968).

1234. Etude Cristallographique et Magnetique de la Serie $Mn_xFe_{1-x}HoO_3$, Andre Apostolov and Paul Pataud, *Mat. Res. Bull.* **4**, 1-6 (1969).

1235. Structures et Propriétés Magnetiques de Quelques Composés de Formule $MFeF_3$ (M Na, K, Rb, Cs, NH_4, Tl), J. Portier, A. Tressaud, J.-L. Dupin, and R. de Pape, *Mat. Res. Bull.* **4**, 45-50 (1969).

1236. Preparation and Magnetic Properties of Bi Substituted Single Crystal Rare Earth Orthoferrites, J. P. Remeika, E. M. Gyorgy, and D. L. Wood, *Mat. Res. Bull.* **4**, 51-56 (1969).

1237. Invar Behavior of Face-Centered Cubic FeCo Alloys Precipitated from Copper, Yoji Nakamura, Masayuki Shiga, and Sumio Santa, *J. Phys. Soc. Japan* **26**, 210 (1969).

1238. Ferromagnetic Resonance in $Ga_{0.85}Fe_{1.15}O_3$, M. P. Petrov, S. A. Kizhaev, and G. A. Smolenskii, *Phys. Stat. Sol.* **30**, 871 (1968).

1239. Thermal Conductivity of Europium Oxide (EuO) across the Curie Temperature, Robert G. Morris and John L. Cason, Jr., *Helv. Phys. Acta* **41**, No. 6/7, 1045-51 (1968).

1240. Propriétés Magnetiques du Composé $CsFeF_3$, Josik Portier, Alain Tressaud, Rene Pauthenet, and Paul Hagenmuller, *Compt. Rend.* **267**, 1329-31 (1968).

1241. Determination de la Temperature de Neel du Sulfure de Manganese par Mesure Calorimetrique de la Capacite Calorifique Molaire dans le Domaine 20-300°K, Jean Bousquet, Michel Diot and Marc Roubin, *Compt. Rend.* **267**, 861-62 (1968).

1242. Magnetic Structure of $NaNiF_3$, A. Epstein, J. Makovsky, M. Melamud, and H. Shaked, *Phys. Rev.* **174**, 560-61 (1968).

1243. Etude de Cr_2S_3 Rhomboedrique par Diffraction Neutronique et Mesures Magnetiques, E. F. Bertaut, J. Cohen, B. Lambert-Andron, and P. Mollard, *J. Phys.* **29**, 813-24 (1968).

1244. Magnetization, Ferromagnetic Curie Temperature and Lattice Parameter of $(Zr_{1-x}Ti_x)Zn_2$ Compounds, H. J. Blythe and J. Crangle, *Phil. Mag.* **18**, 1143-48 (1968).

1245. Magnetic Properties of $Ga_{2-x}Fe_xO_3$, B. F. Levine, C. H. Nowlin, and R. V. Jones, *Phys. Rev.* **174**, 571-82 (1968).

1246. Superconductivity of Cu-Sb Phases and Absence of Antiferromagnetism in Cu_2Sb, K. Andres, E. Bucher, J. P. Maita, and A. S. Cooper, *Phys. Letters* **28A**, 67-8 (1968).

1247. NMR Evidence to the Absence of Antiferromagnetism in Cu_2Sb, L. C. Gupta, S. M. Malik, and R. Vijayaraghavan, *Phys. Letters* **28A**, 255 (1968).

1248. Crystal Structure of Manganese at 77-300°K, V. A. Finkel', *Zh. Eksper. Teor. Fiz. (USSR)* **54**, No. 6, 1697-9 (1968).

1249. Transformation to the Fan Structure in MnP Single Crystal, Takemi Komatsubara, Takashi Suzuki, and Eiji Hirahara, *J. Phys. Soc. Japan* **26**, 208 (1969).

1250. NMR in Paramagnetic $NdBr_3$ and UI_3 and in Antiferromagnetic UI_3, Stephen Ira Parks, thesis, Florida State Univ., 1967.

1251. Magnetic Properties of the Intermetallic Compound AuMn, William Bindloss, Ph.D. thesis, Cal. Univ., Berkeley (available from University Microfilms, Ann Arbor, Mich., Order No. 68-33).

1252. Investigation of Magnetic Susceptibility and Magnetostriction of the Compound MnSe, I. G. Kerimov and T. A. Mamedov, *Fiz. Metallov Metallovedenie (USSR)* **26**, No. 1, 188-91 (July 1968) (in Russian); English translation in *Phys. Metals Metallography (GB)*.

1253. Magnetic Properties of $Ni(CN)_2 \cdot NH_3 \cdot C_6H_6$, T. Watanabe, S. Takayanagi, T. Maruyamauchi, *J. Phys. Soc. Japan* **25**, No. 2, 346-51 (1968).

1254. Transport Properties of Thulium Single Crystals, L. Roger Edwards and Sam Legvold, *Phys. Rev.* **176**, 753-60 (1968).

1255. The Growth of Large Single Crystals of Lithium Ferrite, A. J. Pointon and J. M. Robertson, *J. Mat. Sci.* **2**, 293-94 (1967).

1256. Elementary Magnetic Properties of Some U-Ga Compounds, V. Ansorge and A. Menovsky, *Phys. Stat. Sol.* **30**, K31 (1968).

1257. Electric and Magnetic Properties of $Li_xV_{2-x}O_2$, K. Kobayashi, K. Kosuge, and S. Kachi, *Mat. Res. Bull.* **4**, 95-106 (1969).

1258. Structural and Magnetic Phase Transitions of Chromium Sulfides $Cr_{1-x}S$ with $0 > x < 0.12$, T. J. A. Popma and C. F. van Bruggen, *J. Inorg. Nucl. Chem.* **31**, 73-80 (1969).

1259. Characterization of Ferrous Ions in Hafnium-Doped Yttrium Iron Garnet, J. M. Robertson and D. Elwell, *J. Phys. C (Solid St. Phys.), Ser. 2*, **2**, 14-17 (1969).

1260. Magnetocrystalline Anisotropy of Europium Selenide, R. F. Brown, A. W. Lawson, and Glen E. Everett, *Phys. Rev.* **172**, 559-64 (1968).

REFERENCES

1261. Magnetic Properties of Some Co-Rich Erbium Cobalt Intermetallic Compounds, K. H. J. Buschow, J. F. Fast, and A. S. van der Goot, *Phys. Stat. Sol.* **29**, 719 (1968).

1262. Magnetic and Crystallographic Transitions in the Alpha-Mn_2O_3-Fe_2O_3 System, R. W. Grant, S. Geller, J. A. Cape, and G. P. Espinosa, *Phys. Rev.* **175**, 686-695 (1968).

1263. Temperature Dependence of the Sublattice Magnetization in Cobaltous Oxide, D. C. Khan and R. A. Erickson, *J. Phys. Chem. Solids* **29**, 2087-90 (1968).

1264. Magnetic Structures and Exchange Interactions in the Mn-Pt System, E. Kren, G. Kadar, L. Pal, J. Solyom, P. Szabo, and T. Tarnoczi, *Phys. Rev.* **171**, 574-85 (1968).

1265. Nature of the Low Temperature Phase Transition in Alpha-Uranium, G. T. Meaden and N. H. Sze, *Cryogenics* **8**, 396-97 (1968).

1266. Magnetic Structures of Er_2O_3 and Yb_2O_3, R. M. Moon, W. C. Koehler, H. R. Child, and L. J. Raubenheimer, *Phys. Rev.* **176**, 722-31 (1968).

1267. The Effects of Non-Stoichiometry on the Magnetic Properties of Cadmium Chromium Chalcogenide Spinels, H. L. Pinch and S. B. Berger, *J. Phys. Chem. Solids* **29**, 2091-99 (1968).

1268. Ferromagnetisme Faible dans les Alliages Pd-Cr, Jacques Rault and Jean-Paul Burger, *Compt. Rend.* **267**, 750-52 (1968).

1269. Elastic Moduli and Ultrasonic Attenuation of Polycrystalline Uranium from 4.2 to 300°K, M. Rosen, *Phys. Letters* **28A**, 438-39 (1968).

1270. Induced Weak Ferromagnetism in Antiferromagnetic Hematite, R. A. Beyerlein and I. S. Jacobs, *Bull. Am. Phys. Soc.* **14**, 349 (1969).

1271. Magnetic Order of $Mn_2P_2O_7$, David C. Fowlis and Carl V. Stager, *Bull. Am. Phys. Soc.* **14**, 387 (1969).

1272. Magnetic Transitions in Ti_3O_5 and Ti_4O_7, W. J. Danley and L. N. Mulay, *Bull. Am. Phys. Soc.* **14**, 350 (1969).

1273. Low-Temperature Magnetic Susceptibilities of Some Tetrachloroferrate Compounds, Edwin R. Jones, Oliver B. Morton, and M. Elton Hendricks, *Bull. Am. Phys. Soc.* **14**, 350 (1969).

1274. Magnetic Characteristics of Lanthanide-Silver Compounds Having the CsCl Structure, R. E. Walline and W. E. Wallace, *J. Chem. Phys.* **41**, 3285 (1964).

1275. Magnetic Properties of $DyPt_3$ and $DyIn_3$, G. Arnold and N. Nereson, *J. Chem. Phys.* **51**, 1495-99 (1969).

1276. Magnetic Properties of Plutonium Monophosphide, D. J. Lam, F. Y. Fradin, and O. L. Kruger, *Bull. Am. Phys. Soc.* **14**, 387 (1969).

1277. Antiferromagnetism and Electrical Transport Properties of Dilute Cr-Fe Alloys, Sigurds Arajs, Elmer E. Anderson, and Emerson E. Ebert, *Bull. Am. Phys. Soc.* **14**, 349 (1969).

1278. Magnetic Structures in $UP_{0.75}S_{0.25}$, G. H. Lander, M. Kuznietz, J. Crangle, and Y. Baskin, *Bull. Am. Phys. Soc.* **14**, 387 (1969).

1279. The Variation of T_c with Composition and Annealing Temperature in Ferromagnetic $ZrZn_{2-x}$ Alloys, G. S. Knapp and E. Corenzwit, *Bull. Am. Phys. Soc.* **14**, 348 (1969).

1280. Chalkogenides of the Transition Elements. VI. X-Ray, Neutron, and Magnetic Investigation of the Spinels Co_3O_4, $NiCo_2O_4$, Co_3S_4, and $NiCo_2S_4$, Osvald Knop, K. I. G. Reid, Sutarno, and Yasuaki Nakagawa, *Can. J. Chem.* **46**, 3463-76 (1968).

1281. Mixed Magnetic Ordering: Magnetic Susceptibility and Mossbauer Studies on Iron(III) Tellurate (Fe_2TeO_6), J. T. Dehn, R. E. Newnham, and L. N. Mulay, *J. Chem. Phys.* **49**, 3201-03 (1968).

1282. Magnetic Susceptibility Measurements on Cubic FeGe, L. Lundgren, K. A. Blom, and O. Beckman, *Phys. Letters* **28A**, 175-76 (1968).

1283. The Shift of the Ferromagnetic Curie Temperature in EuS by Hydrostatic Pressure, P. Schwob and O. Vogt, *Phys. Letters* **24A**, 242-44 (1967).

1284. Chaleurs Specifiques des Composes Perovskites $GdCrO_3$, $SmCrO_3$ et $SmAlO_3$, Andre de Combarieu, Jean Mareschal, Jean-Claude Michel, Jean Peyrard, and Jean Sivardiere, *Compt. Rend., Ser. B*, **267**, 1169-72 (1968).

1285. Magnetic Properties of Fused Mixtures of $MnCl_2$ and Alkali Chlorides, C. N. Owston, *Brit. J. Appl. Phys. (J. Phys. D), Ser. 2*, **1**, 1839-40 (1968).

1286. Propriétés Metamagnetiques des Orthoferrites de Terres Rares, Andre Apostolov and Jean Sivardiere, *Compt. Rend., Ser. B*, **267**, 1315-18 (1968).

1287. Fero- und paramagnetisches Verhalten in den Mischkristallen des Nickels mit Vanadium, Rhodium und Platin im Temperaturbereich von 14° bis 1000°K, Fritz Bolling, *Phys. kondens. Materie* **7**, 162-84 (1968).

1288. Chaleurs Specifiques à Basse Temperature de Quelques Orthoferrites de Terres Rares, J. Peyrard and J. Sivardiere, *Solid State Commun.* **7**, 605-09 (1969).

1289. Structures Magnetiques des Orthoferrites de Terres Rares, J. Mareschal and J. Sivardiere (Centre d'Etudes Nucleaires de Grenoble, France), to be published in *J. Phys.* (received at RMIC March 1969).

1290. Metamagnetic Behavior of Europium Selenide (EuSe), C. Kuznia and G. Kneer, *Phys. Letters* **27A**, No. 10, 664-5 (1968).

1291. Magnetic Susceptibility and ^{51}P Nuclear Resonance in Mn_2P, Satish K. Malik and R. Vijayaraghavan, *Phys. Letters* **28A**, 648 (1969).

1292. Neutron Diffraction Study of the Hexagonal Ferrite $Ba_2Zn_2Al_{2.5}Fe_{9.5}O_{22}$, N. N. Aganova, V. A. Sizov, and I. I. Yamzin, *Soviet Phys.-Solid State* **10**, 2258 (1969).

1293. Lattice Changes in Spinel-Type Iron Chromites, M. H. Francombe, *J. Phys. Chem. Solids* **3**, 37-43 (1957).

1294. Magnetic Properties of Gold-Rich Gold-Vanadium Alloys, L. Creveling, Jr., and H. L. Luo, *Phys. Rev.* **176**, 614-30 (1968).

1295. Magnetic Studies of Rare Earth Zinc Compounds with CsCl Structure, K. Kanematsu, G. T. Alfieri, and E. Banks, *J. Phys. Soc. Japan* **26**, 244 (1969).

1296. Magnetometrie und Magnetokalorischer Effekt des Kobalts in der Umgebung der Curie-Temperatur, Heinrich Lange, Rudolf Kohlhaas, and Werner Rocker (Institut fur Theoretische Physik der Universitat Koln, Abteilung fur Metallphysik), Forschungsberichte des Landes Nordrhein-Westfalen Nr. 1992 (Westdeutscher Verlag, Koln and Opladen, 1969).

1297. Magnetic Properties of Uranium Selenides and Sulfides, V. I. Chechernikov, A. V. Pechennikov, E. I. Yarembash, L. F. Martynova, and V. K. Slavyanskikh, *Soviet Phys. JETP* **26**, 328-30 (1969).

1298. Spin Re-Orientation in $SmFeO_3$, G. Gorodetsky and Lionel M. Levinson, *Solid State Commun.* **7**, 67-70 (1969).

1299. Electromagnetic-Sound Conversion by Linear Magnetostriction in $TlFeF_3$, L. R. Testardi, H. J. Levinstein, E. M. Gyorgy, and H. J. Guggenheim, *Solid State Commun.* **7**, 241-43 (1969).

1300. Propriétés Magnetiques des Alliages $TbCu_{1-x}Zn_x$, Jacques Pierre, *Compt. Rend., Ser. B*, **265**, 1169-72 (1967).

1301. Pressure Effect of Curie Point of CoS_2 and Ferromagnetic Properties of $Co(S_xSe_{1-x})_2$, Kiyoo Sato, Kengo Adachi, Tetsuhiko Okamoto, and Eiji Tatsumoto, *J. Phys. Soc. Japan* **26**, 639 (1969).

1302. Magnetic Properties of Cobalt and Nickel Dichalcogenide Compounds with Pyrite Structure, Kengo Adachi, Kiyoo Sato, and Motohiko Takeda, *J. Phys. Soc. Japan* **26**, 631 (1969).

1303. Temperature Dependence of the High Field Susceptibility of $CoCr_2O_4$, Kiiti Siratori and Masayoshi Ohashi, *J. Phys. Soc. Japan* **26**, 856 (1969).

1304. Magnetic Properties of Iron Phosphide Fe_2P, S. Chiba, *J. Phys. Soc. Japan* **15**, 581 (1960).

1305. Etude de $FeSb_2O_4$ par Effet Mossbauer, F. Varret, P. Imbert, A. Gerard, and F. Hartmann-Boutron, *Solid State Commun.* **6**, 889-92 (1968).

1306. Magnetic Structures in $UP_{0.95}S_{0.05}$, G. H. Lander, Moshe Kuznietz, and Y. Baskin, *Solid State Commun.* **6**, 877-79 (1968).

1307. Weak-Ferromagnetism in $EuCrO_3$, K. Tsushima, I. Takemura, and S. Osaka, *Solid State Commun.* **7**, 71-73 (1969).

1308. Exchange Striction of Ferromagnetic Compound CrTe, Hideaki Ido, Kiwamu Shirakawa, Takanobu Suzuki, and Takejiro Kaneko, *J. Phys. Soc. Japan* **26**, 663 (1969).

1309. Magnetic and Structural Characteristics of Lanthanide-Copper Compounds, R. E. Walline and W. E. Wallace, *J. Chem. Phys.* **42**, 604-7 (1965).

REFERENCES

1310. The Ferromagnetism of High Purity $ZrZn_2$, H. J. Blythe, *J. Phys. Soc. (Proc. Phys. Soc), Ser. 2*, **1**, 1604-07 (1968).
1311. Spin Quenching in the System $MnAs_{1-x}P_x$ and $MnAs_{1-y}Sb_y$, J. B. Goodenough, D. H. Ridgley, and W. A. Newman, *Proc. Intern. Conf. on Magnetism, Nottingham, 1964* (Proc. Phys. Soc., London, 1965).
1312. The Magnetic Susceptibilities of Some Rare Earth Silicides and Germanides with the $D8_8$ Structure, K. S. V. L. Narasimhan, H. Steinfink, and E. V. Ganapathy, *J. Appl. Phys.* **40**, 51-54 (1969).
1313. Magnetic Characteristics of $TbAu_2$, L. R. Sill, A. J. Fedro, and C. W. Kimball (Northern Illinois Univ., DeKalb, Ill.), CONF-681020-(Vol. 1), pp. 117-22.
1314. Magnetic Properties of Lanthanide Ferrite Garnets near the Curie Point, K. P. Belov, E. V. Talalaeva, and G. A. Yarkho, *Zh. Eksperim. i Teor. Fiz.* **52**, No. 6, 1489-1494 (1967).
1315. Properties of the Compounds Rubidium Manganese Trifluoride, Potassium Manganese Trifluoride, and Cesium Manganese Trifluoride, Joan S. Friebely and William J. Ince (Lexington Labs, Inc., Cambridge, Mass.), Contract AF 19(628) -5167, AD-677285, ESD-TR-68-240, August 8, 1968, 25 pages.
1316. Preparation and Properties of $NiCrO_3$, B. L. Chamberland and W. H. Cloud, *J. Appl. Phys.* **40**, 434 (1969).
1317. Interactions Magnetoelastiques dans le Sulfure de Manganese, Roland Georges, *Compt. Rend., Ser. B*, **268**, 16-19 (1969).
1318. Magnetic Susceptibility of Some Transition Metal Chalcogenides Having the Cr_3S_4 Structure, B. L. Morris, R. H. Plovnick, and A. Wold, *Solid State Commun.* **7**, 291-93 (1969).
1319. Etude par Diffraction des Neutrons du Compose Spinelle $MgCr_2O_4$, Rene Plumier and Michel Sougi, *Compt. Rend., Ser. B*, **268**, 365-67 (1969).
1320. Magnetic Properties of Er-Based Binary Rare Earth Alloys, A. H. Millhouse and W. C. Koehler (Solid State Division, Oak Ridge National Laboratory, Oak Ridge, Tenn.), for presentation at the *Intern. Colloquium on Rare Earth Elements*, Paris-Grenoble, France, May 5-10, 1969.
1321. Magnetism of Piezoelectric $Ga_{2-x}Fe_xO_3$, C. H. Nowlin and R. V. Jones, *J. Appl. Phys.* **34**, 1262 (1963).
1322. Kristallstruktur und magnetische Ordnung des Hubnerits, $MnWO_4$, H. Dachs, E. Stoll, and H. Weitzel, *Z. Krist.* **125**, 120-29 (1967).
1323. Magnetic Susceptibility of $Mn_2P_2O_7$, D. C. Fowlis and C. V. Stager, *Can. J. Phys.* **47**, 371 (1969).
1324. Messung der differentiellen magnetischen Suszeptibilität von $HoCl_3$, $Ho(Y)Cl_3$ und $Ho(La)Cl_3$ bei Heliumtemperaturen. Beobachtung eines Phononenengpasses in $Ho(La)Cl_3$, K. H. Hellwege, J. Kotzler, and G. Weber, *Z. Physik* **217**, 373-85 (1968). (149)
1325. Antiferromagnetism in Au_5Mn_2, J. H. Smith and P. Wells, *J. Phys. C (Solid State Phys.), Ser. 2*, **2**, 356-60 (1969). (149)
1326. Antiferromagnetism of $Cs_2MnCl_4 \cdot 2H_2O$, T. Smith and S. A. Friedberg, *Phys. Rev.* **177**, 1012-16 (1969). (149)
1327. Magnetic States of Cerium Antimonide with Sodium Chloride Structure, Takashi Tsuchida and Yoji Nakamura, *J. Phys. Soc. Japan* **25**, 284 (1968). (129)
1328. Magnetic Susceptibility and Electrical Conductivity in a Higher Chromium Germanide, V. L. Zagryazhskii, P. V. Gel'd, and A. K. Shtol'ts, *Izv. VUZ Fiz. USSR* No. 5, 46-9 (1968) (translated into English in *Soviet Phys. J.*).
1329. An Investigation of $CuFeO_2$ by the Mossbauer Effect, A. H. Muir, Jr., and H. Wiedersich, *J. Phys. Chem. Solids* **28**, 65-72 (1967).
1330. A Neutron Diffraction Investigation of Cr_2Te_3 and Cr_5Te_6, Arne F. Andersen, *Acta Chem. Scand.* **17**, 1335-42 (1963).
1331. A Neutron Diffraction Study of Fe_7Se_8, A. F. Andersen and J. Leciejewicz, *J. Phys.* **25**, 574-8 (1964).
1332. Magnetic First-Order Phase Transition in Single-Crystal MnAs, R. W. DeBlois and D. S. Rodbell, *Phys. Rev.* **130**, 1347-60 (1963).

1333. Magnetische Suszeptibilität des Mangans zwischen Helium-Temperatur und 2000°K, Rudolf Kohlhaas and Wolf Dieter Weiss, *Z. Naturforschung* **24a**, 287-88 (1969).

1334. Neutron Diffraction Study on Manganese Telluride, Nobuhiko Kunitomi, Yoshikazu Hamaguchi, and Shuichiro Anzai, *J. Phys.* **25**, 568-74 (1964).

1335. Etude de l'Antiferromagnetisme dans $Mn(DCOO)_2 \cdot 2H_2O$ par Diffraction Neutronique, Erwin F. Bertaut, Paulette Burlet, and Paul Burlet, *Solid State Commun.* **7**, 343-49 (1969).

1336. Solid-State Reaction between Zinc Oxide and Ferric Oxide, R. Parker, C. J. Rigden, and C. J. Tinsley, *Trans. Faraday Soc.* No. 553, **65**, Part 1, 219-24 (1969).

1337. Nuclear Magnetic Resonance in Cr_3As_2, Takeshi Shinohara and Hiroshi Watanabe, *J. Phys. Soc. Japan* **21**, 2076 (1966).

1338. Influence of s–f Exchange Interaction on Electrical Conduction in Rare-Earth Dialuminides, H. J. van Daal and K. H. J. Buschow, *Solid State Commun.* **7**, 217-21 (1969).

1339. Magnetic Properties of Jarosites, $RFe_3(OH)_6(SO_4)_2$ (R = NH_4, Na or K), Mikio Takano, Teruya Shinjo, Masao Kiyama, and Toshio Takada, *J. Phys. Soc. Japan* **25**, 902 (1968).

1340. Etude par Diffraction de Neutrons a 0.31°K de la Structure Antiferromagnetique des Grenats d'Aluminium-Terbium et d'Aluminium-Holmium, J. Hammann (Service de Physique du Solide et de Resonance Magnetique, Centre d'Etudes Nucleaires de Saclay, BP no 2, 91, Gif-sur-Yvette, France), DPh-SRM/DOC/68-754/EC, Oct. 1968. Also in *Acta Cryst.* **25B**, 1853-56 (1969).

1341. Effect of Ferromagnetic Ordering on the Thermoelectric Power of $GdAl_2$, M. P. Kawatra and J. A. Mydosh, *Phys. Letters* **28A**, 182-83 (1968).

1342. Solid State Reactions in Lithium Ferrite, A. J. Pointon and R. C. Saull, *J. Am. Ceram. Soc.* **52**, 157-60 (1969).

1343. Metamagnetism in the Perovskite Compound Gd_2CoMnO_6, A. Marsh and C. C. Clark, *Phil. Mag.* **19**, 449-63 (1969).

1344. Magnetic Phase Transitions in the Two-Dimensional Antiferromagnet Rb_2MnF_4, R. J. Birgeneau, H. J. Guggenheim, and G. Shirane, *Bull. Am. Phys. Soc.* **14**, 738 (1969).

1345. The Magnetic and Crystallographic Properties of MnBi Studied by Neutron Diffraction, A. F. Andresen, W. Halg, P. Fischer, and E. Stoll, *Acta Chem. Scand.* **21**, 1543-54 (1967).

1346. Magnetic Susceptibility of Vanadium Sulphides at Elevated Temperatures, G. M. Loginov, *Russ. J. Inorg. Chem.* **6**, No. 2, 133-37 (1961).

1347. Origin of Ferrimagnetism in Compounds, Cr_5S_6 and Cr_2S_3, Motoyoshi Yuzuri and Yoji Nakamura, *J. Phys. Soc. Japan* **19**, 1350-54 (1964).

1348. The Crystal Structure of K_3MoCl_6, Z. Amilius, B. van Laar, and H. M. Rietveld, *Acta Cryst.* **25B**, 400-02 (1969).

1349. Magneto-Optical Properties of a Green Room-Temperature Ferromagnet: $FeBO_3$, A. J. Kurtzig, R. Wolfe, R. C. LeCraw, and J. W. Nielsen, *Appl. Phys. Letters* **14**, 350-52 (1969).

1350. Anomalies in Electrical Conductivity and Magnetic Susceptibility of Chromium Selenides, Kan-ichi Masumoto, Tadamiki Hihara, and Takahiko Kamigaichi, *J. Phys. Soc. Japan* **17**, 1209-10 (1962).

1351. On the Influence of the Oxygen Content on the Magnetocrystalline Anisotropy and on the Magnetization of Ni-Fe and Mg-Fe Ferrites, G. Elbinger, *Phys. Stat. Sol.* **31**, K127 (1969).

1352. Magnetic Properties of $CrSb_2$, Kengo Adachi, Kiyoo Sato, and Makoto Matsuura, *J. Phys. Soc. Japan* **26**, 906 (1969).

1353. Pressure Dependence of the Curie Temperature of Ferromagnetic Laves Phase Alloys, G. T. Alfieri, E. Banks, and K. Kanematsu, *J. Appl. Phys.* **40**, 1322 (1969).

1354. Magnetic Structures and Properties of $UFeO_4$, M. Bacmann, E. F. Bertaut, A. Blaise, R. Chevalier, and G. Roult, *J. Appl. Phys.* **40**, 1131 (1969).

1355. Magnetic Studies of Annealed and Alloyed Chromium by Neutron Diffraction, G. E. Bacon and N. Cowlam, *J. Phys. C (Solid St. Phys.)*, Ser. 2, **2**, 238-51 (1969).

1356. Neutron Scattering from K_2NiF_4: a Two-Dimensional Heisenberg Antiferromagnet, R. J. Birgeneau, H. J. Guggenheim, and G. Shirane, *Phys. Rev. Letters* **22**, 720-23 (1969).

1357. Magnetic Properties of the Oxygen Molecule in Solid Oxygen-Argon Mixtures, T. G. Blocker III, C. L. Simmons, and F. G. West, *J. Appl. Phys.* **40**, 1154 (1969).

REFERENCES

1358. Magnetic Structure of Iron Manganite by Neutron Diffraction, B. Boucher, R. Buhl, and M. Perrin, *J. Appl. Phys.* **40**, 1126 (1969).
1359. Manganites Spinelles Purs d'Elements de Transition Preparations et Structures Cristallographiques, Robert Buhl, *J. Phys. Chem. Solids* **30**, 805-12 (1969).
1360. Magnetic Properties of the System $Zn_{1-x}Cd_xCr_2Se_4$, G. Busch, B. Magyar and O. Vogt, *Solid State Commun.* **7**, 509-10 (1969).
1361. Crystalline Spark Splitting in $TmAl_3$, K. H. J. Buschow, *Z. Phys. Chem. Neue Folge* **59**, 21-26 (1968).
1362. Magnetothermal Studies of Phase Transitions in $CuCl_2 \cdot 2H_2O$, G. J. Butterworth and V. S. Zidell, *J. Appl. Phys.* **40**, 1033 (1969).
1363. Raman Scattering by Magnetic Excitations in $RbNiF_3$, S. R. Chinn, H. J. Zeiger, and J. R. O'Connor, *J. Appl. Phys.* **40**, 1603 (1969).
1364. Exchange-Enhanced Paramagnetism and Weak Ferromagnetism in the Ni_3Al and Ni_3Ga Phases; Giant Moment Inducement in Fe-Doped Ni_3Ga, F. R. de Boer, C. J. Schinkel, J. Biesterbos, and S. Proost, *J. Appl. Phys.* **40**, 1049 (1969).
1365. Magnetization and Hyperfine Structure of ^{57}Fe in $CsFeF_3$, M. Eibschutz, L. Holmes, H. J. Guggenheim, and H. J. Levinstein, *J. Appl. Phys.* **40**, 1312 (1969).
1366. Polarized Optical Reflectivity of Ferromagnetic EuO, J. Feinleib, p. 8 in Massachusetts Institute of Technology, Francis Bitter National Laboratory, Quarterly Technical Status Report, January 1, 1969-March 31, 1969, QTSR 35, Air Force Contract F44620-67-C-0047, March 31, 1969.
1367. Mossbauer Investigation of Metamagnetic $FeCO_3$, D. W. Forester and N. C. Koon, *J. Appl. Phys.* **40**, 1316 (1969).
1368. Structural and Magnetic Properties of Nonstoichiometric Praseodymium Monophosphide, Enrico Franceschi and Giorgio L. Olcese, *J. Phys. Chem. Solids* **30**, 903-07 (1969).
1369. Crystallographic and Magnetic Properties of Solid Solutions of the Phosphides M_2P, M = Cr, Mn, Fe, Co, and Ni, R. Fruchart, A. Roger and J. P. Senateur, *J. Appl. Phys.* **40**, 1250 (1969).
1370. Cubic-Magnetic and Optical Behavior of Orthorhombic $RbFeF_3$, E. M. Gyorgy, H. J. Levinstein, J. F. Dillon, Jr., and H. J. Guggenheim, *J. Appl. Phys.* **40**, 1599 (1969).
1371. Divergences of the Magnetic Properties of $CrBr_3$ near the Critical Point, John T. Ho and J. D. Litster, *J. Appl. Phys.* **40**, 1270 (1969).
1372. Total Intensities of Some Crystal Field Transitions in MnO and MnS Related to the Antiferromagnetism, Donald R. Huffman, *J. Appl. Phys.* **40**, 1334 (1969).
1373. Magnetic Properties of $CoSe_xS_{2-x}$, Vancliff Johnson and Aaron Wold, *J. Appl. Phys.* **40**, 1287 (1969).
1374. Scattering of Light by Spin-Ordering Fluctuations Close to the Neel Point in Antiferromagnetic $KNiF_3$, Z. M. Khashkhozev, R. V. Pisarev, and G. A. Smolenskii, *Soviet Phys.-Solid State* **10**, 2772 (1969).
1375. Magnetic Properties of Mixed Rare Earth Antimonide Crystals, J. P. Kopp, *Solid State Commun.* **7**, 505-07 (1969).
1376. Neutron Diffraction Study of NpC, G. H. Lander, L. Heaton, M. H. Mueller and K. D. Anderson, *J. Phys. Chem. Solids* **30**, 733-37 (1969).
1377. Magnetic Properties, Conductivity and Ionic Ordering in $Fe_{1-x}Cu_xCr_2S_4$, F. K. Lotgering, R. P. van Stapele, G. H. A. M. van der Steen and J. S. van Wieringen, *J. Phys. Chem. Solids* **30**, 799-804 (1969).
1378. Chemical and Magnetic Study of Layered Strontium Lanthanum Manganate Structures, J. B. MacChesney, J. F. Potter, and R. C. Sherwood, *J. Appl. Phys.* **40**, 1243 (1969).
1379. Magnetic Structure and Spin-Density Distribution in UFe_2, M. Yessik, *J. Appl. Phys.* **40**, 1133 (1969).
1380. Magnetic Structure and Exchange Interactions in Cubic Gadolinium Compounds, T. R. McGuire, R. J. Gambino, S. J. Pickart, and H. A. Alperin, *J. Appl. Phys.* **40**, 1009 (1969).
1381. Magnetic and Semiconducting Properties of SmB_6, A. Menth, E. Buehler, and T. H. Geballe, *Phys. Rev. Letters* **22**, 295-97 (1969).

1382. Magnetic Properties of $FeTi_2S_4$, Bernard L. Morris, Vancliff Johnson, Ross H. Plovnick, and Aaron Wold, *J. Appl. Phys.* **40**, 1299 (1969).
1383. Neutron-Diffraction Investigations on USb, UBi, and U_3Sb_4, Clayton E. Olsen and W. C. Koehler, *J. Appl. Phys.* **40**, 1135 (1969).
1384. Thermal Expansion in Dysprosium Aluminum Garnet, J. W. Philip, R. Gonano, and E. D. Adams, *J. Appl. Phys.* **40**, 1275 (1969).
1385. Spin Reorientation in Mixed Samarium-Dysprosium Orthoferrites, R. D. Pierce, R. Wolfe, and L. G. van Uitert, *J. Appl. Phys.* **40**, 1241 (1969).
1386. Anomalous and Ordinary Hall Effect in Terbium, J. J. Rhyne, *J. Appl. Phys.* **40**, 1001 (1969).
1387. Antiferromagnetism in Chromium and its Alloys, T. M. Rice, A. S. Barker, Jr., B. I. Halperin, and D. B. McWhan, *J. Appl. Phys.* **40**, 1337 (1969).
1388. Optical Measurements on Ferromagnetic $Tb(OH)_3$, P. D. Scott and W. P. Wolf, *J. Appl. Phys.* **40**, 1031 (1969).
1389. Magnetic and Optical Properties of Ferrimagnets in the $RbMgF_3$-$RbCoF_3$ System, M. W. Shafer, *J. Appl. Phys.* **40**, 1601 (1969).
1390. Magnetic Behavior of the $FeSiO_3$-$MgSiO_3$ Orthopyroxene System from NGR in ^{57}Fe, G. K. Shenoy, G. M. Kalvius, and S. S. Hafner, *J. Appl. Phys.* **40**, 1314 (1969).
1391. Temperature Dependence of M_s and K_1 of $BaFe_{12}O_{19}$ and $SrFe_{12}O_{19}$ Single Crystals, B. T. Shirk and W. R. Buessem, *J. Appl. Phys.* **40**, 1294 (1969).
1392. Pressure Dependence of Ferromagnetic Phase Transitions of Chromium Chalcogenide Spinels, Vishnu C. Srivastava, *J. Appl. Phys.* **40**, 1017 (1969).
1393. Magnetic and Magneto-Optic Properties of FeRh and CrO_2, A. M. Stoffel, *J. Appl. Phys.* **40**, 1238 (1969).
1394. Magnetic Structure of CrAs and Mn-Substituted CrAs, H. Watanabe, N. Kazama, Y. Yamaguchi, and M. Ohashi, *J. Appl. Phys.* **40**, 1128 (1969).
1395. Exchange Interactions in $Mn(Sc_2)S_4$, Peter J. Wojtowicz, Lynne Darcy, and Martin Rayl, *J. Appl. Phys.* **40**, 1023 (1969).
1396. Ferromagnetism in VAu_4 and Related Compounds, J. R. Asik, R. J. Gambino, and T. R. McGuire, *J. Appl. Phys.* **40**, 1385 (1969).
1397. Mossbauer Study of Perovskites of Composition $SrFe_{1-x}Cr_xO_{3-y}$, E. Banks and M. Mizushima, *J. Appl. Phys.* **40**, 1408 (1969).
1398. Intrinsic Magnetization of Fe-Ni-Mn Alloys, D. A. Colling, *J. Appl. Phys.* **40**, 1379 (1969).
1399. Ferromagnetic to Canted-Ferrimagnetic Transition in $Fe(Pd,Pt)_3$, J. S. Kouvel and J. B. Forsyth, *J. Appl. Phys.* **40**, 1359 (1969).
1400. First-Order Magnetic Transitions in $(Fe,Mn)_{\geq 2}As$, R. M. Rosenberg, W. H. Cloud, F. J. Darnell, R. B. Flippen, and S. R. Butler, *J. Appl. Phys.* **40**, 1361 (1969).
1401. Mossbauer Studies of Co_3O_4; Bulk Material and Ultrafine Particles, Walter Kundig, M. Kobelt, H. Appel, G. Constabaris, and R. H. Lindquist, *J. Phys. Chem. Solids* **30**, 819-26 (1969).
1402. Etude par Diffraction Neutronique de Fe_3Se_4, B. Lamber-Andron and G. Berodias, *Solid State Commun.* **7**, 623-29 (1969).
1403. Thermal and Magnetic Properties of Cerium Magnesium Nitrate below 1°K, K. W. Mess, J. Lubbers, L. Niesen, and W. J. Huiskamp, *Physica* **41**, 260-68 (1969).
1404. Magnetic Behavior of the Pt + Fe System near Pt_3Fe, D. Palaith, C. W. Kimball, R. S. Preston, and J. Crangle, *Phys. Rev.* **178**, 795-99 (1969).
1405. Ferrimagnetic Structure of Mn_2Co_2C, N. S. Satya Murthy, R. J. Begum, C. S. Somanathan, B. S. Srinivasan, and M. R. L. N. Murthy, *J. Phys. Chem. Solids* **30**, 939-45 (1969).
1406. Effet Hall et Magnetoresistance Transverse dans Fe_7Se_8 (Structure NiAs4c), Jacques Serre and Gerard Villers, *Compt. Rend., Ser. B*, **268**, 162-65 (1969).
1407. Low Temperature Minimum of the Electrical Resistance in Ni-Cu and Ni-Cu-H Alloys, T. Skoskiewicz and B. Baranowski, *Solid State Commun.* **7**, 647-49 (1969).
1408. Studies of Au_4X-Ordered Alloys: Electron and Neutron Diffraction, Resistivity and Specific Heat, R. S. Toth, A. Arrott, S. S. Shinozaki, S. A. Werner, and H. Sato, *J. Appl. Phys.* **40**, 1373 (1969).

1409. Mechanism for the First-Order Magnetic Transition in the FeRh System, P. Tu, A. J. Heeger, J. S. Kouvel, and J. B. Comly, *J. Appl. Phys.* **40**, 1368 (1969).

1410. The Effect of Deuteration on the Neel Temperature of $MnCl_2 \cdot 4H_2O$, B. G. Turrell, C. L. Yue, and D. S. Sahri, *Phys. Letters* **28A**, 680-81 (1969).

1411. The Low Temperature Specific Heats of a Number of Ferromagnetic Pd-Co and Pt-Co Alloys, J. C. G. Wheeler, *J. Phys. C (Solid St. Phys.), Ser. 2,* **2**, 135-46 (1969).

1412. Ferromagnetic Transitions in Dilute Solutions of Cobalt in Palladium, B. D. Dunlap and J. C. Dash, *Phys. Rev.* **155**, 460-67 (1967).

1413. Electrical Conductivity and Thermal emf of a $Li_xMn_{1-x}Se$, A. G. Rustamov, I. G. Kerimov, and L. M. Valiev, *Izv. Akad. Nauk Azerb. SSR, Ser. Fiz.-Tekh. Mat. Nauk* No. 1, 28-32 (1968). (in Russian)

1414. Neutron Diffraction Study of NiS under Pressure, F. A. Smith and J. T. Sparks, *J. Appl. Phys.* **40**, 1332 (1969).

1415. W. Trzebiatowski, A. Sliwa, and B. Stalinski, *Roczniki Chem.* **26**, 110 (1952); *Roczniki Chem.* **28**, 12 (1954).

1416. Ferromagnetic Structure of Uranium Hydride, M. K. Wilkinson, C. G. Shull, and R. E. Rundle, *Phys. Rev.* **99**, 627 (1955).

1417. Specific Heat of Antiferromagnetic $KMnCl_3 \cdot 2H_2O$, H. Forstat, J. N. McElearney, and P. T. Bailey, *Phys. Letters* **27A**, 549-50 (1968).

1418. Relaxation Effects in Antiferromagnetic $FeCO_3$, Hang Nam Ok, *Bull. Am. Phys. Soc.* **14**, 729 (1969).

1419. Etude Cristallographique et Magnetique d'un Fluorure Inedit de Type Trirutile, J. Portier, A. Tressaud, R. de Pape, and P. Hagenmuller, *Compt. Rend., Ser. C*, **267**, 1711-13 (1968).

1420. Exchange Enhancement of Conduction Electron Spin Lifetimes in $ZrZn_{1.9}$, W. M. Walsh, Jr., G. S. Knapp, L. W. Rupp, Jr., and P. H. Schmidt, *J. Appl. Phys.* **41**, 1081-82 (1970).

1421. Magnetic Susceptibilities of Rare-Earth-Indium Compounds: RIn_3, K. H. J. Buschow, H. W. de Wijn, and A. M. van Diepen, *J. Chem. Phys.* **50**, 137-41 (1969).

1422. Mossbauer Study of β- and δ-FeOOH and Their Disintegration Products, I. Dezsi, L. Keszthelyi, D. Kulgawczuk, B. Molnar, and N. A. Eissa, *Phys. Stat. Sol.* **22**, 617 (1967).

1423. Sound Propagation near the Magnetic Phase Transition in EuO, B. Luthi and R. J. Pollina, *Phys. Rev. Letters* **22**, 717-20 (1969).

1424. Untersuchungen an magnetischen Phasen intermetallischer Erbiumverbindungen, Georg Petrich, *Z. Physik* **221**, 431-50 (1969).

1425. Interaction of Magnetic Sublattices in Gadolinium Iron Garnet, V. P. Polyakov, *Soviet Phys.-Solid State* **10**, 2891-93 (1969).

1426. Magnetic Properties of U_3P_4 and U_3As_4 Single Crystals, T. Slotwinski and J. Trivisonno, *J. Phys. Chem. Solids* **30**, 1273-76 (1969).

1427. Magnetic Properties of Certain Magnetically Ordered Fluorides in a Strong Pulsed Magnetic Field, A. T. Starovoitov, V. I. Ozhogin, V. A. Bokov, and P. P. Syrnikov, *Soviet Phys.-Solid State* **10**, 2862-66 (1969).

1428. A Neutron Diffraction Determination of the Coherent Scattering Amplitude of Np and the Possible Antiferromagnetism of Neptunium Dioxide, L. Heaton, M. H. Mueller, and J. M. Williams, *J. Phys. Chem. Solids* **28**, 1651-54 (1967).

1429. Magnetic Properties of $A^{2+}Cr_2S_4$ Compounds with the NiAs Structure, R. E. Tressler and V. S. Stubican, *J. Am. Ceram. Soc.* **51**, 391-93 (1968).

1430. Low Temperature Crystallographic and Magnetic Study of $LaCoO_3$, N. Menyuk, K. Dwight, and P. M. Raccah, *J. Phys. Chem. Solids* **28**, 549-56 (1967).

1431. The Influence of Exchange Interaction on the Ordering of Holmium Ions in Orthoferrites of the $Ho_xSm_{1-x}FeO_3$ System, K. P. Belov, A. M. Kadomtseva, T. M. Ledneva, T. L. Ovchinnikova and V. A. Timofeeva, *Soviet Phys.-Solid State* **10**, 2997 (1969).

1432. Magnetic Properties of hcp Fe-Ru Alloys, Hiroyasu Fujimori and Hideo Saito, *J. Phys. Soc. Japan* **26**, 1115 (1969).

1433. Precision Measurements of the F^{19} NMR in FeF_2 from 4.2°K to the Critical Region, Stanley M. Kulpa, *J. Appl. Phys.* **40**, 2274 (1969).

1434. The Para Process near the Curie Point in Polycrystalline Cobalt-Zinc Ferrite, S. A. Poltinnikov, *Soviet Phys.-Solid State* **10**, 2928 (1969).

1435. Transition from Ferromagnetism to Paramagnetism in Ni-Cu Alloys, C. G. Robbins, Helmut Claus, and Paul A. Beck, *J. Appl. Phys.* **40**, 2269 (1969).

1436. Propriétés Electriques, Optiques et Magnetiques des Tellurures de Cobalt et de Nickel (Phase de Structure NiAs), Georges Saut, thesis, Faculte des Sciences de l'Universite de Paris, 1967, p. 36.

1437. Propriétés Physiques de Fe_7Se_8, Jacques Serre, Pierre Gibart, and Jacques Bonnerot, *J. Phys. (France)* **30**, 93-96 (1969).

1438. Crystal and Magnetic Structures of $Sr_2Fe_2O_5$, Takayoshi Takeda, Yasuo Yamaguchi, Hiroshi Watanabe, Shoichi Tomiyoshi, and Hisao Yamamoto, *J. Phys. Soc. Japan* **26**, 1320 (1969).

1439. Zur Temperaturabhängigkeit der Molwärme und der magnetischen Suszeptibilität von $Fe_2O_3-Cr_2O_3$-mischkristallen, A. Knappwost, H. Lechert, and W. Gunsser, *Z. Phys. Chem. Neue Folge* **47**, 207-223 (1965).

1440. Magnetic Properties of σ-Phase of Some 3d-Transition Alloys, Nobuo Mori and Tadayasu Mitsui, *J. Phys. Soc. Japan* **26**, 1087-93 (1969).

1441. Spin Waves around the Neel Temperature in MnO, N. Kroo and L. Bata, p. 111-15 in Neutron Inelastic Scattering, Vol. 2, (IAEA, 1968).

1442. Revue des Proprietes Magnetiques et Electroniques des Composes des Terres Rares avec les Anions du 5^{ieme} Groupe du Systeme Periodique, P. Junod, A. Menth, and O. Vogt, *Phys. kondens. Materie* **8**, 323-370 (1969).

1443. Preparation of Ferromagnetic Chromium Dioxide, Joseph H. Balthis, Jr. (du Pont de Nemours, E.I., and Co.), U.S. Patent 3, 423, 320, Jan. 21, 1969.

1444. Magnetic Excitations in Cobaltous Oxide, W. J. L. Buyers, G. Dolling, J. Sakurai, and R. A. Cowley (Atomic Energy of Canada, Ltd., Chalk River, Ontario), p. 123-31 in IAEA Neutron Inelastic Scattering, Vol. 2, 1968.

1445. Magnetic Susceptibility of Yttrium-Cobalt Intermetallic Compounds, W. G. D. Frederick, K. J. Strnat, and P. P. Yaney, *Bull. Am. Phys. Soc.* **14**, 798 (1969).

1446. Contribution à la Connaissance des Ferrites de Cuivre et de Certaines de leurs Solutions Solides, J. Mexmain, Theses Doct. Etat. Sci. Paris, 1968; Arch. orig. Centre Document C.N.R.S., No. 2292, April 1968, 91 p.

1447. Sur les Vanadates. XV. La Susceptibilité Magnetique des Composés $Na_x V^{IV}_x V^V_{2-x} O_5$ ($0 < x \leq 1$), C. Mirel, D. Lupu, I. Lukacs, C. Strusievici, and C. Liteanu, *Rev. Roumaine Chim.* **13**, No. 8, 1035-41 (1968).

1448. Determination of T_c in Ferrimagnetic Garnet Materials, Benjamin R. Capone, *IEEE Trans.* **MAG-3**, No. 4, 705 (1967).

1449. Ultrasonic Propagation in EuSe, B. Golding and E. Buehler, *Solid State Commun.* **7**, 747-50 (1969). (156)

1450. Susceptibilités Magnetiques des Mononitrures et Sesquicarbures de Thorium, Uranium et Plutonium, Georges Raphael and Charles de Novion, *Solid State Commun.* **7**, 791-93 (1969).

1451. Crystal Chemistry of Metal Dioxides with Rutile-Related Structures, D. B. Rogers, R. D. Shannon, A. W. Sleight, and J. L. Gillson, *Inorg. Chem.* **8**, 841 (1969).

1452. Magnetic Properties of Uranium Compounds, Wojciech Suski and Bohdan Stalinski, Fizkochem. Ciala Stalego, Ewa Modrakowa, ed. (Panst. Wydawnictwo Nauk, Warsaw, 1967), p. 257-79.

1453. Magnetic Studies on $CuCr_2O_4$ Catalysts, H. P. Walter, I. Schulz, and J. Scheve, *Z. Anorg. Allgem. Chem.* **352**, No. 5-6, 241-45 (1967).

1454. Hyperfine Field of ^{237}Np in NpO_2, B. D. Dunlap, G. M. Kalvius, D. J. Lam, and M. B. Brodsky, *J. Phys. Chem. Solids* **29**, 1365-67 (1968).

1455. Mossbauer Study of Fe^{6+} in Potassium Ferrate, K_2FeO_4, Atsuko Ito and Kazuo Ono, *J. Phys. Soc. Japan* **26**, 1548 (1969).

1456. NMR study in MnB_2, Mitsuo Kasaya, Tadamiki Hihara, and Yoshitaka Koi, *J. Phys. Soc. Japan* **26**, 1549 (1969).
1457. Antiferromagnetic Structures of USb and UBi, M. Kuznietz, G. H. Lander, and F. P. Campos, *J. Phys. Chem. Solids* **30**, 1642-44 (1969).
1458. Electrical and Magnetic Properties of Cu_3Se_2 and Some Related Compounds, Kimihiko Okamoto, Shichio Kawai, and Ryoiti Kiriyama, *Japan. J. Appl. Phys.* **8**, 718-24 (1969).
1459. Susceptibilité Magnetique de Quelques Sulfures et Oxydes de Plutonium, Georges Raphael and Charles de Novion, *J. Phys. (Fr.)* **30**, 261-66 (1969).
1460. Antiferromagnetism in $Cu(NH_3)_4SO_4 \cdot H_2O$, Shinhachiro Saito, *J. Phys. Soc. Japan* **26**, 1388-95 (1969).
1461. Magnetic Properties of Ferrite-Chromite Series of Nickel and Cobalt, Tachiro Tsushima, *J. Phys. Soc. Japan* **18**, 1162-66 (1963).
1462. Spin-Structure Determination in Single Crystals of Some Chromites, T. Tsushima, Y. Kino, and S. Funahashi, *J. Appl. Phys.* **39**, 626-28 (1968).
1463. Antiferromagnetic Resonance in $LiMnPO_4$, P. R. Elliston, J. G. Creer, and G. J. Troup, *J. Phys. Chem. Solids* **30**, 1335-40 (1969).
1464. Magnetic Properties of Er-Dy Alloys, A. H. Millhouse, W. C. Koehler, and H. R. Child, *J. Appl. Phys.* **40**, 1006 (1969).
1465. Magnetic and Electrical Properties of $Co_{1-x}Cu_xRh_2S_4$, F. K. Lotgering, *J. Phys. Chem. Solids* **30**, 1429-34 (1969).
1466. Elastic Moduli and Ultrasonic Attenuation of Praseodymium, Neodymium, and Samarium from 4.2 to 300°K, M. Rosen, *Phys. Rev.* **180**, 540-44 (1969).
1467. Contribution to the Magnetic Properties of Some B8-Type Phases (CoSb, $CoSn_{0.7}$, NiSb, $Co_xCr_{1-x}Sb$ and $Cr_{0.75}Fe_{0.25}Sb$), Hans Schmid, *Cobalt* **7**, 26-32 (1960).
1468. A Mossbauer Effect Study of the Metallic Compounds U_6Fe and Pu_6Fe, and the Relevance of the Results to Theories of the Behaviour of Actinide Metals, S. Blow, *J. Phys. Chem. Solids* **30**, 1549-59 (1969).
1469. Magnetism of Synthetic and Natural $MnCO_3$, I. Maartense (Submitted to *Phys. Rev.* 1969). (158)
1470. Heat Capacity of Manganese Titanate from 30° to 300°K, C. C. Stephenson and David Smith, *J. Chem. Phys.* **49**, 1814-18 (1968).
1471. The Magnetic Properties of the Compound Dy_3Al_2, K. H. J. Buschow, *Phys. Letters* **29A**, 12-13 (1969).
1472. Magnetic Susceptivility Studies of the Cubic Laves Phases $Sc(Ni_{1-x}Co_x)_2$, E. W. Collings, R. D. Smith, and R. G. Lecander, *J. Less-Common Metals* **18**, 251-66 (1969).
1473. Mossbauer Effect in a Cubic Antiferromagnet near the Neel Point, G. K. Wertheim, H. J. Guggenheim, H. J. Williams, and D. N. E. Buchanan, *Phys. Rev.* **158**, 446-50 (1967).
1474. Absence of Antiferromagnetism in Ti_2O_3, R. M. Moon, T. Riste, W. C. Koehler, and S. C. Abrahams, *J. Appl. Phys.* **40**, 1445-47 (1969).
1475. Propriétés Magnetiques et Thermodynamiques de FeP et Propriétés Magnetiques des Solutions Solides $Mn_{1-x}Fe_xP$, J. Bonnerot, R. Fruchart, and A. Roger, *Phys. Letters* **26A**, 536-37 (1968).
1476. Magnetic Properties of the Intermetallic Compound $Co_{3+x}Ti_{1-x}$, Yoshihira Aoki, *J. Phys. Soc. Japan* **27**, 258 (1969).
1477. Electronic Properties of Two New Elemental Ferromagnets: fcc Pr and Nd, E. Bucher, C. W. Chu, J. P. Maita, K. Andres, A. S. Cooper, E. Buchler, and K. Nassau, *Phys. Rev. Letters* **22**, 1260-63 (1969).
1478. Magnetic Properties of $Mn_2Ge_ySb_{1-y}$ Alloys, A. A. Galkin, E. A. Zavadskii, and E. M. Morozov, *Soviet Phys.-Solid State* **11**, 76-80 (1969). (Transl. from *Fiz. Tverd. Tela.* **11**, 106-12, 1969).
1479. Magnetische Eigenschaften von Verbindungen der Seltenen Erdmetalle mit Mangan und Eisen, H. R. Kirchmayr, *Z. Angew. Phys.* **27**, 18-27 (1969).
1480. Magnetic Properties of $KNiF_3$ Containing Impurities, T. Miyashita, H. Kondo, and S. Miyahara, *J. Phys. Soc. Japan* **27**, 256 (1969).

1481. Magnetocrystalline Anisotropy of Single Crystals of Lithium Gallium Ferrites, G. A. Petrakovskii, V. N. Seleznev, K. A. Sablina, and L. M. Protopopova, *Soviet Phys.-Solid State* **11**, 7-11 (1969). (Transl. from *Fiz. Tverd. Tela* **11**, 11-16, 1969).

1482. An X-Ray Diffraction Investigation of the Magnetic and Electric Properties of the System $BiFeO_3$-$LaFeO_3$, Yu. E. Roginskaya, Yu. N. Venevtsev, S. A. Fedulov, and G. S. Zhdanov, *Soviet Phys.-Cryst.* **8**, 490-94 (1964). (Transl. from *Kristallografiya* **8**, 610-16, 1963).

1483. Chromium-Like Antiferromagnetic Behavior of CrB_2, R. G. Barnes and R. B. Creel, *Phys. Letters* **29A**, 203-04 (1969).

1484. Magnetic Phase Diagram of $CsMnCl_3 \cdot 2H_2O$, G. J. Butterworth and J. A. Woolam, *Phys. Letters* **29A**, 259 (1969).

1485. Specific Heat of Antiferromagnetic $Cs_3Cu_2Cl_7 \cdot 2H_2O$, G. J. Butterworth and V. S. Zidell, *J. Phys. Soc. Japan* **28**, 526-27 (1970).

1486. Magnetic Properties of Gadolinium Ortho-Vanadate, J. D. Cashion, A. H. Cooke, L. A. Hoel, D. M. Martin, and M. R. Wells (Clarendon Laboratory, Oxford, England). (Rec'd RMIC Aug. 1969; to appear in the *Proc. Colloque International du C.N.R.S. sur les Elements des Terres Rares*).

1487. Magnetic Transitions in Uranium and Plutonium Mononitrides, Monocarbides and Sesquicarbides, P. Costa, R. Lallement, F. Anselin, and D. Rossignol, Met. Soc. AIME, Petroleum Eng., Inst. Metals Div. Special Report Ser. 10, No. 13, 83-91 (1964).

1488. Ordre Antiferromagnetique dans les Alliages $Cr_{2-x}Fe_xAl$, Ahmed Kallel, *Compt. Rend.* **268B**, 455-58 (1969).

1489. Study of the Sublattices Magnetisations in $Mn_{0.6}Zn_{0.4}Fe_2O_4$ by Neutron Diffractometry and Mossbauer Spectrometry, U. Konig, Y. Gros, and G. Chol, *Phys. Stat. Sol.* **33**, 811 (1969).

1490. Some Experiments on Synthetic Titanomagnetites, M. Lewis, *Geophys. J.* **16**, 295-310 (1968).

1491. Preparation and Properties of Polycrystalline Cerium Orthoferrite ($CeFeO_3$), M. Robbins, G. K. Wertheim, A. Menth, and R. C. Sherwood, *J. Phys. Chem. Solids* **30**, 1823-25 (1969).

1492. A New High Pressure Phase of $MnTiO_3$ and its Magnetic Property, Y. Syono, S. Akimoto, Y. Ishikawa, and Y. Endoh, *J. Phys. Chem. Solids* **30**, 1665-72 (1969).

1493. Low Temperature Electrical Resistivities of Some Dilute Pd Fe Alloys, Gwyn Williams and J. W. Loram, *J. Phys. Chem. Solids* **30**, 1827-33 (1969).

1494. Ferrimagnetism in Gadolinium Bromide, F. Varsanyi, K. Andres, and M. Marezzio, *Bull. Am. Phys. Soc.* **13**, 460 (1968).

1495. Low Temperature Crystallographic and Magnetic Study of $LaCoO_3$, N. Menyuk, K. Dwight, and P. M. Raccah, *J. Phys. Chem. Solids* **28**, 549-56 (1967).

1496. Experimental Evidence for Spontaneous Nuclear Ordering in Paramagnetic PrBi, K. Andres and E. Bucher, *Phys. Rev. Letters* **22**, 600-03 (1969).

1497. Superconductors with Magnetic Impurities in a Singlet Ground State, E. Bucher, K. Andres, J. P. Maita, and G. W. Hull, Jr., *Helv. Phys. Acta* **41**, 723-30 (1968).

1498. Low-Temperature Magnetic Transitions in Some Rare-Earth Trichlorides, J. C. Eisenstein, R. P. Hudson, and B. W. Mangum, *Phys. Rev.* **137**, A 1886-95 (1965).

1499. Magnetische Fernordnung bei Silberdifluorid AgF_2, P. Fischer and G. Roult, *Helv. Phys. Acta* **41**, 416 (1968).

1500. $Co_{16}Fe_4Al_3B_6$ a New Permeability Material for Video Heads, H. Hirota, Y. Tawara, and Y. Komatsu, *Japan. J. Appl. Phys.* **8**, 962 (1969).

1501. Magnetic and Crystallographic Studies of Compounds $Mn_{1-x}Cr_xAs$ ($0.3 \geq x > 0$), Hideaki Ido, *J. Phys. Soc. Japan* **27**, 318-21 (1969).

1502. Magnetic Behavior of $CoCr_2O_4$ near its Curie Point, H. Isaji, K. Iwahashi, and Y. Masuda, *J. Phys. Soc. Japan* **27**, 503 (1969).

1503. Magnetoconductance of $ZnMn_2As_2$, Mitsuo Kasaya, *J. Phys. Soc. Japan* **27**, 507 (1969). (161)

1504. The Resistance Maximum in Copper Manganese Dilute Alloys, A. Nakamura and N. Kinoshita, *J. Phys. Soc. Japan* **27**, 382-96 (1969).

1505. Thermal Conductivities and Lorenz Functions of Gadolinium, Terbium, and Holmium Single Crystals, W. J. Nellis and S. Legvold, *Phys. Rev.* **180**, 581-90 (1969).

REFERENCES

1506. Electrical Resistivity of $SrRuO_3$, Y. Noro and S. Miyahara, *J. Phys. Soc. Japan* **27**, 518 (1969).
1507. Curie Temperatures and Residual Resitivities of Nickel-Carbon Solid Solutions, F. C. Schwerer, *J. Appl. Phys.* **40**, 2705-7 (1969).
1508. Sur l'Aimantation des Couches Minces de Ni_3Mn, V. Tutovan, M. Sorohan, and H. Chiriac, *Thin Solid Films* **3**, 287-92 (1969).
1509. Magnetization and Mossbauer Effect in Single Crystal Fe_3BO_6, R. Wolfe, R. D. Pierce, M. Eibschutz, and J. W. Nielsen, *Solid State Commun.* **7**, 949-52 (1969).
1510. Structure Magnetique de l'Oxysulfure d'Ytterbium, R. Ballestracci and J. Rossat-Mignod, *Solid State Commun.* **7**, 1011-14 (1969).
1511. Low-Temperature Magnetic Properties of Some Hexagonal Rare-Earth Trihalides, J. H. Colwell, B. W. Mangum, and D. B. Utton, *Phys. Rev.* **181**, 842-54 (1969).
1512. Mossbauer and Magnetic Studies of Dicalcium Ferrite ($Ca_2Fe_2O_5$), M. Eibschutz, U. Ganiel, and S. Shtrikman, *J. Mater. Sci.* **4**, 574-80 (1969).
1513. Spin-Flop Transition in $BaMnF_4$, L. Holmes, M. Eibschutz, and H. J. Guggenheim, *Solid State Commun.* **7**, 973-76 (1969). (162)
1514. Metamagnetism of Siderite ($FeCO_3$), I. S. Jacobs, *J. Appl. Phys.* **34**, 1106-07 (1963).
1515. Low-Temperature Magnetic Susceptibility of $FeCl_3$, Edwin R. Jones, Jr., O. B. Morton, L. Cathey, Theo Auel, and E. L. Amma, *J. Chem. Phys.* **50**, 4755-57 (1969).
1516. Preparation and Properties of Ferrimagnets in the $RbMgF_3$-$RbCoF_3$ System, M. W. Shafer and T. R. McGuire, *J. Phys. Chem. Solids* **30**, 1989-97 (1969).
1517. Magnetic-Phase Diagram of MnF_2 from Ultrasonic and Differential Magnetization Measurements, Y. Shapira, S. Foner, and A. Misetich, *Phys. Rev. Letters* **23**, 98-101 (1969).
1518. Suszeptibilitäten, Neutronenbeugung mit Verdampferkryostat und Mössbauereffekt in Mischkristallen (Mn, Fe) WO_4, Wolframit, H. Weitzel, Dissertation der Universitat Tubingen (1969).
1519. Neutron Diffraction Studies of Ferrites and Selenates, H. Fuess, Ulrich Koenig, Georg Will, and Erich Woelfel, Diskussionstag. Neutronenphys. Forschungsreaktoren, Vortragskurzfassungen, 3rd, p. 144-46 (1967) (in German).
1520. Thermal Expansion and Magnetostriction of Manganese Chalcogenides, T. A. Mamedov, I. G. Kerimov, and N. G. Aliev, *Dokl. Akad. Nauk Azerb. SSR* **24**, No. 10, 15-20 (1968).
1521. Phase Transitions of NH_4FeF_3, a Perovskite Fluoride, D. N. E. Buchanan and W. J. Ordille, *Mat. Res. Bull.* **4**, 627-32 (1969).
1522. Ultrasonic Attenuation near the Antiferromagnetic Critical Point of CoO, A. Ikushima, *Phys. Letters* **29A**, 417-18 (1969).
1523. Sound Velocity near the Neel Point of MnF_2, A. Ikushima, *Phys. Letters* **29A**, 364-65 (1969).
1524. Effect of the Molecular Field on the Electrical Resistivity near the Magnetic Transition: $GdNi_2$, M. P. Kawatra, S. Skalski, J. A. Mydosh, and J. I. Budnick, *Phys. Rev. Letters* **23**, 83-86 (1969). (163)
1525. Crystal and Magnetic Structures in the Mn-Pd System near $MnPd_3$, E. Kren and G. Kadar, *Phys. Letters* **29A**, 340-41 (1969).
1526. Polarized-Neutron Study of the Magnetic Moment Density in Antiferromagnetic $CuSO_4$, F. Menzinger, D. E. Cox, B. C. Frazer, and H. Umebayashi, *Phys. Rev.* **181**, 936-945 (1969).
1527. Etude du Systeme $Fe_{2(1-s)}Ge_sCu_{1+s}O_4$, Jean-Claude Tellier and Michel Lenglet, *Compt. Rend., Ser. C*, **268**, 1593-95 (1969).
1528. Canted Antiferromagnetism in Gadolinium Tribromide, F. Varsanyi, K. Andres, and M. Marezio, *J. Chem. Phys.* **50**, 5027-30 (1969).
1529. Magnetic Characteristics of Intermetallic Compounds Containing Two Lanthanides Combined with Nickel, W. E. Wallace, H. P. Hopkins, Jr., and K. Lehman, in Thermal, Structural and Magnetic Studies of Metals and Intermetallic Compounds, Annual Report to the U.S. Atomic Energy Commission, May 15, 1968 to May 15, 1969, by W. E. Wallace and R. S. Craig (University of Pittsburgh, Penn. 15213), Contract AT(30-1) 3454, NYO-3454-27.

1530. Crystallographic and Magnetic Characteristics of the $Y_2Ni_{17-x}Cu_x$ System, P. D. Carfagna, W. E. Wallace and R. S. Craig, in Thermal, Structural and Magnetic Studies of Metals and Intermetallic Compounds, Annual Report to the U.S. Atomic Energy Commission, May 15, 1968 to May 15, 1969, by W. E. Wallace and R. S. Craig (University of Pittsburgh, Penn. 15213), Contract AT(30-1) 3454, NYO-3454-27.

1531. Heat Capacity Measurements on Rb_3CoCl_5 Compared with the Ising Model, H. W. J. Blote and W. J. Huiskamp, *Phys. Letters* **29A**, 304-05 (1969).

1532. Thermomagnetic Measurements of Chromium Alloys of Platinum Metals, Albrecht Kussmann, Karin Muller, and Ernst Raub, ORNL-tr-2179, translated from *Z. Metallkde.* **59**, 859-60 (1968).

1533. Magnetic and Caloric Study of the Phase Transitions of Copper, Nickel, Manganese and Cobalt Lanthanum Double Nitrate, K. W. Mess, E. Lagendijk, N. J. Zimmerman, A. J. van Duyneveldt, J. J. Giesen, and W. J. Huiskamp, *Physica* **43**, 165-208 (1969).

1534. Application of Adiabatic Calorimetry to Metal Systems, E. E. Stansbury and C. R. Brooks (Dept. of Chemical and Metallurgical Engineering, University of Tennessee, Knoxville, Tenn.), AT-(40-1)-3291, Progr. Rept. ORO-3291-13, Sept. 1, 1968 to July 30, 1969.

1535. Specific Heat of $Tb_xY_{1-x}Sb$ Mixed Crystals, W. Stutius (Laboratorium fur Festkorperphysik ETH, Zurich, Switzerland), to be published in Physics of Condensed Matter (Springer-Verlag, Heidelberg, Germany, 1969).

1536. Transition Metal Moment Collapse in $Gd(Co_{1-x}Ni_x)_2$, K. N. R. Taylor, *Phys. Letters* **29A**, 372-73 (1969).

1537. Properties of $(SnTe)_{1-x}(MnTe)_x$ Alloys, Jerrold Cohen, A. Globa, P. Mollard, H. Rodot, and M. Rodot, *J. Phys. (Paris), Colloq.* **29**, No. 4, 142-44 (1968).

1538. Contribution a l'Etude des Systemes VO_2-NbVO_4 et VO_2-TaVO_4, H. Trarieux, J. C. Bernier, and A. Michel, *Ann. Chim.* **4**, 183-94 (1969).

1539. Low-Temperature Phase Transition in Cerium Magnesium Nitrate, D. J. Abeshouse, George O. Zimmerman, D. R. Kelland, and E. Maxwell, *Phys. Rev. Letters* **23**, 308-10 (1969).

1540. Magnetocaloric Effect in Rare-Earth Iron Garnets, K. Belov, E. V. Talalaeva, L. A. Chernikova, V. I. Ivanovskii, and T. V. Kudryavtseva, *JETP Letters* **9**, 416-19 (1969) translated from *Zh. ETF Pis. Red.* **9**, No. 12, 671-75 (June 20, 1969).

1541. Effect of the Number of Fe^{3+} Exchange Couples on the Effective Magnetic Fields at the Nuclei of These Ions in $Y_{3-2x}Ca_{2x}Fe_{5-x}V_xO_{12}$ Garnets, V. A. Bokov, G. V. Popov, and S. I. Yushchuk, *Soviet Phys.-Solid State* **11**, 479-82 (1969).

1542. Vanadium Chalcogenides and the Nickel-Arsenide Structure, F. M. A. Carpay, Phillips Res. Repts. Suppl. No. 10 (1968).

1543. Observation d'une Transition Antiferro-Paramagnetique du Premier Ordre dans les Chromites de Zinc et de Magnesium, Francoise Hartmann-Boutron, Andre Gerard, Pierre Imbert, Romain Kleinberger, and Francois Varret, *Compt. Rend., Ser. B*, **268**, 906-08 (1969).

1544. Effect of Inhomogeneity on Magnetic Properties, X-Ray Diffraction and ^{31}P NMR Linewidths in $UP_{1-x}S_x$ Powders, Moshe Kuznietz, F. P. Campos, and Y. Baskin, *J. Appl. Phys.* **40**, 3621-25 (1969).

1545. Die Absorptionskante von $HgCr_2S_4$ im metamagnetischen Gebiet, H. W. Lehmann and G. Harbeke, Bericht uber die Tagung der Schweizerischen Physikalischen Gesellschaft, p. 612 (1969). (to be published in *Phys. Rev.*)

1546. Les Grenats de Neodyme-Fer-Scandium; Conditions d'Existence, Etude Cristographie et Propriétés Magnetiques, Jean Loriers, Gerard Villers, Francoise Clerc, and Francine Lacour, *Compt. Rend., Ser. B*, **268**, 1553-56 (1969).

1547. Direct Observation of Coherent Exchange Scattering by Low-Energy Electron Diffraction from Antiferromagnetic NiO, P. W. Palmberg, R. E. DeWames, and L. A. Vredevoe, *Phys. Rev. Letters* **21**, 682-84 (1968).

1548. Electrolytic Growth and Properties of Transition Metal Compound Single Crystals, Summary Technical Report, May 1, 1967–April 30, 1969, Aaron Wold (Brown University, Providence, Rhode Island 02912), Contract AF 33(615)-3844, AFML-TR-69-191, July 1969, 34 p.

REFERENCES

1549. MgO Containing Fe^{++}: Magnetic Properties and Clustering, K. N. Woods and M. E. Fine, *J. Appl. Phys.* **40**, 3425-33 (1969).
1550. Structure Antiferromagnetique de FeOF, J. Chappert and J. Portier, *Solid State Commun.* **4**, 395-98 (1966).
1551. Magnetic Structure of the ζ-Phase of Pd_3Mn_2 Alloy, J. A. Gonzalo and M. I. Kay, *J. Phys. Soc. Japan* **21**, 1626-27 (1966).
1552. Atomic and Magnetic Order in Mn_2Pd_3, E. Kren, G. Kadar, and T. Tarnoczi, *Phys. Letters* **25A**, 56-57 (1967).
1553. The Antiferromagnetic Structure of Mn_5Si_3, G. H. Lander, P. J. Brown, and J. B. Forsyth, *Proc. Phys. Soc.* **91**, 332-40 (1967).
1554. Neutron Diffraction Studies of Mn-Zn Alloys, Yasuaki Nakagawa and Tomiei Hori, *J. Phys. Soc. Japan* **19**, 2082-87 (1964).
1555. The Magnetic Structure of Trigonal Cr_2S_3, B. van Laar, *Phys. Letters* **25A**, 27-29 (1967).
1556. Magnetic Properties of Mixed Ferrites. I. Behavior near the Curie Point, J. G. Booth and J. Crangle, *Proc. Phys. Soc.* **79**, 1271 (1962).
1557. On the Vanadium Selenides, Erling Rost and Liv Gjertsen, *Z. Anorg. Allgem. Chem.* **328**, 299-308 (1964).
1558. A New Type of Spin Ordering in the System of Hexagonal $Ba_{2-x}Sr_xZn_2Fe_{12}O_{22}$(Y) Ferrites, V. A. Sizov, R. A. Sizov, and I. I. Yamzin, *Soviet Phys. JETP* **26**, 736-41 (1968).
1559. A New Type of Spin Ordering in the Hexagonal Ferrite $(Sr,Ba)_2Zn_2Fe_{12}O_{22}$(Y), V. A. Sizov, R. A. Sizov, and I. I. Yamzin, *Soviet Phys. JETP Letters* **6**, 176-77 (1967).
1560. Triangular Spin Configurations in the Antiferromagnetic Intermetallic Compounds Mn_3Sn, Mn_3Ge and Mn_3Rh, J. S. Kouvel and J. S. Kasper, *Proc. Internat. Conf. on Magnetism, Nottingham, 1964*, pp. 169-70.
1561. A Neutron Diffraction Study of the Influence of Temperature on Spin Configuration in Spinel MnV_2O_4, R. Plumier, *Proc. Internat. Conf. on Magnetism, Nottingham, 1964*, pp. 295-98.
1562. Magnetic Structure of Cobalt Manganite by Neutron Diffraction, B. Boucher, R. Buhl, and M. Perrin, *J. Appl. Phys.* **39**, 632-34 (1968).
1563. Magnetic Structure of Neodymium, R. M. Moon, J. W. Cable, and W. C. Koehler, *J. Appl. Phys.* **35**, 1041-42 (1964).
1564. The Paramagnetic Form Factor of Gadolinium, H. R. Child, R. M. Moon, L. J. Raubenheimer, and W. C. Koehler, *J. Appl. Phys.* **38**, 1381-83 (1967).
1565. Three-Lattice Ferrimagnetic Structure, B. Y. Boucher, R. Buhl, and M. Perrin, *J. Appl. Phys.* **38**, 1109-1110 (1967).
1566. Magnetic Structures in CrTe–CrSb Solid Solutions, W. J. Takei, D. E. Cox, and G. Shirane, *J. Appl. Phys.* **37**, 973-74 (1966).
1567. Etude par Effet Mossbauer du Systeme $Mn_{2-x}Fe_xO_3$ et Transitions Magnetiques dans Mn_2O_3 par Diffraction Neutronique, R. R. Chevalier, G. Rould, and E. F. Bertaut, *Solid State Commun.* **5**, 7-11 (1967).
1568. The Spontaneous Magnetization of the bcc Heisenberg Ferromagnet $Cu(NH_4)_2Br_4 \cdot 2H_2O$, R. F. Wielinga and W. J. Huiskamp, *Physica* **40**, 602-24 (1969).
1569. Propriétés Magnetiques de l'Europium à l'Etat de Sesquioxyde et de Metal, Francois Kern and Nicolas Perakis, *Compt. Rend., Ser. B,* **269**, 241-43 (1969).
1570. Magnetische Eigenschaften der Kobalt-Stickstoff-Phase Co_3N, K. H. Mader, F. Thieme, and A. Knappwost, *Z. Anorg. Allgem. Chem.* **366**, 274-79 (1969).
1571. Ultrasonic Propagation in $RbMnF_3$. I. Elastic Properties, R. L. Melcher and D. I. Bolef, *Phys. Rev.* **178**, 864-73 (1969).
1572. The Magnetic Behavior of CuMn Dilute Alloys, Yoshihito Miyako and Takashi Watanabe (Dept. of Physics, Faculty of Science, Hokkaido University, Sapporo, Japan) and Masao Watanabe (Research Institute for Catalysis, Hokkaido University, Sapporo, Japan), preprint received Oct. 1969.
1573. Sur la Structure et le Comportement Magnetique de l'Oxyde de Nickel Additionne d'Ions Lithium, Nicolas Perakis and Francois Kern, *Compt. Rend., Ser. B,* **269**, 281-84 (1969).

1574. Magnetic Properties of UP-US Solid Solutions, W. Trzebiatowski and T. Palewski, *Phys. Stat. Sol.* **34**, K51-K54 (1969).

1575. Possible Nuclear Ferromagnetism of High Density Hydrogen Solids, T. Ishizuka, T. Hamada, and Y. Nakamura, *Publ. Astron. Soc. Japan* **20**, No. 3, 300-02 (1968).

1576. Tellurium-125 Moessbauer Effect in Paramagnetic and Antiferromagnetic Manganese Telluride, M. Pasternak and A. L. Spijkervet, *Phys. Rev.* **181**, No. 2, 574-9 (1969).

1577. Electronic Structure and Magnetism in Transition Metal Diborides, J. Castaing, R. Caudron, G. Toupance, and P. Costa, in *Third Intern. Conf. on Solid Compounds of Transition Elements, Oslo, June 16-20, 1969.* Also in *Solid State Commun.* **7**, 1453-56 (1969).

1578. Ferromagnetism and Antiferromagnetism in Non-Ferrous Metals and Alloys, J. Crangle, *Met. Rev.* **7**, No. 26, 133 (1962).

1579. Magnetic Properties of Solid Solutions of the Heavy Rare Earths with Each Other, R. M. Bozorth and R. J. Gambino, *Phys. Rev.* **147**, 487-94 (1966).

1580. Heat Capacities of Cr_5Te_6, Cr_3Te_4 and Cr_2Te_3 from 5 to 350°K, Fredrik Gronvold and Edgar F. Westrum, Jr., *Z. Anorg. Allgem. Chem.* **328**, 272-82 (1964).

1581. Effect of High Pressure on Antiferromagnetism in Cr Alloys, A. Jayaraman, T. M. Rice and E. Bucher, *J. Appl. Phys.* **41**, 869-70 (1970).

1582. Magnetic Measurements on Compounds with K_2NiF_4 Structure, E. Legrand and A. Van den Bosch, *Solid State Commun.* **7**, 1191-94 (1969).

1583. Two Antiferromagnetic Phases in Mixed Crystals $(Mn_{1-x},Fe_x)WO_4$, H. Weitzel, *Solid State Commun.* **7**, 1249-52 (1969).

1584. Electron-Magnon Scattering in Dilute PdMn at Low Temperatures, Gwyn Williams and John W. Loram, *Solid State Commun.* **7**, 1261-65 (1969).

1585. Calcium-Vanadium-Indium Substituted Yttrium-Iron-Garnets with Very Low Linewidths of Ferrimagnetic Resonance, G. Winkler and P. Hansen, *Mat. Res. Bull.* **4**, 825-38 (1969).

1586. Anhydrous Ruthenium Chlorides, J. M. Fletcher, W. E. Gardner, E. W. Hooper, K. R. Hyde, F. H. Moore, and J. L. Woodhead, *Nature* **199**, 1089-90 (1963).

1587. Antiferromagnetism of $CoCO_3$, R. A. Alikhanov, *Soviet Phys. JETP* **12**, 1029-30 (1961).

1588. Magnetic Structure Studies at Brookhaven National Laboratory, L. M. Corliss and J. M. Hastings, *J. Phys.* **25**, 557-62 (1964).

1589. Magnetic Structure of Manganese Chromite, L. M. Corliss and J. M. Hastings, *J. Appl. Phys.* **33**, Suppl., 1138 (1962).

1590. The Magnetic Structure of $MnCr_2O_4$, J. M. Hastings and L. M. Corliss, *J. Phys. Soc. Japan* **17**, Suppl. B-III, 43-45 (1962).

1591. Magnetic Characteristics of Hydrogenated Holmium, Y. Kubota and W. E. Wallace, *J. Appl. Phys.* **33**, Suppl., 1348-49 (1962).

1592. High Temperature Vapour Phase Growth of Europium Sulphide, Selenide and Telluride, E. Kaldis, *J. Crystal Growth* **3**, 4, 146-49 (1968).

1593. Magnetic and Optical Properties of the High and Low Pressure Forms of $CsCoF_3$, J. M. Longo, J. A. Kafalas, J. R. O'Connor, T. W. Hilton, and D. A. Batson, pp. 41-43 in Solid State Research Quarterly Technical Summary, May 1 through July 31, 1969, Peter E. Tannenwald (Lincoln Laboratory, Mass. Inst. of Technology, Cambridge, Mass.), Contract No. AF 19(628)-5167, ESD-TR-69-211, August 15, 1969, 76 p.

1594. Magnetic Scattering of Phonons in the Thermal Conductivity of $MnCl_2 \cdot 4H_2O$, J. E. Rives, D. Walton, and G. S. Dixon, *J. Appl. Phys.* **41**, 1435-36 (1970) (presented at *15th Annual Conf. Magnetism and Magnetic Materials, Philadelphia, Nov. 18-21, 1969*).

1595. Magnetic Properties of Rare Earth Oxysulfides, J. Rossat-Mignod, R. Ballestracci, G. Quezel, and F. Tcheou, (Commissariat a l'Energie Atomique, Centre d'Etudes Nucleaires de Grenoble, France), CEA-CONF-1349 (*Colloque Internat. du CNRS sur les Elements des Terres Rares, Grenoble, 5-10 Mai 1969*).

1596. Magnetic Properties of Tb-Ho Single-Crystal Alloys. I. Magnetization Measurements, F. H. Spedding, R. G. Jordan, and R. W. Williams, *J. Chem. Phys.* **51**, 509-20 (1969).

REFERENCES

1597. The Effect of Magnetic and Configurational Disordering on the High Temperature Heat Capacity of a Ni-50 at. % Pd Alloy, R. E. Bingham and C. R. Brooks, *J. Phys. Chem. Solids* **30**, 2365-70 (1969).
1598. Lorenz-Function Enhancement due to Inelastic Processes near the Neel Point of Chromium, G. T. Meaden, K. V. Rao, and H. Y. Loo, *Phys. Rev. Letters* **23**, 475-77 (1969).
1599. Direct Thermal Evidence for a First-Order Change at the Spin-Flip Transition of Chromium, G. T. Meaden and N. H. Sze, *Phys. Rev. Letters* **23**, 1242-44 (1969).
1600. The Pressure Dependence of the Lattice Parameters of CrTe and CrSb, Hiroshi Nagasaki, Ippei Wakabayashi, and Shigeru Minomura, *J. Phys. Chem. Solids* **30**, 2405-08 (1969).
1601. Production of a γ-Oxygen Single Crystal, R. A. Alikhanov and V. A. Pavlov, p. 202 in Growth of Crystals, Volume 7, N. N. Sheftal', ed. (Consultants Bureau, New York, 1969).
1602. The Magnetization of Face-Centered Cubic and Body-Centered Cubic Iron-Nickel Alloys, J. Crangle and G. C. Hallam, *Proc. Roy. Soc. (London)* **272-A**, 119-32 (1963).
1603. Magnetic Properties of the Pd_3Mn_2 Alloy, Hiroshi Yamauchi, *J. Phys. Soc. Japan* **19**, 652-57 (1964).
1604. Lattice Parameters and Curie-Point Anomalies of Iron-Cobalt Alloys, H. Stuart and N. Ridley, *Brit. J. Appl. Phys. (J. Phys. D)*, Ser. 2, **2**, 485-91 (1969).
1605. Thermal and Magnetic Study of Bonding in $CsCuCl_3$, F. J. Rioux and B. C. Gerstein, *J. Chem. Phys.* **50**, No. 2, 758-64 (1969).
1606. Nuclear Magnetic Resonance in Ferrous Fluoride Below the Neel Temperature, Stanely M. Kulpa, Ph. D. thesis, Brandeis University, Waltham, Mass., 1968, 93 p. (Available from University Microfilms, Ann Arbor, Mich., Order No. 69-2056).
1607. The Effect of Composition and Purity on the Neel Temperature of CrN, T. Mills (Aeronautical Research Labs., Melbourne, Australia), ARL-MET-74, March 1962, 21 p.
1608. Phase Relations, Crystal Structures, and Magnetic Properties of Erbium-Iron Compounds, K. H. J. Buschow and A. S. van der Goot, *Phys. Stat. Sol.* **35**, 515-22 (1969).
1609. On the Demagnetized States in a Small Single Crystal of $SrO \cdot 3.6Fe_2O_3 \cdot 2.4Al_2O_3$, R. Carey and C. Tanasiou, *Phys. Stat. Sol.* **35**, K 115-118 (1969).
1610. Magnetic Properties of Cr_7Te_8, Takasu Hashimoto and Masuhiro Yamaguchi, *J. Phys. Soc. Japan* **27**, 1121-26 (1969).
1611. Anomalous Thermal Expansion and Specific Heat of the Ferromagnetic Compound Au_4Mn, Takejiro Kaneko and Minoru Matsumoto, *J. Phys. Soc. Japan* **27**, 1141-43 (1969).
1612. Les Propriétés Cristallographiques et Magnetiques des Solutions Solides entre les Phosphures $Ni_2P-Co_2P-Fe_2P-Mn_2P$ et Cr_2P, A. Roger, J.-P. Senateur, and R. Fruchart, *Ann. Chim.* **4**, 79-91 (1969).
1613. Contribution of s-d Interaction to the Internal Magnetic Field in Heusler Alloys, Takeshi Shinohara, *J. Phys. Soc. Japan* **27**, 1127-35 (1969).
1614. Spezifische Wärme der Nitride Seltener Erden, W. Stutius, *Phys. kondens. Materie* **10**, 152-185 (1969).
1615. Heat Capacity Studies of Intermetallic Compounds Containing Rare Earth Elements, W. E. Wallace, R. S. Craig, A. Thompson, C. Deenadas, M. Dixon, M. Aoyagi, and N. Marzouk, pp. 57-58 in *Colloques Internationaux du Centre National de la Recherche Scientifique*, No. 180, *Les Elements des Terres Rares, Paris-Grenoble, 5-10 Mai, 1969* (Centre National de la Recherche Scientifique).
1616. Magnetic Ordering in $Cu_2Cs_3Cl_7 \cdot 2H_2O$ Single Crystals Studied by Nuclear Magnetic Resonance, T. O. Klaassen, S. Wittekoek, J. J. van der Klink, and N. J. Poulis, *Physica* **41**, No. 3, 523-35 (1969).
1617. Linear Chain Antiferromagnetism in $CsMnCl_3 \cdot 2H_2O$, T. Smith and S. A. Friedberg, *Phys. Rev.* **176**, 660-65 (1968).
1618. Heat Capacity Measurements on Rare-Earth Double Oxides $R_2M_2O_7$, H. W. J. Blote, R. F. Wielinga, and W. J. Huiskamp, *Physica* **43**, 549-68 (1969).

1619. Ferromagnetism of Ce and Nd Chalcogenides with a Th_3P_4 Structure, A. T. Starovoitov, V. I. Ozhogin, G. M. Loginov, and V. M. Sergeyeva, *Zh. Eksper. Teor. Fiz.* **57**, 791-93 (1969).

1620. New Magnetic Transformation in the System $Mn_2Ge_ySb_{1-y}$, A. A. Galkin, E. A. Zavadskii, and E. M. Morozov, *JETP Letters* **8**, 400-02 (1968).

1621. Structural and Magnetic Properties of Pseudobinary System $(Zr_{1-x}Nb_x)Fe_2$, Kazuo Kanematsu, *J. Phys. Soc. Japan.* **27**, 849-56 (1969).

1622. Electrical Properties and Mossbauer Effect in the System $Sr_2(FeMo_xW_{1-x})O_6$, Takehiko Nakagawa, Kazuo Yoshikawa, and Shoichiro Nomura, *J. Phys. Soc. Japan* **27**, 880-86 (1969).

1623. Characteristic Features of the Magnetic Behavior of Orthoferrites Belonging to the $YFe_{1-x-y}Al_xMn_yO_3$ System, V. P. Polyakov, M. V. Bystrov, and G. K. Seliber, *Soviet Phys.-Solid State* **11**, 1147-49 (1969).

1624. High Curie Points of Calcium-Vanadium Garnets, V. M. Yudin, S. A. Poltinnikov, O. B. Proskuryakov, and N. N. Parfenova, *Phys. Letters* **28A**, 483-84 (1969).

1625. Low Temperature Magnetic Susceptibility of the Nickel Hexamine Halides, N. F. deOliveira, Jr., and C. Quadros, *Solid State Commun.* **7**, 1531-33 (1969).

1626. Propriétés Metamagnetiques de $LaErO_3$, J. Mareschal, J. M. Moreau, G. Ollivier, P. Pataud, and J. Sivardiere, *Solid State Commun.* **7**, 1669-72 (1969).

1627. Thermal Expansion and Magnetostriction of $ZrZn_2$, P. P. M. Meincke, E. Fawcett, and G. S. Knapp, *Solid State Commun.* **7**, 1643-45 (1969).

1628. Microhardness of Gadolinium near the Curie Point, O. A. Nabutovskaya, *Soviet Phys.-Solid State* **11**, 1172-73 (1969).

1629. Kondo Effect in Amorphous Palladium-Silicon Alloys Containing Transition Metals, C. C. Tsuei and R. Hasagawa, *Solid State Commun.* **7**, 1581-85 (1969).

1630. Magnetic Structures in $FeCr_2S_4$ and $FeCr_2O_4$, G. Shirane, D. E. Cox, and S. J. Pickart, *J. Appl. Phys.* **35**, 954-55 (1964).

1631. Detection of Magnetic Order in Ferroelectric $BeFeO_3$ by Neutron Diffraction, S. V. Kiselev, R. P. Ozerov, and G. S. Zhdanov, *Soviet Phys.-Dokl.* **7**, 742-44 (1963).

1632. Study of $(La_{1-x}Ca_x)MnO_3$, G. Matsumoto and K. Tsushima, *Bull. Am. Phys. Soc.* **15**, 156 (1970).

1633. Optical Investigation of Metamagnetic $DyAlO_3$, H. Schuchert, S. Hufner, and R. Faulhaber, *Z. Phys. (Germany)* **222**, No. 2, 105-27 (1969).

1634. Die Magnetstrukturen von $NiSeO_4$ und $CoSeO_4$ und die Bestimmung der Neel-Punkte von $MnSeO_4$, $CoSeO_4$ und $NiSeO_4$, H. Fuess, *Z. Angew. Phys.* **27**, 311-18 (1969).

1635. Magnetic Structure of Manganese Chromite, J. M. Hastings and L. M. Corliss, *Phys. Rev.* **126**, 556-65 (1962).

1636. On the Preparation of Very Pure Fe_xS, C. B. van den Berg and R. C. Thiel, *Z. Anorg. Allgem. Chem.* **368**, 106-12 (1969).

1637. Observation of Cr^{3+} Absorption Bands in $CdIn_2S_4(Cr)$ and the Consequences for the Interpretation of the Absorption Edge of $CdCr_3S_4$, S. Wittekoek and P. F. Bongers, *Solid State Commun.* **7**, 1719-22 (1969).

1638. Magnetic Structure of $HoAg_2$, Masao Atoji, *J. Chem. Phys.* **51**, 3882-85 (1969).

1639. Magnetic Structures of $DyAu_2$ and $DyAg_2$, Masao Atoji, *J. Chem. Phys.* **51**, 3877-82 (1969).

1640. Magnetic and Crystal Structures of the Trigonal Tb_2C, Masao Atoji, *J. Chem. Phys.* **51**, 3872-76 (1969).

1641. The Magnetic Phase Diagram of Deuterated Manganese Chloride, H. Forstat, P. T. Bailey, and J. R. Ricks, *Phys. Letters* **30A**, 52-53 (1969).

1642. Magnetic Properties of $FeCr_2S_4$ and $CoCr_2S_4$, P. Gibart, J. L. Dormann, and Y. Pellerin, *Phys. Stat. Sol.* **36**, 187 (1969).

1643. Susceptibilités Magnetiques de Monocristaux de $CoCl_2 \cdot 6H_2O$ et $CoCl_2 \cdot 6H_2O$ à Basse Temperature, Bernard Lecuyer and Jean-Pierre Renard, *Compt. Rend., Ser. B*, **269**, 78-81 (1969).

1644. Helical Spin Arrangement in Cubic FeGe, L. Lundgren, O. Beckman, V. Attia, S. P. Bhattacherjee, and M. Richardson, *Physica Scripta* **1**, 1-4 (1970).

REFERENCES

1645. Thermal Behavior of the Antiferromagnet $MnBr_2 \cdot 4H_2O$ in Applied Magnetic Fields, J. H. Schelleng and S. A. Friedberg, *Phys. Rev.* **185**, 728-34 (1969).

1646. Temperature Dependence of Hyperfine Interactions in Near-Stoichiometric FeS, I. Experiment, R. C. Thiel and C. B. van den Berg, *Phys. Stat. Sol.* **29**, 837-846 (1968).

1647. Optical Constants Versus Purity in EuTe and EuSe, R. Verreault, *Solid State Commun.* **7**, 1653-56 (1969).

1648. Heusler Alloys, Peter J. Webster, *Contemp. Phys.* **10**, 559-77 (1969).

1649. Magnetic Properties of K_2CoF_4 and $RbCoF_4$; Two-Dimensional Ising Antiferromagnets, D. J. Breed, K. Gilijamse, and A. R. Miedema, *Physica* **45**, 205-16 (1969).

1650. Studies of the Ionic Ferromagnet $(LaPb)MnO_3$. II. Static Magnetization Properties from 0 to 800°K, L. K. Leung, A. H. Morrish, and C. W. Searle, *Can. J. Phys.* **47**, 2697 (1969).

1651. Tracing Phase Transitions by Means of High Frequency ac Measurements Using a Q-Meter, Mats Nygren and Arne Magneli, *Arkiv for Kemi* **28**, 217-21 (1967).

1652. Studies of the Ionic Ferromagnet $(LaPb)MnO_3$. III. Ferromagnetic Resonance Studies, C. W. Searle and S. T. Wang, *Can. J. Phys.* **47**, 2703 (1969).

1653. Magnetic Resistivity and Magnetic Transitions in Rare-Earth Alloys, Tien-Tsai Yang and Lawrence Baylor Robinson, *Phys. Rev.* **185**, 743-51 (1969).

1654. Preparation and Magnetic Properties of Ruthenium Spinels, H. M. Kasper, p. 18 in Solid State Research, Lincoln Laboratory (Mass. Inst. Tech., Lexington, Mass.), Contract AF 19(628)-5167, ESD-TR-69-336, January 13, 1970.

1655. Structure and Magnetic Properties of $VOSO_4$, J. M. Longo, R. J. Arnott, and D. A. Batson, p. 18-25 in Solid State Research, Lincoln Laboratory (Mass. Inst. Tech., Lexington, Mass.), Contract AF 19(628)-5167, ESD-TR-69-336, January 13, 1970.

1656. Magnetic Properties of Cr_5S_6, N. Menyuk, K. Dwight, and J. A. Kafalas, p. 38-40 in Solid State Research, Lincoln Laboratory (Mass. Inst. Tech., Lexington, Mass.), Contract AF19(628)-5167, ESD-TR-69-336, January 13, 1970.

1657. Mossbauer Measurement of Curie Temperatures and X-Ray Measurement of Lattice Parameters of Some Iron-Palladium-Hydrogen Alloys, J. S. Carlow and R. E. Meads, *J. Phys. C (Solid State Phys.)*, Ser. 2, **2**, 2120-27 (1969).

1658. Magnetic Properties of Amorphous Fe-Pd-Si Alloys, R. Hasegawa, p. 13 in Annual Summary Report, Dec. 1, 1968 to Dec. 1, 1969 (Division of Engineering, California Institute of Technology, Pasadena, Calif.), Contract AT(04-3)-221, C.I.T. Project Engineering 91, CALT-221-82.

1659. Uber die Temperaturabhängigkeit der paramagnetischen Suszeptibilität von Scandium, Titan, Vanadium, Chrom und Mangan, R. Kohlhaas and W. D. Weiss, *Z. Angew. Physik* **28**, 16-20 (1969).

1660. Structures Magnetiques des Chromates de Cobalt et de Nickel, M. Pernet, G. Quezel, Jean Coing-Boyat, and Erwin-F. Bertaut, *Bull. Soc. Fr. Mineral. Cristallogr.* **92**, 264-73 (1969).

1661. The Magnetic Properties of the Synthetic Iron Oxyhydrates, N. Yamamoto, *Bull. Inst. Chem. Res. Kyoto Univ. (Japan)* **46**, No. 6, 275-82 (Nov. 1968).

1662. Synthesis and Properties of $EuNb_2S_4$, O. Berkooz and E. Hermon, *Mat. Res. Bull.* **5**, 173-78 (1970).

1663. Antiferromagnetic Phase Diagram and Magnetic Band Gap Shift of $NaCrS_2$, K. W. Blazey and H. Rohrer, *Phys. Rev.* **185**, 712-19 (1969).

1664. Low-Temperature Magnetic Properties of $DyPO_4$: an Ideal Three-Dimensional Ising Antiferromagnet, J. H. Colwell, B. W. Mangum, D. D. Thornton, J. C. Wright, and H. W. Moos, *Phys. Rev. Letters* **23**, 1245-47 (1969).

1665. Lattice Distortion, Specific Heat and Antiferromagnetic Ordering in AuMn, D. C. Finbow, P. Makhurane, and P. Gaunt, *J. Phys. Chem. Solids* **31**, 179-85 (1970).

1666. Antiferromagnetism of γ FeOOH: a Mossbauer Effect Study, C. E. Johnson, *J. Phys. C (Solid St. Phys.)*, Ser. 2, **2**, 1996-2002 (1969).

1667. Pressure-Induced Pyrochlore to Perovskite Transformations in the $Sr_{1-x}Pb_xRuO_3$ System, J. A. Kafalas and J. M. Longo, *Mat. Res. Bull.* **5**, 193-98 (1970).

1668. Spin Waves around the Neel Temperature in MnO, N. Kroo and L. Bata (Central Research Institute for Physics, Hungarian Academy of Sciences, Budapest, Hungary), SM-104/8 (preprint received at RMIC Feb. 1970).

1669. The Magnetic Ordering of Beta-Ni(IO$_3$)$_2$·2H$_2$O, H. C. Meijer, L. M. W. A. Pimmelaar, and J. van den Handel, *Physica* **44**, 626-28 (1969).

1670. Optical Properties of Layer Antiferromagnets with K$_2$NiF$_4$ Structure, J. B. Parkinson, *J. Phys. C (Solid St. Phys.), Ser. 2*, **2**, 2012-21 (1969).

1671. The Effect of Deuteration on the Magnetic Critical Temperatures of Hydrogen-Bonded Materials, B. G. Turrell and C. L. Yue, *Can. J. Phys.* **47**, 2575 (1969).

1672. EPR Study of Antiferromagnetic Ordering in CrVO$_4$, P. R. Elliston, *Can. J. Phys.* **47**, 1865 (1969).

1673. On the Magnetic Properties of CoSe$_2$, NiS$_2$, and NiSe$_2$, Sigrid Furuseth, Arne Kjekshus, and Arne F. Andresen, *Acta Chem. Scand.* **23**, 2325-34 (1969).

1674. Thermal Conductivity, Electrical Resistivity, and Seebeck Coefficient of Uranium Mononitride, J. P. Moore, W. Fulkerson, and D. L. McElroy, *J. Am. Ceram. Soc.* **53**, No. 2, 76-82 (1979).

1675. Magnetic Neutron Scattering in Dysprosium Aluminum Garnet. I. Long-Range Order, J. C. Norvell, W. P. Wolf, L. M. Corliss, J. M. Hastings, and R. Nathans, *Phys. Rev.* **186**, 557-66 (1969).

1676. On the Two-Dimensional Antiferromagnetic Character of MnTiO$_3$, J. Akimitsu, Y. Ishikawa, and Y. Endoh, *Solid State Commun.* **8**, 87-90 (1970).

1677. Antiferromagnetic Resonance in K$_2$NiF$_4$, R. J. Birgeneau, F. DeRosa, and H. J. Guggenheim, *Solid State Commun.* **8**, 13-16 (1970).

1678. Ferromagnetic VBO$_3$ and Antiferromagnetic CrBO$_3$, T. A. Bither, Carol G. Frederick, T. E. Gier, J. F. Weiher, and H. S. Young, *Solid State Commun.* **8**, 109-112 (1970).

1679. Some Magnetic Properties of Platinum-Rich Pt-Fe Alloys, J. Crangle, *J. Phys. Rad.* **20**, 435-37 (1959).

1680. Critical Magnetic Fluctuations in MnF$_2$, O. W. Dietrich, *J. Phys. C (Solid State Phys.), Ser. 2*, **2**, 2022-36 (1969).

1681. Temperature Dependence of the Elastic Moduli of the Metamagnetic Compound MnAu$_2$, I. G. Fakidov and E. V. Kleimenov, *Soviet Phys.-Solid State* **11**, No. 8, 1865-67 (1970).

1682. Structure Magnetique de Mn$_3$GaC, D. Fruchart, E. F. Bertaut, F. Sayetat, M. Nasr Eddine, R. Fruchart, and J. P. Senateur, *Solid State Commun.* **8**, 91-99 (1970).

1683. Magnetic Properties of Manganese Arsenide in Strong Pulsed Magnetic Fields, N. P. Grazhdankina, E. A. Zavadskii, and I. G. Fakidov, *Soviet Phys.-Solid State* **11**, 1879-85 (1970).

1684. Uber Perowskit-Verbindungen vom Typ Ba$_3$MeU$_2$O$_9$, Sibylle Kemmler-Sack, *Z. Naturforsch.* **24b**, 1398-1401 (1969).

1685. NMR of ^{55}Mn and $^{(69)(71)}$Ga in Mn$_3$GaC and Mn$_3$GaC$_{0.95}$, Le Dang Khoi, Mme E. Fruchart, and M. R. Fruchart, *Solid State Commun.* **8**, 49-51 (1970).

1686. Magnetic Properties of Plutonium Monophosphide, D. J. Lam, F. Y. Fradin, and O. L. Kruger, *Phys. Rev.* **187**, 606-10 (1969).

1687. Etude Thermomagnetique de Quelques Composés Intermetalliques Fer-Titane de Composition Voisine de Fe$_2$Ti, Mototaka Okazaki, *Compt. Rend., Ser. B*, **270**, 254-56 (1970).

1688. Etude de FeP par Spectrometrie Mossbauer. Effet du Manganese sur le Point de Neel, Jean-Pierre Senateur, Alain Roger, Robert Fruchart, and Jean Chappert, *Compt. Rend., Ser. C*, **269**, 1385-87 (1969).

1689. Paramagnetic to Antiferromagnetic Phase Boundaries of FeF$_2$ from Ultrasonic Measurements, Y. Shapira, *Phys. Letters* **30A**, 388-89 (1969).

1690. Magnetic Properties in Strong Pulsed Magnetic Fields of Some Manganites Having the Perovskite Structure, A. T. Starovoitov, V. I. Ozhogin, and V. A. Bokov, *Soviet Phys.-Solid State* **11**, 1740-44 (1970).

1691. Structure et Proprietes Magnetiques du Monosiliciure de Terbium TbSi, Nguyen van Nhung, J. Laforest and J. Sivardiere, *Solid State Commun.* **8**, 23-30 (1970).

REFERENCES

1692. Kinetic Phenomena in Neodymium Chalcogenides in the Magnetic Ordering Region, V. P. Zhuze, V. I. Novikov, V. M. Sergeeva, and S. S. Shalyt, *Soviet Phys.-Solid State* 11, 1770-72 (1970).

1693. Propriétés Magnetiques de $FeCuO_2$ et $CrCuO_2$, A. Apostolov, *Compt. Rend. Acad. Bulg. Sci.* 22, No. 1, 11-13 (1969).

1694. The Synthesis of New Ilmenite-Type Derivatives, $CuVO_3$ and $CoVO_3$, B. L. Chamberland, *J. Solid State Chem.* 1, 138-42 (1970).

1695. On the Magnetic Behavior of New $2H-NbS_2$-Type Derivatives, Fritz Hulliger and Eva Pobitschka, *J. Solid State Chem.* 1, 117-19 (1970).

1696. Mossbauer Study of Some 2-17 Lanthanide-Iron Compounds, Lionel M. Levinson, E. Rosenberg, A. Shalulov, S. Shtrikman, and K. Strnat, *J. Appl. Phys.* 41, 910-12 (1970).

1697. Magnetic Properties of the System $Au_4(VFe)$, L. R. Sill, C. W. Kimball, R. H. Clark, W. J. Mass, and J. B. Darby, Jr., *J. Appl. Phys.* 41, 865-66 (1970).

1698. Antiferromagnetism in V_2O_3, R. M. Moon, *J. Appl. Phys.* 41, 883 (1970).

1699. Temperature Dependence of Magnon-Pair Modes in Antiferromagnets and Paramagnets, Paul A. Fleury, *J. Appl. Phys.* 41, 886-88 (1970).

1700. Magnetic Studies of the Alloy System SnTe-MnTe, M. P. Mathur, D. W. Deis, C. K. Jones, A. Patterson, W. J. Carr, Jr., and R. C. Miller, *J. Appl. Phys.* 41, 1005-07 (1970).

1701. Neutron Diffraction Study of the Magnetic Structure of $BaNiF_4$, D. E. Cox, M. Eibschutz, H. J. Guggenheim, and L. Holmes, *J. Appl. Phys.* 41, 943-45 (1970).

1702. Effect of Atomic Ordering on the Magnetic Structure of the $MnPd_3$ Phase, E. Kren, G. Kadar, and L. Pal, *J. Appl. Phys.* 41, 941-42 (1970).

1703. Magnetic and Optical Properties of the High- and Low-Pressure Forms of $CsCoF_3$, J. M. Longo, J. A. Kafalas, J. R. O'Connor, and J. B. Goodenough, *J. Appl. Phys.* 41, 935-36 (1970).

1704. Ferrimagnetic Compositions in the System $Fe_{1+x}Cr_{2-x}S_4$, M. Robbins, R. Wolff, A. J. Kurtzig, R. C. Sherwood, and M. A. Miksovsky, *J. Appl. Phys.* 41, 1086-87 (1970).

1705. Optical Constants of the Eu Chalcogenides Above and Below the Magnetic Ordering Temperatures, G. Guntherodt, J. Schoenes, and P. Wachter, *J. Appl. Phys.* 41, 1083-84 (1970).

1706. Electron Spin Resonance in Ferromagnetic $Sc_{3+x}In$ Alloys, G. L. Dunifer, G. S. Knapp, and E. Corenzwit, *J. Appl. Phys.* 41, 1075-76 (1970).

1707. Field-Induced Spin Reorientation in $ErCrO_3$, L. Holmes, M. Eibschutz, and L. G. Van Uitert, *J. Appl. Phys.* 41, 1184-85 (1970).

1708. Magnetic and Crystallographic Properties of Pr-Gd Alloys, L. Tissot and A. Blaise, *J. Appl. Phys.* 41, 1180-82 (1970).

1709. Further Magnetic and Thermal Studies of Cast Rare-Earth Permanent Magnets, E. A. Nesbitt, G. Y. Chin, P. K. Gallagher, R. C. Sherwood, and J. H. Wernick, *J. Appl. Phys.* 41, 1107-08 (1970).

1710. Etude Magnetique et Structurale des Phases MSn_2 et MSn (M = Fe,Co) et des Solutions Solides $(Fe_xM_{1-x})Sn$, $(Fe_xMn_{1-x})Sn_2$ (M = Co,Ni), C. Djega-Mariadassou, P. Lecocq and A. Michel, *Ann. Chim.* 4, 175-82 (1969).

1711. Antiferromagnetism and Electrical Resistivity of Chromium Alloys Containing Ruthenium and Osmium, Sigurds Arajs, Tice F. De Young, and Elmer E. Anderson, *J. Appl. Phys.* 41, 1426-28 (1970).

1712. Magnetic Properties of $Ba_2Zn_2Fe_{28}O_{46}$ and $Ba_2Co_2Fe_{28}O_{46}$ Single Crystals, A. Tauber, J. S. Megill, and J. R. Shappirio, *J. Appl. Phys.* 41, 1353-54 (1970).

1713. Magnetic and Structural Properties of Some Rare-Earth-Sn_3 Compounds, G. K. Shenoy, B. D. Dunlap, G. M. Kalvius, A. M. Toxen, and R. J. Gambino, *J. Appl. Phys.* 41, 1317-18 (1970).

1714. Phase Transitions and Magnetic Correlations in Two-Dimensional Antiferromagnets, R. J. Birgeneau, J. Skalyo, Jr., and G. Shirane, *J. Appl. Phys.* 41, 1303 (1970).

1715. Critical-Point Behavior of $FeBO_3$ Single Crystals by Mossbauer Effect, M. Eibschutz, L. Pfeiffer, and J. W. Nielsen, *J. Appl. Phys.* 41, 1276-77 (1970).

1716. Ordering in Two- and Three-Dimensional Diluted Heisenberg Antiferromagnets, D. J. Breed, K. Gilijamse, J. W. E. Sterkenburg, and A. R. Miedema, *J. Appl. Phys.* 41, 1267-68 (1970).

1717. Relationship Between Stoichiometry and Properties of EuO Films, K. Y. Ahn and M. W. Shafer, *J. Appl. Phys.* **41**, 1260-62 (1970).

1718. Optical Study of Ordering in $DyPO_4$, $DyAsO_4$, and $DyVO_4$, John C. Wright and H. W. Moos, *J. Appl. Phys.* **41**, 1244-45 (1970).

1719. Magnetic Properties of the Pseudobinary Alloys $Au_4(Mn_{1-x}Cr_x)$, M. Yessik, J. Noakes, and H. Sato, *J. Appl. Phys.* **41**, 1234-35 (1970).

1720. Evidence for Impurity Clustering and Matrix Polarization in Nickel-Palladium Alloys, W. A. Ferrando, R. Segnan, and A. I. Schindler, *J. Appl. Phys.* **41**, 1236-37 (1970).

1721. Magnetic Interactions in Gadolinium Orthoferrite, J. D. Cashion, A. H. Cooke, D. M. Martin, and M. R. Wells, *J. Appl. Phys.* **41**, 1193-94 (1970).

1722. Effect of Pressure on the Neel Temperature of $MnCO_3$, $CoCO_3$, and $FeCO_3$, Vishnu C. Srivastava, *J. Appl. Phys.* **41**, 1190-91 (1970).

1723. Room-Temperature Ferromagnetic Materials Transparent in the Visible, R. Wolfe, A. J. Kurtzig, and R. C. LeCraw, *J. Appl. Phys.* **41**, 1218-24 (1970).

1724. Mossbauer Study of Ordered and Disordered Alloy $Au_{80}Cr_{20}$, Masahumi Kohgi, Takemi Yamada, Nobuhiko Kunitomi, and Yutaka Maeda, *J. Phys. Soc. Japan* **28**, 793-94 (1970).

1725. Ferromagnetism and Antiferromagnetism in Co-Mn Alloy, Masaaki Matsui, Tadashi Ido, Kiyoo Sato, and Kengo Adachi, *J. Phys. Soc. Japan* **28**, 791 (1970).

1726. Cubic-Tetragonal Transition and Antiferromagnetism of γ-Phase Mn-Ni Alloys, Hidema Uchishiba, Tomiei Hori, and Yasuaki Nakagawa, *J. Phys. Soc. Japan* **28**, 792 (1970).

1727. Ferromagnetic Properties of the Intermetallic Compound with the Hexagonal Laves-Phase Structure in Cobalt-Titanium System, Takuro Nakamichi, Yoshihira Aoki and Mikio Yamamoto, *J. Phys. Soc. Japan* **28**, 590-95 (1970).

1728. The Magnetic Symmetry of α-Mn and the Parasitic Ferromagnetism from Magnetic Torque Measurements, Takemi Yamada and Shuichi Tazawa, *J. Phys. Soc. Japan* **28**, 609-14 (1970).

1729. Magnetic Characteristics of Lanthanide Zinc ($LnZn_2$) Intermetallic Compounds, D. K. DebRay, W. E. Wallace, and E. Ryba (DebRay and Wallace: Dept. of Chemistry, University of Pittsburgh, Pa. 15213; Ryba: Dept. of Material Science, Penn. State Univ., University Park, Pa. 16802), preprint received April 1970.

1730. Neutron-Diffraction Determination of the Antiferromagnetic Structure of $KCuF_3$, M. T. Hutchings, E. J. Samuelsen, G. Shirane, and K. Hirakawa, *Phys. Rev.* **188**, 919-23 (1969).

1731. Neutron-Diffraction Study of Magnetic Ordering in $UP_{0.75}S_{0.25}$, G. H. Lander, Moshe Kuznietz, and D. E. Cox, *Phys. Rev.* **188**, 963-66 (1969).

1732. Effect of a Magnetic Field on the Electrical and Thermal Conductivities of Holmium, G. S. Nikol'skii, N. M. Zvyagina, and V. V. Eremenko, *Soviet Phys.-Solid State* **11**, 2009-13 (1970).

1733. Magnetic Properties of $Ba_{2-x}Sr_xZn_2Fe_{12}O_{22}$ Hexaferrites with a Helicoidal Magnetic Structure, I. I. Petrova and M. A. Vinnik, *Soviet Phys.-Solid State* **11**, 2177-78 (1970).

1734. Nuclear Resonance in Ferromagnetic Chromium Tribromide: Applications to High-Precision Thermometry in the Range $2°-25°K$, Stephen D. Senturia, *J. Appl. Phys.* **41**, 644-51 (1970).

1735. Structure and Magnetic Properties of Rapidly Quenched Samarium-Type Alloys, J. D. Speight, *J. Less-Common Metals* **20**, 251-62 (1970).

1736. Effect of Pressure on the Neel Temperature of Rare Earth Intermetallic Compounds, Kazuko Sekizawa, Hisashi Sekizawa, and C. T. Tomizuka, *J. Phys. Chem. Solids* **31**, 215-18 (1970).

1737. Antiferromagnetism of $Zn_{0.87}Mn_{0.13}Fe_2O_4$, M. K. Fayek, J. Leciejewicz, A. Murasik, and I. I. Yamzin, *Phys. Stat. Sol.* **37**, 843-50 (1970).

1738. Ultrasonic Attenuation in MnF_2 near the Neel Temperature, Akira Ikushima, *J. Phys. Chem. Solids* **31**, 283-89 (1970).

1739. Magnetic Characteristics of $CuFeCl_4$, $AgFeCl_4$, and $TlFeCl_4$ near the Neel Temperatures, E. R. Jones, M. E. Hendricks, S. L. Finklea III, L. Cathey, Theo Auel, and E. L. Amma, *J. Chem. Phys.* **52**, 1922-1926 (1970).

1740. Studies of the Ionic Ferromagnet (LaPb) MnO_3. IV. Lattice Distortion and Ferromagnetism, M. J. Oretzki, and P. Gaunt, *Can. J. Phys.* **48**, 346-48 (1970).

REFERENCES

1741. Elastic Moduli and the Order of the Antiferromagnetic Transition in MnO, M. F. Cracknell and R. G. Evans, *Solid State Commun.* **8**, 359-61 (1970).
1742. Pressure Effect on the Antiferromagnetism of Chromium-Vanadium-Manganese Alloys, Yuzi Furuya, Nobuo Mori, and Tadayasu Mitsui, *J. Phys. Soc. Japan* **28**, 257 (1970).
1743. The Temperature Dependence of the Electrical Resistivity of Cr-V Alloys, Shoji Kashida, Yorihiko Tsunoda, and Nobuhiko Kunitomi, *J. Phys. Soc. Japan* **28**, 261 (1970).
1744. Magnetic Ordering in Iron Gel, Z. Mathalone, M. Ron, and A. Biran, *Solid State Commun.* **8**, 333-36 (1970).
1745. The Critical Phenomena of the Heisenberg Ferromagnet $CdCr_2Se_4$, Kazuo Miyatani, *J. Phys. Soc. Japan* **28**, 259-60 (1970).
1746. Magnetic Properties of Several Phases of Barium Orthoferrate, $BaFeO_x$, Saburo Mori, *J. Phys. Soc. Japan* **28**, 44-50 (1970).
1747. Magnetic and Crystallographic Evidence for Localization to Delocalization of V d-Electron Levels in $CuCr_{2-x}V_xS_4$ Spinels, M. Robbins, A. Menth, M. A. Miksovsky, and R. C. Sherwood, *J. Phys. Chem. Solids* **31**, 423-30 (1970).
1748. Magnetic Phase Transition for Dilute Heisenberg Spin System in Two-Dimensional Lattice, Kazuyoshi Takeda, Motohiro Matsuura, and Taiichiro Haseda, *J. Phys. Soc. Japan* **28**, 29-35 (1970).
1749. Paramagnetic Diffuse Scattering of Neutrons by a 90 at. % Manganese Copper Alloy, E. R. Vance and J. H. Smith, *J. Phys. Chem. Solids* **31**, 485-88 (1970).
1750. Electrical Resistivity of the Dilute PdCo System, Gwyn Williams, *J. Phys. Chem. Solids* **31**, 529-39 (1970).
1751. Mossbauer Effect for Fe^{57} Nuclei in $Bi_2Fe_4O_9$, V. A. Bokov, G. V. Novikov, V. A. Trukhtanov, and S. I. Yushchuk, *Soviet Phys.-Solid State* **11**, 2324-26 (1970).
1752. Electrical Resistivity and Thermal Expansion of CeAs, Takashi Tsuchida, Masahiko Kawai, and Yoki Nakamura, *J. Phys. Soc. Japan* **28**, 528 (1970).
1753. Temperature Dependence of Domain Structure in Strontium Alumino-Ferrites Single Crystals, V. Florescu and M. Rosenberg, *Japan. J. Appl. Phys.* **9**, 217-23 (1970).
1754. Hydrostatic Pressure Effect on the Neel Point of Chromium-Molybdenum Alloys, Nobuo Mori, Yuji Furuya, and Tadayasu Mitsui, *J. Phys. Soc. Japan* **28**, 531 (1970).
1755. Magnetic Properties of fcc Fe-Ir Alloys, Tadashi Mizoguchi and Terufumi Sasaki, *J. Phys. Soc. Japan* **28**, 532 (1970).
1756. Nuclear Magnetic Resonance in Heusler Alloys: Ni_2MnSn, Co_2MnSn and Ni_2MnSb, Takeshi Shinohara, *J. Phys. Soc. Japan* **28**, 313-17 (1970).
1757. Magnetic Investigation of Single Crystal of $Ni(NH_3)_2 \cdot Ni(CN)_4 \cdot 2C_6H_6$, Sigeru Takayanagi and Takashi Watanabe, *J. Phys. Soc. Japan* **28**, 296-301 (1970).
1758. On New Ferromagnetic Intermetallic Compounds PtMnSn and PtMnSb, Kiyoshi Watanabe, *J. Phys. Soc. Japan* **28**, 302-07 (1970).
1759. Anomalous Temperature Dependence of Exchange Interactions in $K_2CuCl_4 2H_2O$ and $K_2CuCl_4 2D_2O$, Takashi Okuda and Muneyuki Date, *J. Phys. Soc. Japan* **28**, 308-12 (1970).
1760. Exchange Interaction in Europium Chalcogenides, A. A. Berdyshev and B. M. Letfulov, *Soviet Phys.-Solid State* **11**, 2234-37 (1970).
1761. Critical Phonon Scattering near a Ferromagnetic Transition and its Effect on the Thermal Conductivity of $CuK_2Cl_4 \cdot 2H_2O$, G. S. Dixon and D. Walton, *Phys. Rev.* **185**, 735-38 (1969).
1762. Synthesis of a New Perovskite $CaFeO_3$, F. Kanamaru, H. Miyamoto, Y. Mimura, M. Koizumi, M. Shimada, S. Kume, and S. Shin, *Mat. Res. Bull.* **5**, 257-62 (1970).
1763. Linear Antiferromagnetic Chains in Hexagonal $ABCl_3$-Type Compounds (A: Cs, or Rb, B: Cu, Ni, Co, or Fe), Norio Achiwa, *J. Phys. Soc. Japan* **27**, 561-74 (1969).
1764. Magnetism of γ Fe-Ni Invar Alloys with Low Nickel Concentration, Hajime Asano, *J. Phys. Soc. Japan* **27**, 542-53 (1969).
1765. Magneto-Optic Properties of Quenched Thin Films of MnBi and Optical Memory Experiments, D. Chen, R. L. Aagard, and T. S. Liu, *J. Appl. Phys.* **41**, 1395-6 (1970).

1766. Magneto-Optic Properties of CrTe Films Prepared by Sequential Evaporation, R. L. Comstock and P. H. Lissberger, *J. Appl. Phys.* **41**, 1397-98 (1970).

1767. Magnetic Studies of Cl_b-Compounds $Ni_xCu_{1-x}MnSb$, Keizo Endo, Yuko Fujita, Ritsuo Inaba, Tetuo Ohoyama, and Ren'iti Kimura, *J. Phys. Soc. Japan.* **27**, 785 (1969).

1768. Effect of Pressure on the Curie Temperatures of PdCo Alloys, W. B. Holzapfel, J. A. Cohen, and H. G. Drickamer, *Phys. Rev.* **187**, 657-59 (1969).

1769. Ultrasonic Measurements in MnF_2 near the Néel Temperature, R. G. Leisure and R. W. Moss, *Phys. Rev.* **188**, 840-44 (1969).

1770. Critical-Region Thermal Expansion in $MnCl_2 \cdot 4H_2O$ and $MnBr_2 \cdot 4H_2O$, J. W. Philip, R. Gonano, and E. D. Adams, *Phys. Rev.* **188**, 973-82 (1969).

1771. Cubic-Tetragonal Transition and Antiferromagnetism of a'-$MnZn_3$, Hidema Uchishiba, Tomiei Hori, and Yasuaki Nakagawa, *J. Phys. Soc. Japan.* **27**, 600-04 (1969).

1772. Transport in Gd-doped EuO, S. von Molnar and M. W. Shafer, *J. Appl. Phys.* **41**, 1093-94 (1970).

1773. Thermal Expansion Anomaly in $ZrZn_2$, Shinji Ogawa and Naoko Kasai, *J. Phys. Soc. Japan.* **27**, 789 (1969).

1774. Nuclear Magnetic Resonance and Magnetic Structure of Antiferromagnetic $Mn_2P_2O_7$, Sung Ho Choh and C. V. Stager, *Can. J. Phys.* **48**, 521-30 (1970).

1775. Application of Stoner Theory to Ferromagnetic $ZrZn_{2-x}$ Alloys, G. S. Knapp, *J. Appl. Phys.* **41**, 1073-74 (1970).

1776. Effect of Pressure on the Curie Point of the Pseudo-NiAs-Type Cr_3Te_4, K. Ozawa, S. Yanagisawa, T. Yoshimi, M. Ogawa, and S. Anzai, *Phys. Stat. Sol.* **38**, 385 (1970).

1777. Magnetic Properties of RXO_4 System, Haruo Saji, Tokio Yamadaya, and Mitsuru Asanuma, *J. Phys. Soc. Japan.* **28**, 913 (1970).

1778. The Temperature Dependence of the Rhombohedral Distortion in NiO, M. W. Vernon, *Phys. Stat. Sol.* **37**, K1-3 (1970).

1779. Mössbauer Isomer Shifts, Hyperfine Interactions, and Magnetic Hyperfine Anomalies in Compounds of Iridium, F. Wagner and Ursel Zahn, *Z. Physik* **233**, 1-20 (1970).

1780. Etude de l'Aimantation sur l'Orthoferrite d'Holium $HoFeO_3$, Yves Allain, Jean-Eric Bouree, Jean Denis, Germain Jehanno, and Serge Legrand, *Compt. Rend. Ser. B*, **269**, 1024-26 (1969).

1781. Investigations on the Resistivity of the Compound $CeAl_3$, K. H. J. Buschow and H. J. van Daal, *Solid State Commun.* **8**, 363-65 (1970).

1782. The Low-Temperature Electrical and Magnetic Properties of $TaSe_2$ and $NbSe_2$, H. N. S. Lee, M. Garcia, H. McKinzie, and A. Wold, *J. Solid State Chem.* **1**, 190-94 (1970).

1783. Antiferromagnetic Interactions between Fe^{3+} Ions at a Large Distance in $Fe_{1/2}Cu_{1/2}Rh_2S_4$, R. Plumier and F. K. Lotgering, *Solid State Commun.* **8**, 477-80 (1970).

1784. Propriétés Magnétiques des Oxysulfures de Terres Rares, G. Quezel, R. Ballestracci, and J. Rossat-Mignod, *J. Phys. Chem. Solids* **31**, 669-84 (1970).

1785. Magnetic Properties of MnP, Nobuo Iwata, *J. Sci. Hiroshima Univ., Ser. A-II*, **33**, No. 1, 1-21 (June 1969).

1786. Growth of V_5O_9 Single Crystals, Koichi Nagasawa, Yoshichika Bando, and Toshio Takada, *Japan. J. Appl. Phys.* **9**, 407 (1970).

1787. Electron Magnetic Resonance in Antiferromagnetic $FeCl_2$-$4H_2O$ over the Temperature Range 0.38-1.1°K, Donald David Thornton, Ph.D. thesis, Syracuse University, New York, 1968, 92 p. (Available from University Microfilms, Ann Arbor, Mich., Order No. 69-8654).

1788. Electrical Conductivity of the Antiferromagnetic Semiconductor MnTe Below its Néel Temperature, A. I. Zvyagin, L. V. Povstyanyi, and R. M. Aref'eva, *Soviet Phys.-Solid State* **11**, 2759-60 (1970).

1789. The Magnetic Properties of Single Crystals of Erbium and Some Yttrium-Erbium and Lutetium-Erbium Alloys between 1.2 and 300°K, W. J. Gray and F. J. Spedding (Ames Lab., Iowa State Univ. of Science and Technology, Ames, Iowa), IS-2044 (Feb., 1968).

1790. Semiconducting Ferromagnetic Materials, Peter J. Wojtowicz, Stuart B. Berger, and Harry L. Pinch (RCA Labs, Princeton, N. J. 08540), AFCRL-70-0045 (Jan., 1970) (Final Report, Dec. 16, 1966-Dec. 15, 1969).

1791. Pressure-Induced Pyrochlore to Perovskite Transformations in $Sr_{1-x}Pb_xRuO_3$ System, J. A. Kafalas, J. M. Longo, and D. A. Batson, pp. 17-21 in ESD-TR-70-33, Lincoln Laboratory, Mass. Institute of Technology (May 1970).

REFERENCES

1792. Attempt to Measure the Isotope Effect in the Weak Itinerant Ferromagnet $ZrZn_2$, G. S. Knapp, E. Corenzwit, and C. W. Chu, *Solid State Commun.* **8**, 577-80 (1970).
1793. Temperature Dependent Microwave Absorption in Terbium Orthoaluminate, R. C. LeCraw and K. P. Sinha, *Solid State Commun.* **8**, 577-80 (1970).
1794. Sublattice Magnetization in $YbFeO_3$ and $YFeO_3$ as Obtained by Neutron Diffraction and its Relation to the Hyperfine Field, H. Pinto, G. Shachar, and H. Shaked, *Solid State Commun.* **8**, 597-99 (1970).
1795. Preparation and Properties of New Ferrous Basic Sulfate, Toshio Takada, Toshihiko Ariyama, Naochi Yamamoto, and Masao Kiyama, *Bull. Inst. Chem. Res.*, Kyoto Univ., **46**, No. 6, 295-99 (1968).
1796. Magnetic Structures in the $Ca_2Fe_{2-x}Al_xO_5$ System, S. Geller, R. W. Grant, and L. D. Fullmer, *J. Phys. Chem. Solids* **31**, 793-803 (1970).
1797. Etude des Structures Magnétiques des Composés Er_3Co et Er_3Ni par Diffraction Neutronique, D. Gignoux, R. Lemaire, and D. Paccard, *Solid State Commun.* **8**, 391-400 (1970).
1798. Observation of a Magnetically Controllable Jahn Teller Distortion in Dysprosium Vanadate at Low Temperatures, A. H. Cooke, C. J. Ellis, K. A. Gehring, M. J. M. Leask, D. M. Martin, B. M. Wanklyn, M. R. Wells, and R. L. White, *Solid State Commun.* **8**, 689-92 (1970).
1799. Anisotrophy of the Ultrasonic Attenuation in MnF_2 near the Néel Temperature, Akira Ikushima, *J. Phys. Chem. Solids* **31**, 939-46 (1970).
1800. Magnetic and Structural Characteristics of Ternary Intermetallic Systems Containing Lanthanides, Burke Leon and W. E. Wallace, *J. Less-Common Metals* **22**, 1-10 (1970).
1801. Magnetic Properties of ACr_2Se_4 (A = Fe, Co, Ni) and $NiCr_2S_4$, B. L. Morris, P. Russo, and A. Wold, *J. Phys. Chem. Solids* **31**, 635-38 (1970).
1802. Magneto-Mössbauer Study of $Cu_{0.5}Fe_{2.5}O_4$, A. Nagarajan and A. H. Agajanian, *J. Appl. Phys.* **41**, 1642-47 (1970).
1803. Some New Mixed A-site Chromium Chalcogenide Spinels, H. L. Pinch, M. J. Woods, and E. Lopatin, *Mat. Res. Bull.* **5**, 425-30 (1970).
1804. Hexagonal Ferrites for Bubble Domain Devices, L. G. van Uitert, D. H. Smith, W. A. Bonner, W. H. Grodkiewicz, and G. J. Zydzik, *Mat. Res. Bull.* **5**, 455-64 (1970).
1805. Structures and Magnetic Properties of the Pseudo-Binary Systems $CeNi_2$-$CeCu_2$ and $GdNi_2$-$GdCu_2$, W. E. Wallace, T. V. Volkman, and R. S. Craig, *J. Phys. Chem. Solids* **31**, 2185-90 (1970).
1806. Etude par des Mesures de Diffraction de Neutrons et de Magnétisme des Propriétés Cristallines et Magnétiques de Composés Cubiques Spinelles $Co_{3-x}Mn_xO_4$ ($0.6 \leqslant X \leqslant 1.2$), B. Boucher, R. Buhl, R. di Bella, and M. Perrin, *J. Phys.* **31**, 113-19 (1970).
1807. Crystallographic and Magnetic Characteristics of the $Y_2Ni_{17-x}Cu_x$ System, P. D. Carfagna, W. E. Wallace, and R. S. Craig, *Solid State Chemistry* **2**, 1-5 (1970).
1808. Preparation and Characterization of $BaMnO_3$ and $SrMnO_3$ Polytypes, B. L. Chamberland, A. W. Sleight, and J. F. Weiher, *J. Solid State Chem.* **1**, 506-11 (1970).
1809. Conductivité Electrique et Susceptibilité Magnetique des Lames Minces de Nickel-Phosphore, Jean Flechon and Maurice Viard, *Compt. Rend.* **270**, 556-59 (1970).
1810. Mössbauer Study of Several Ferrimagnetic Spinels, G. A. Sawatzky, F. van der Woude, and A. H. Morrish, *Phys. Rev.* **187**, 747-57 (1969).
1811. Magnetic Properties of MTa_2O_6 (M = Fe, Co, or Ni), Mikio Takano and Toshio Takada, *Mat. Res. Bull.* **5**, 449-54 (1970).
1812. Magnetic Characteristics of Intermetallic Compounds Containing Two Lanthanides Combined with Nickel, W. E. Wallace, H. P. Hopkins, Jr., and K. Lehman, *J. Solid State Chem.* **1**, 39-44 (1969).
1813. Low-Temperature Phase Transition of Vanadium Sesquioxide, Mitsuoki Nakahira, Shigeo Horiuchi, and Hirotoshi Ooshima, *J. Appl. Phys.* **41**, 836-38 (1970).
1814. Magnetic Properties of $(Cr_{1-x}Fe_x)O_2$, Eiichi Hirota, Toshihiro Mihara, Tadashi Kawamata, and Mitsuru Asanuma, *Japan. J. Appl. Phys.* **9**, 647-51 (1970).
1815. Antiferromagnetism of a γ Iron-Nickel-Chromium Alloy (Non-Magnetic Stainless Steel), Yoshikazu Ishikawa, Yasuo Endoh, and Teiji Takimoto, *J. Phys. Chem. Solids* **31**, 1225-34 (1970).
1816. Antiferromagnetism in Disordered Cr-Pd Alloys, Takejiro Kaneko and Hiroyasu Fujimori, *J. Phys. Soc. Japan.* **28**, 1373 (1970).

1817. Investigation of the Dynamics of a Magnetic Transition in Mn_2Sn, A. I. Kaplienko, B. N. Leonov, and V. V. Chekin, *Soviet Phys.-Solid State* **11**, 3053-54 (1970).

1818. Magnetic Properties of Cerium Monochalcogenides, G. M. Loginov, A. T. Starovoitov, and A. V. Golubkov, *Soviet Phys.-Solid State* **11**, 3053-54 (1970).

1819. Magnetic Susceptibility of Sulfides of Cerium, Praseodymium, and Neodymium Having the Th_3P_4 Type Structure, G. M. Logunov and V. M. Sergeeva, *Soviet Phys.-Solid State* **11**, 3061-63 (1970).

1820. Metamagnetism of Ni $(NO_3)_2 \cdot 2H_2O$, V. A. Schmidt and S. A. Friedberg, *Phys. Rev. B* **1**, 2250-56 (1970).

1821. The Dysprosium-Iron System: Structural and Magnetic Properties of Dysprosium-Iron Compounds, A. S. van der Goot and K. H. J. Buschow, *J. Less-Common Metals* **21**, 151-57 (1970).

1822. Barium-Strontium Ferrite with Admixtures of Elements of the III-rd Group, Wlodzimierz Wolski and Jadwiga Kowalewska, *Japan. J. Appl. Phys.* **9**, 711-15 (1970).

1823. Nuclear Magnetic Resonances of ^{95}Mo, ^{97}Mo and ^{57}Fe in $Ba_2(FeMo)O_6$, $Sr_2(FeMo)O_6$, and $Ca_2(FeMo)O_6$, Hirotaka Yokoyama and Takehiko Nakagawa, *J. Phys. Soc. Japan.* **28**, 1197-1201 (1970).

1824. Magnetic and Optical Properties of Single-Crystal $CsFeF_3$, T. R. McGuire, V. L. Moruzzi, and M. W. Shafer, *J. Appl. Phys.* **41**, 956 (1970).

1825. NMR of Magnetic Ordering in Two-Dimensional Antiferromagnet, $Mn(HCOO)_2 \cdot 2H_2O$, Yoshitami Ajiro, Norio Terata, Motohiro Matsuura, and Taiichiro Haseda, *J. Phys. Soc. Japan.* **28**, 1587-88 (1970).

1826. Hall Effect of NiAs-Type $Ni_{1-x}S$, Tsukio Ohtani, Koji Kosuge, and Sukeji Kachi, *J. Phys. Soc. Japan.* **28**, 1588 (1970).

1827. Magnetic Properties of the Intermetallic Compound with the Cu_3Au-Type Structure in Cobalt-Titanium Alloy System, Yoshihira Aoki, *J. Phys. Soc. Japan.* **28**, 1451-56 (1970).

1828. Optical Properties of Single-Crystal Films of $CdCr_2S_4$, S. B. Berger and L. Ekstrom, *Phys. Rev. Letters* **23**, 1499-1503 (1969).

1829. Temperature Dependence of the Periodicity of the Magnetic Structure of Thulium Metal, T. O. Brun, S. K. Sinha, N. Wakabayashi, G. H. Lander, L. R. Edwards, and F. H. Spedding, *Phys. Rev. B* **1**, 1251-53 (1970).

1830. Preparation and Characterization of $MgMnO_3$ and $ZnMnO_3$, B. L. Chamberland, A. W. Sleight, and J. F. Weiher, *J. Solid State Chem.* **1**, 512-14 (1970).

1831. Models of the Magnetic Structure of Zinc Ferrite, U. Konig, E. F. Bertaut, Y. Gros, M. Mitrikov, and G. Chol, *Solid State Commun.* **8**, 759-64 (1970).

1832. New Ferromagnetic Europium Compounds, F. Hulliger and O. Vogt, *Solid State Commun.* **8**, 771-72 (1970).

1833. Structure and Magnetic Properties of $VOSO_4$, J. M. Longo and R. J. Arnott, *J. Solid State Chem.* **1**, 394-98 (1970).

1834. Les Propriétés Magnetostatiques des Manganites de Terres Rares, René Pauthenet and Claudine Veyret, *J. Phys. (Fr.)* **31**, 65-72 (1970).

1835. Sur une Nouvelle Phase Oxyfluorée de Type Grenat: $Gd_3Fe_5O_{12-x}F_x$ ($0 \leq x \leq 0.60$), Josik Portier, Bernard Tanguy, Antoinette Morell, René Pauthenet, Roger Olazcuaga, and Paul Hagenmuller, *Compt. Rend., Ser. C,* **270**, 821-24 (1970).

1836. Determination of Curie, Néel, or Crystallographic Transition Temperatures via Differential Scanning Calorimetry, Henry W. Williams and B. L. Chamberland, *Anal. Chem.* **41**, 2084-86 (1969).

1837. Ferromagnetism in K_2CuF_4, Isao Yamada, *J. Phys. Soc. Japan.* **28**, 1585 (1970).

1838. Specific Heat of $MnBr_2$ near the Critical Point, Alan J. Friedman, Ph.D. thesis, Florida State University, 1970, 142 pp.

1839. Analysis of the Specific Heat Cusp of Two Dilute Ferromagnetic Pd-Mn Alloys, B. M. Boerstoel and R. F. Wielinga, *Phys. Letters* **31a**, No. 7, 359-60 (1970).

1840. Iron-Holmium Binary System and Related Magnetic Properties, Gerald John Roe, thesis, University of Missouri, Rolla, Missouri, 1969, 119 pp.

1841. Neutron Scattering Investigation of Phase Transitions and Magnetic Correlations in the Two-Dimensional Antiferromagnets K_2NiF_4, Rb_2MnF_4, Rb_2FeF_4, R. J. Birgeneau, H. J. Guggenheim, and G. Shirane, *Phys. Rev. B* **1**, 2211-30 (1970).

1842. Transition from Van Vleck Paramagnetism to Induced Antiferromagnetism in $Tb_xY_{1-x}Sb$, Bernard R. Cooper and Oscar Vogt, *Phys. Rev. B* **1**, 1218-26 (1970).

1843. Effet Mössbauer en Fonction de la Temperature du ^{57}Fe en Impureté dans le Nickel, Christian Janot and Hubert Scherrer, *Compt. Rend., Ser. B*, **270**, 1293-96 (1970).

1844. Specific Heat Measurement of $ErCl_3 \cdot 6H_2O$ Having Strongly Anisotropic Dipolar Interactions, E. Lagendijk, R. F. Wielinga, and W. J. Huiskamp, *Phys. Letters* **31a**, 375-76 (1970).

1845. $DyPO_4$: A Three-Dimensional Ising Antiferromagnet, J. C. Wright, H. E. Moos, J. H. Colwell, B. W. Mangum, and D. D. Thornton, *Phys. Rev. B* **3**, 843-58 (1971).

1846. Ferromagnetische Halbleiter mit Austauschwechselwirkung über die Leitungselektronen, S. V. Vonsovskii, A. A. Samokokhvalov, and A. A. Berdyshev, *Helv. Phys. Acta* **43**, 9-16 (1970).

1847. Effect of Magnetic Fields on the Crystal Structure of Dysprosium, V. A. Finkel and V. S. Belovol, *Zh. Eksper. Teor. Fiz.* **57**, No. 3, 774-80 (1969).

1848. Method of Producing High Curie Temperature EuO Single Crystals, F. Holtzberg and M. W. Shafer, IBM Corp., U. S. Patent 3488286, Aug. 1, 1968.

1849. Superconductivity and Magnetism of Lanthanum-Rare Earth Dialuminides, Merrill Brian Maple, thesis, University of California, San Diego, California, 1969, 126 pp.

1850. On the Thermal Expansion Coefficient and the Temperature Coefficient of Young's Modulus of Cobalt-Palladium Alloys, Hakaru Masumoto and Shohachi Sawaya, *J. Japan. Inst. Metals* **33**, No. 6, 685-87 (1969).

1851. Nuclear Resonance Determination of the Magnetic Space Group of $MnCl_2 \cdot 2H_2O$, R. D. Spence and K. V. S. Rama Rama Rao, *J. Chem. Phys.* **52**, 2740-44 (1970).

1852. Investigation of Ferro-Antiferromagnetic Transformations in Iron-Platinum Alloys by a Method Based on the Mössbauer Effect, L. I. Vinokurova, I. N. Nikolaev, E. V. Mel'inkov, I. K. Adic'evich, and Yu. B. Reutov, *Fiz. Metallov Metalloved.* **28**, No. 6, 1098-1102 (1969).

1853. Effect of Pressure on the Curie Temperature of $ZrZn_2$, R. C. Wayne and L. R. Edwards, *Phys. Rev.* **188**, 1042-44 (1969).

1854. Weak Ferromagnetism in $NiCO_3$, N. M. Kreines and T. A. Shalnikova, *Zh. Eksper. Teor. Fiz.* **58**, No. 2, 522-27 (1970) (in Russian).

1855. Behavior of Europium Oxide, M. Crisan, *Stud. Cercet. Fiz.* **22**, 25-27 (1970) (in Rumanian).

1856. Mössbauer Studies of Pt-Fe Alloys, S. H. Devare and H. G. Devare, abstract only, in *Nuclear Physics and Solid State Physics Symp.*, Roorkee, India, Dec. 28-31, 1969 (Department of Atomic Energy, New Delhi, India, 1969).

1857. NMR in Fe-Mn-Ge, R. Fatchally, N. P. Sastry, R. Nagarajan, and M. S. Multani, abstract only, in *Nuclear Physics and Solid State Physics Symp.*, Roorkee, India, Dec. 28-31, 1969 (Department of Atomic Energy, New Delhi, India, 1969).

1858. The Thermal Conductivity of Europium Sulfide across the Curie Temperature, Rees D. Morgan (South Dakota School of Mines and Technology, Rapid City, So. Dak.), TR-26; AD-704 498 (Nov. 15, 1969).

1859. Magnetic Transitions in FeMnGe, M.R.L.N. Murthy, N. G. Natera, R. J. Begum, and N.S.S. Murthy, abstract only, in *Nuclear Physics and Solid State Physics Symp.*, Roorkee, India, Dec. 28-31, 1969 (Department of Atomic Energy, New Delhi, India, 1969).

1860. Influence of the Magnetic Transition of the Heat Capacity Behavior of Chromium Telluride, S. G. Sankar, V. G. Gunjikar, and A. B. Biswas, *13th Proc. Nucl. Phys. Solid State Phys. Symp.*, 1968 (publ. 1969), **3**, 179-82 (in English).

1861. Anomalies of Magnetic and Electrical Properties in Li-Al Ferrite, K. P. Belov, A. N. Goryaga, T. Ya. Gridasova, and O. I. Lavroskaya, *Soviet Phys.-Solid State* **12**, 221-22 (1970).

1862. Magnetic Behavior of $Mn(NH_4)_2(SO_4)_2 \cdot 6H_2O$ in its Curie Region in Fields Directed Along the b-axis, J. J. Fritz, M. M. Cesarano, and J. T. Clark, *J. Low Temp. Phys.* **2**, 87-102 (1970).

1863. Possible Coexistence of Superconductivity and Ferromagnetism in Mixed Crystals of the Type $Ce_{1-x}Tb_xRu_2$, B. Hillenbrand and M. Wilhelm, *Phys. Letters* **31A**, 448-49 (1970).

1864. The Magnetic Transition, Thermal Capacity, and Thermodynamic Properties of Uranium Dioxide from 5 to 350°K, James J. Huntzicker and Edgar F. Westrum, Jr., *J. Chem. Thermodynamics* **3**, 61-76 (1971).

1865. The Magnetic Structures of $RbNiCl_3$ and $CsNiCl_3$, V. J. Minkiewicz, D. E. Cox, and G. Shirane, *Solid State Commun.* **8**, 1001-05 (1970).

1866. Magnetostriction of Single Crystals of Lithium Ferrite-Aluminate Systems, G. A. Petrakovskii, E. M. Smokotin, L. M. Protopopova, and K. A. Sablina, *Soviet Phys.-Solid State* **12**, 135-37 (1970).

1867. Mössbauer ^{57}Fe Spectra from the Intermetallic Compounds $PuFe_2$, $NpFe_2$, and UFe_2, S. Blow, *J. Phys. C. Proc. Phys. Soc. (Solid State Phys.)* **3**, No. 4, 835-45 (1970).

1868. Exchange Narrowing of the Paramagnetic Resonance Line in Ferromagnetic Compounds of Divalent Europium, A. A. Samokhvalov and V. S. Babushkin, *Soviet Phys.-Solid State* **12**, 9-11 (1970).

1869. Magnetic Resonance in Antiferromagnetic $FeCl_2 \cdot 4H_2O$, D. D. Thornton, *Phys. Rev. B* **1**, 3193-3203 (1970).

1870. Etude de l'Ordre dans les Solutions Solides $(CuO)_x(NiO)_{1-x}$, R. Zilber, E. F. Bertaut, and P. Burlet, *Solid State Commun.* **8**, 935-41 (1970).

1871. Magnetische Bereiche in Chromtrihalogeniden, O. Bostanjoglo and W. Vieweger, *Phys. Stat. Sol.* **39**, 471 (1970).

1872. Effect of Strong Pulsed Magnetic Fields on Magnetic Transitions in Mn_3B_4, I. G. Fakidov and V. I. Timoshchuk, *Soviet Phys.-Solid State* **12**, 218-20 (1970).

1873. Effect of Hydrostatic Pressure on the Magnetic Transition Temperature of the System $Cd_{1-x}Zn_xCr_2Se_4$, Hironobu Fujii, Takahiko Kamigaichi, Yasuharu Hidaka, and Tetsuhito Okamoto, *J. Phys. Soc. Japan.* **29**, 244 (1970).

1874. Studies of Nickelous and Cobaltous Oxide and Effects of n-Type and p-Type Dopants, William Rives Helms (Dept. of Phys., Purdue University, Lafayette, Indiana), COO-AT-(11-1)-1616-14 (1970).

1875. The Specific Heats of $GdCl_3$, $GdBr_3$, and GdI_3 at Low Temperatures, V. Hovi, R. Vuola, and L. Salmenpera, *J. Low Temp. Phys.* **2**, 283-87 (1970).

1876. Some Magnetic Properties of $MnCO_3$ at Low Field Strengths, H. C. Meijer, L.M.W.A. Pimmelaar, S. R. Brouwer, and J. van den Handel, *Physica* **46**, 279-90 (1970).

1877. Neutron Diffraction Study of the Magnetic Structure of Alpha-Phase Chromia-Alumina Solid Solutions, T. M. Sabine and E. R. Vance, *J. Solid State Chem.* **1**, 554-56 (1970).

1878. Temperature Effects on Spin Waves in Cr_2O_3 Studied by Means of Inelastic Neutron Scattering, E. J. Samuelsen, *Physica* **45**, 12-28 (1969).

1879. Fe^{57} Mössbauer Effect and Magnetic Susceptibility of Hexavalent Iron Compounds; K_2FeO_4, $SrFeO_4$ and $BaFeO_4$, Teruya Shinjo, Toshio Ichida, and Toshio Takada, *J. Phys. Soc. Japan.* **29**, 111-16 (1970).

1880. Contribution a l'Etude de l'Influence de l'Ordre sur les Propriétés Magnétiques des Alliages Fer-Nickel, Francois Barruel, thesis, University of Grenoble, France, April 25, 1968.

1881. Magnetic Properties of U_3P_4-U_3As_4 Solid Solutions, W. Trzebiatowski and A. Misiuk, *Roczniki Chemii Ann. Soc. Chim. Polonorum* **44**, 695-97 (1970).

1882. Sur l'Aimantation de l'Alliage Ni_3Mn, V. Tutovan and J. Feron, Analele Stiintifice ale Universitatii "Al. I. Cuza" Din Iasi (Serie Noua), Sectiunea I, b. Fizica, Vol. XIII, 17-24 (1967) (in French).

1883. Double Peaks in the Specific Heats of ErAs and ErSb at Low Temperatures, V. Hovi, M. Lehtinen and R. Vuola, *Phys. Letters* **31A**, 451-52 (1970).

1884. The Preparation of Beta-Molybdenum Nitride, Ronald Karam and Roland Ward, *Inorg. Chem.* **9**, 1385-87 (1970).

1885. Electrical Resistance Measurements on Pd Base Alloys, A. M. Schiffrin, *Phys. Stat. Sol.* **39**, K81-83 (1970).

1886. Manganese Disulfide (Hauerite) and Manganese Ditelluride. Thermal Properties from 5 to 350°K and Antiferromagnetic Transitions, Edgar F. Westrum, Jr., and Fredrik Gronvold, *J. Chem. Phys.* **52**, 3820-26 (1970).

1887. Structure Magnetique du Spinelle Antiferromagnetique $ZnFe_2O_4$, B. Boucher, R. Buhl, and M. Perrin, *Phys. Stat. Sol.* **40**, 171-82 (1970).

1888. Etude par Diffraction de Neutrons de la Structure Cristalline et Magnétique de AgF_2, Pierrette Charpin and Pierre Meriel, *Bull. Soc. Fran. Mineral. Cristallogr.* **93**, 7-13 (1970).

1889. NMR in UH_3, J. Grunzweig and M. Kuznietz, in Magnetic Resonance and Relaxation, Proc. XIVth Colloque Ampere, Ljubljana, Sept. 6-11, 1966 (North Holland Publishing Company, Amsterdam, 1967), p. 1224-28.

1890. Review on the Single Crystal Growth of Chalcogenide Spinel, F. Okamoto, T. Takahashi, and Y. Wada, *Oyo Butsuri* **39**, 471-79 (1970).

1891. Magnetic Structure of Magnesium Chromite, H. Shaked, J. M. Hastings, and L. M. Corliss, *Phys. Rev. B* **1**, 3116-24 (1970).

1892. Magnetic Phase Diagram of MnF_2 from Ultrasonic and Differential Magnetization Measurements, Y. Shapira and S. Foner, *Phys. Rev. B* **1**, 3083-96 (1970).

REFERENCES

1893. Magnetic Properties of Materials by Ultrasonic Studies. Paramagnetic to Antiferromagnetic Phase Boundary of CoF_2 from Ultrasonic Measurements, Y. Shapira (MIT), in *Massachusetts Institute of Technology Francis Bitter National Magnetic Lab. Quart. Tech. Status Report*, April 1, 1970-June 30, 1970, QTSR 40, pp. 6-7 (June 30, 1970).

1894. ^{63}Cu Nuclear Magnetic Relaxation in Cu-Ni Alloys Near the Critical Concentration, F. Y. Fradin and T. J. Rowland, *Solid State Commun.* **8**, 1047-50 (1970).

1895. Magneto-photoconductivity of Europous Oxide near the Curie Temperature, Koji Kajita and Taizo Masumi, *Solid State Commun.* **8**, 1039-42 (1970).

1896. Knight Shifts of ^{31}P in the Paramagnetic State of the UP-US Solid Solutions, M. Kuznietz, G. A. Matzkanin, and Y. Baskin, *Phys. Letters* **28A**, 122-23 (1968).

1897. Magnetism and Bonding in a $D8_8$ Structure: Mössbauer and Magnetic Investigation of the System $Mn_5Si_3-Ge_5Si_3$, K.S.V.L. Narasimhan, W. M. Reiff, H. Steinfink, and R. L. Collins, *J. Phys. Chem. Solids* **31**, 1511-24 (1970).

1898. Magnetic Susceptibility of Uranium Carbonitride, Toshihiko Ohmichi and Shoichi Nasu, *J. Nucl. Sci. Technol. (Tokyo)* **7**, 268-70 (1970).

1899. Trivalent and Tetravalent Chromium in the Seleno-Spinel Series $CuCr_{0.3}Rh_{1.7-x}Sn_xSe_4$, R.P. van Stapele and F. K. Lotgering, *J. Phys. Chem. Solids* **31**, 1547-51 (1970).

1900. Propriétés Structurales et Magnétiques de Quelques Composés du Type Stannite, J. Allemand and Micheline Wintenberger, *Bull. Soc. Fr. Mineral. Cristallogr.* **93**, 14-17 (1970).

1901. Temperature Dependence of the Specific Heat and Electrical Resistivity of Dysprosium in the Vicinity of the Néel Point, E. B. Amitin, Yu. A. Kovalevskaya, F. S. Rakhmenkulov, and I. E. Paukov, *Soviet Phys.-Solid State* **12**, 599-604 (1970).

1902. Neutron Diffraction Studies of $TmAu_2$, $YbAu_2$, and $TmAg_2$ at 300–1.7°K, Masao Atoji, *J. Chem. Phys.* **52**, 6433-34 (1970).

1903. Optical Studies of Antiferromagnetism in Chromium and Its Alloys, A.S. Barker, Jr., and J. A. Ditzenberger, *Phys. Rev. B* **1**, 4378-4400 (1970).

1904. Preparation and Properties of Fe_2P, Technical Report, May 1, 1969-April 30, 1970, D. Bellavance, J. Mikkelsen, and A. Wold (Brown University, Providence, R. I.), AFML-TR-70-102 (July 1970).

1905. Nonresonance Absorption of High-Frequency Energy in Yttrium Ferrite Garnet in the Region of the Curie Point, K. P. Belov and N. V. Shebaldin, *Soviet Phys.-Solid State* **12**, 501-02 (1970).

1906. Optical Spectroscopy of Erbium Chromite ($ErCrO_3$), R. Courths, S. Hufner, J. Pelzl, and L. G. van Uitert, *Solid State Commun.* **8**, 1163-65 (1970).

1907. Anisotropy of Thermal Expansion of Iron Digermanide, R. P. Krentsis, A. V. Mikhel'son, and P. V. Gel'd, *Soviet Phys.-Solid State* **12**, 727-728 (1970).

1908. Magnetic and Electrical Properties of $(La_{1-x}Ca_x)MnO_3$, G. Matsumoto, *IBM J. Res. Develop.* **14**, 258-60 (1970).

1909. Magnetism. Critical Magnetic Properties of EuO, N. Menyuk and K. Dwight, in *Solid State Research, Quarterly Technical Summary*, Feb. 1-April 30, 1970, pp. 55-57, Alan L. McWhorter, ed. (Lincoln Lab., MIT, Lexington, Mass.), ESD-TR-70-148 (May 15, 1970), 92 pp.

1910. On the Volume Magnetostriction of Nickel, Iron, and Cobalt, F. Richter and U. Lotter, *Phys. Stat. Sol.* **34**, K149-52 (1969).

1911. Thermomagnetic Properties of Ferrite $Cu_{0.5}Fe_{2.5}O_4$, Z. Simsa and V. Houdek, *Czech. J. Phys. B* **20**, 301-06 (1970).

1912. Ferromagnetic Resonance and Nonlinear Effects in Lithium-Zinc Ferrite Single Crystals, Yu. M. Yakovlev, M. A. Vinnik, E. V. Rubal'skaya, and B. L. Lapovok, *Soviet Phys.-Solid State* **12**, 669-73 (1970).

1913. Magnetic Phase Diagram of $Cs_3Cu_2Cl_7 \cdot 2H_2O$, G. J. Butterworth, V. S. Zidell, and J. A. Woollam, *Phys. Letters* **32A**, 24-25 (1970).

1914. Single Crystal Neutron Diffraction Studies of hcp Rare Earth Thorium Alloys, H. R. Child and W. C. Koehler, Intern. Conf. on Magnetism, Grenoble, France, Sept. 14-17, 1970 (to be published in *J. Phys.*).

1915. Metamagnetism in $UP_{0.75}S_{0.25}$, J. Crangle, M. Kuznietz, G. H. Lander, and Y. Baskin, *J. Phys. C (Solid State Phys.) Ser. 2*, **2**, 925-927 (1969).

1916. Etude par Effet Mössbauer et Diffraction Neutronique des Amas Accidentels dans les Solutions Solides Desordonnées Ba(Fe$_x$Al$_{1-x}$)$_2$O$_4$, C. Do-Dinh, R. Chevalier, P. Burlet, and E. F. Bertaut, *J. Phys.* **31**, 401-08 (1970).

1917. Magnetic and Crystallographic Transitions in Sc^{3+}, Cr^{3+}, and Ga^{3+} Substituted Mn$_2$O$_3$, S. Geller and G. P. Espinosa, *Phys. Rev. B* **1**, 3763-69 (1970).

1918. Thermal Conductivity Measurements in Crystals of CoCs$_3$Cl$_5$ and Cu(NH$_4$)$_2$·2H$_2$O, J. N. Haasbroek and A.S.M. Gieske, *Phys. Letters* **31A**, 351-52 (1970).

1919. A High-Sensitive, Simple Calorimeter for Phase-Change Detection, Kinshiro Hirakawa and Minoru Furukawa, *Japan. J. Appl. Phys.* **9**, 971-75 (1970).

1920. Determination of the Néel Temperature of Face-Centered-Cubic Iron, G. J. Johanson, M. B. McGirr, and D. A. Wheeler, *Phys. Rev. B* **1**, 3208 (1970).

1921. Structures et Propriétés Magnétiques des Orthocobaltites de Terres Rares TCoO$_3$, A. Kappatsch, S. Quezel-Ambrunaz, and J. Sivardiere, *J. Phys.* **31**, 369-76 (1970).

1922. Preparation, Chemistry, and Magnetic Properties of Cobalt Doped BiCa VIG Crystals, R. Krishnan, *Phys. Stat. Sol.* **35**, K63-65 (1969).

1923. Optical Spectra of Ferroelectric-Antiferromagnetic Rare Earth Manganates, K. Kritayakiran, P. Berger, and R. V. Jones, *Optics Commun.* **1**, 95-98 (1969).

1924. Relation between NMR Properties of ^{31}P in the Paramagnetic State of the UP$_{1-x}$S$_x$ System and the Magnetic Ordering, Moshe Kuznietz, Yehuda Baskin, and G. A. Matzkanin, *J. Appl. Phys.* **41**, 1111-13 (1970).

1925. Passage du Ferromagnetisme a l'Antiferromagnetisme dans les Alliages Pd-Mn, Jacques Rault and Jean Paul Burger, *Compt. Rend. Ser. B*, **269**, 1085-88 (1969).

1926. Antiferromagnetism in V$_2$O$_3$, R. M. Moon, *Phys. Rev. Letters* **25**, 527 (1970).

1927. Permeability and Conductivity of Ti-substituted MnZn Ferrites, T.G.W. Stijntjes, J. Klerk, and A. Broese van Groenou, *Philips Res. Repts.* **25**, 95-107 (1970).

1928. Nuclear Magnetic Resonance of ^{31}P in the Antiferromagnetic State of Metallic Uranium Monophosphide, S. L. Carr, C. Long, W. G. Moulton, and M. Kuznietz, *Phys. Rev. Letters* **23**, 785-88 (1969).

1929. Hyperfine Magnetic Fields in USb from Mössbauer Spectroscopy, G. K. Shenoy, G. M. Kalvius, S. L. Ruby, B. D. Dunlap, M. Kuznietz, and F. P. Campos, *Inter. J. Magnetism* **1**, No. 1 (1970).

1930. New Type Metamagnetism in Co(S$_x$Se$_{1-x}$)$_2$: Exchange-compensated Paramagnetism, Kengo Adachi, Kiyoo Sato, Makoto Matsuura, and Masayoshi Ohashi, *J. Phys. Soc. Japan.* **29**, 323-32 (1970).

1931. Critical Exponent β for the Magnetization of YIG, E. E. Anderson, H. J. Munson, S. Arajs, A. A. Stelmach, and B. L. Tehan, *J. Appl. Phys.* **41**, 1274-76 (1970).

1932. Magnetic Phase Diagram of Cs$_3$Cu$_2$Cl$_7$·2H$_2$O, G. J. Butterworth and V. S. Zidell, *Phys. Letters* **32A**, 24-25 (1970).

1933. Effect of Hydrostatic Pressure on Magnetic Properties of Europium Chalcogenides, Yasuharu Hidaka, Hironobu Fujii, Tetsuhiko Okamoto, and Eiji Tatsumoto, *J. Phys. Soc. Japan.* **29**, 515 (1970).

1934. The Effective Magnetic Fields in Cubic Ferricyanides, A. Z. Hrynkiewicz, B. D. Sawicka, and J. A. Sawicki, *Acta Phys. Polon.* **37A**, No. 5, 811-19 (1970).

1935. Effect of Pressure on Curie Temperatures of Chalcogenide Spinels CuCr$_2$X$_4$ (X = S, Se, and Te), Takeshi Kanomata, Hideaki Ido, and Takejiro Kaneko, *J. Phys. Soc. Japan.* **29**, 332-35 (1970).

1936. Magnetic Structure of MnB$_2$, Mitsuo Kasaya and Tâdamiki Kihara, *J. Phys. Soc. Japan.* **29**, 336-42 (1970).

1937. Heat Capacity of Gadolinium near the Curie Point, Edwin A. S. Lewis, *Phys. Rev. B* **1**, 4368-77 (1970).

1938. The Crystal and Magnetic Structures of Mn$_2$GeO$_4$, J. G. Creer and G.J.F. Troup, *Solid State Commun.* **8**, 1183-88 (1970).

1939. Mössbauer Study of Hyperfine Interactions in Divalent Europium Compounds at Ultralow Temperatures, G. J. Ehnholm, T. E. Katila, O. V. Lounasmaa, P. Reivari, G. M. Kalvius, and G. K. Shenoy, *Z. Physik* **235**, 289-307 (1970).

1940. Perpendicular Anisotropy of Electrodeposited Nickel and Nickel-Phosphorus Films, Hiroshi Maeda, *J. Phys. Soc. Japan.* **29**, 311-22 (1970).

1941. Low-Temperature Elasticity and Magnetoelasticity of Dysprosium Single Crystals, M. Rosen and H. Klimker, *Phys. Rev. B* **1**, 3748-56 (1970).

REFERENCES

1942. Magnetic Properties of Amorphous Pd-Si Alloys Containing Iron, Ryusuke Hasegawa, *J. Appl. Phys.* **41**, 4096-4100 (1970).
1943. Ultrasonic Attenuation near the Spin-Alignment Transition of EuTe, Y. Shapira and T. B. Reed, *Phys. Letters* **31A**, 381-82 (1970).
1944. Magnetic Properties of Some Cubic Rare-Earth-Iron Compounds of the Type RFe_2 and $R_xY_{1-x}Fe_2$, K.H.J. Buschow and R. P. van Stapele, *J. Appl. Phys.* **41**, 4066-69 (1970).
1945. W. E. Wallace and A. E. Scrabek, in *Rare Earth Research*, K. Vorres, ed. (Gordon and Breach, New York, 1964), **2**, p. 431.
1946. Low Temperature Melting Curve of He^3, R. T. Johnson, O. V. Lounasmaa, R. Rosenbaum, O. G. Symko, and J. C. Wheatley, *J. Low Temp. Phys.* **2**, 403-21 (1970).
1947. Les Propriétés Magnétiques des Composés Intermétalliques Fer-Titane de Composition Voisine de Fe_2Ti, M. Okazaki (Laboratoire d'Electrostatique et de Physique du Metal, C.N.R.S., Cedex No. 166, Grenoble, France), preprint received RMIC in Oct. 1970.
1948. Structure Relations of Hexagonal Perovskite-Like Compounds ABX_3 at High Pressure and Their Magnetic Properties, Yasuhiko Syono, Syun-iti Akimoto, and Kay Kohn, pp. 415-21 in *Les Propriétés Physiques des Solides sous Pression*, No. 188, Colloques Internationaux du C.N.R.S. (Editions du C.N.R.S., Paris, 1970).
1949. Magnetic Structure of Mn_4N-Type Compounds, W. J. Takei, R. R. Heikes, and G. Shirane, *Phys. Rev.* **125**, 1893-97 (1962).
1950. Magnetic Structure of Rare-Earth Iridium Compounds RIr_2, Gian Piero Felcher and W. C. Koehler, *Phys. Rev.* **131**, 1518-20 (1963).
1951. Neutron Diffraction Studies on Europium Metal, C. E. Olsen, N. G. Nereson, and G. P. Arnold, *J. Appl. Phys.* **33**, Suppl., 1135-36 (1962).
1952. Neutron Study of the Crystal and Magnetic Structures of $MnFe_{2-x}Cr_xO_4$, Stanley J. Pickart and Robert Nathans, *Phys. Rev.* **116**, 317-22 (1959).
1953. Transitions from Ferromagnetism to Antiferromagnetism in Iron-Aluminum Alloys. Experimental Results, Anthony Arrott and Hiroshi Sato, *Phys. Rev.* **114**, 1420-26 (1959).
1954. Magnetic Structures of TbD_2 and HoD_2, D. E. Cox, G. Shirane, W. J. Takei, and W. E. Wallace, *J. Appl. Phys.* **34**, 1352 (1963).
1955. Studies on the Heusler Alloys. II. The Structure of Cu_3Mn_2Al, G. B. Johnston and E. O. Hall, *J. Phys. Chem. Solids* **29**, 201-07 (1968).
1956. Critical Magnetic Behavior of EuO, G. Groll, *Phys. Letters* **32A**, 99-100 (1970).
1957. Ordered Moment of NiS_2, J. M. Hastings and L. M. Corliss, *IBM J. Res. Develop.* **14**, 227-28 (1970).
1958. Magnetic Resonance of $KCuCl_3$, J. E. Hynes, B. B. Garrett, and W. G. Moulton, *J. Chem. Phys.* **52**, 2671-78 (1970).
1959. Sound Propagation near Magnetic Phase Transitions, B. Luthi, T. J. Moran, and R. J. Pollina, *J. Phys. Chem. Solids* **31**, 1741-58 (1970).
1960. Elastic and Magnetoelastic Effects in Rare Earth Metals, T. J. Moran and B. Luthi, *J. Phys. Chem. Solids* **31**, 1735-40 (1970).
1961. The Effect of Deuteration on the Néel Temperature of $MnBr \cdot 4H_2O$, C. L. Yue and B. G. Turrell, *Solid State Commun.* **8**, 1261-64 (1970).
1962. The Effect of Pressure on the Néel Temperature of Chromium-Vanadium Alloys, W. DeDood, J. DeDood, and P. F. DeChatel, *Physica* **47**, 337-44 (1970).
1963. Crystallographic and Magnetic Ordering Studies of Plutonium Carbides Using Neutron Diffraction, J. L. Green, G. P. Arnold, J. A. Leary, and N. G. Nereson, *J. Nucl. Mater.* **34**, 281-89 (1970).
1964. Neutron and X-Ray Diffraction Study of the Low-Temperature Chemical and Magnetic Structure of Alpha-Manganese, J. A. Oberteuffer, J. A. Marcus, L. H. Schwartz, and G. P. Felcher, *Phys. Rev. B* **2**, 670-77 (1970).
1965. Magnetic Properties and Nuclear Magnetic Resonance of Cubic RCu_5 Intermetallic Compounds, K.H.J. Buschow, A. M. van Diepen, and H. W. de Wijn, *J. Appl. Phys.* **41**, 4609-12 (1970).
1966. Magnetic Studies of Cl_b-Compounds CuMnSb, PdMnSb, and Cu_{1-x}(Ni or Pd)$_x$ MnSb, Keizo Endo, *J. Phys. Soc. Japan.* **29**, 643-49 (1970).
1967. Neutron Diffraction of Fe_7Se_8. II., Masaru Kawaminami and Atsushi Okazaki, *J. Phys. Soc. Japan.* **29**, 649-55 (1970).

1968. The Magnetic Structure of Vivianite, $Fe_3(PO_4)_2 \cdot 8H_2O$, J. B. Forsyth, C. E. Johnson, and C. Wilkinson, *J. Phys. C, Solid State Phys.* **3**, 1127-39 (1970).

1969. Ferromagnetism of Europium Amide $Eu(NH_2)_2$, F. Hulliger, *Solid State Commun.* **8**, 1477-78 (1970).

1970. Magnetic Properties of Alpha-$NaFeO_2$, Toshio Ichida, Teruya Shinjo, Yoshichika Bando, and Tashio Takada, *J. Phys. Soc. Japan.* **29**, 795 (1970).

1971. Crystal Fields and the Magnetic Properties of Praseodymium and Neodymium, T. Johansson (Technical Univ., Lyngby, Denmark), B. Lebech, M. Nielsen, H. Bjerrum Moller (AEK Res. Estab., Riso, Denmark), and A. R. Mackintosh (H. C. Orsted Inst., Univ. Copenhagen), preprint received at RMIC Nov. 18, 1970.

1972. Electrical Resistivity near the Magnetic Transition of Some Rare-Earth Laves Phase Compounds, M. P. Kawatra, J. A. Mydosh, and J. I. Budnick, *Phys. Rev. B* **2**, 665-70 (1970).

1973. Torque Measurements on a Single Crystal of $SmFeO_3$ in the Spin-Reorientation Region, Seiichi Kagoshima and Suguru Takayama, *J. Phys. Soc. Japan.* **29**, 793 (1970).

1974. Ferromagnetism in the Laves Phase Compound $Fe_{2+x}Hf_{1-x}$ Annealed at $1000°C$, Takuro Nakamichi, K. Kai, Y. Aoki, K. Ikeda, and M. Yamamoto, *J. Phys. Soc. Japan.* **29**, 794 (1970).

1975. Study of $(La_{1-x}Ca_x)MnO_3$. I. Magnetic Structure of $LaMnO_3$, Gen Matsumoto, *J. Phys. Soc. Japan.* **29**, 606-15 (1970).

1976. Study of $(La_{1-x}Ca_x)MnO_3$. II. Magnetic Properties, Gen Matsumoto, *J. Phys. Soc. Japan.* **29**, 615-22 (1970).

1977. Nuclear Magnetic Resonance of Mn^{55} and P^{31} in Ferromagnetic and Antiferromagnetic MnP, Hiroyuki Nagai, Tadamiki Hihara, and Eiji Hirahara, *J. Phys. Soc. Japan.* **29**, 622-32 (1970).

1978. Magnetoelectric Studies of Critical Behavior in the Ising-Like Antiferromagnet $DyPO_4$, George T. Rado, *Solid State Commun.* **8**, 1349-52 (1970).

1979. Effect of Antiferromagnetic Ordering on the Resistivity of EuTe, Y. Shapira, S. Foner, and N. F. Oliviera, Jr., *Phys. Letters* **32A**, 323-24 (1970).

1980. Cation Distributions in Ferrimagnetic Garnets, M. Shimada, S. Kume, and M. Koizumi, *J. Phys. Chem. Solids* **31**, 2165-67 (1970).

1981. Magnetic Properties of Tb-Ho Single-Crystal Alloys. II. Neutron Diffraction Studies, F. H. Spedding, Y. Ito, and R. G. Jordan, *J. Chem. Phys.* **53**, 1455-65 (1970).

1982. Increase of Curie Temperature in EuO Films by Fe Doping, K. Y. Ahn, *Appl. Phys. Letters* **17**, 347-49 (1970).

1983. Attempt to Find A Magnetic Field Dependence of the Néel Temperature of Chromium, R. L. Melcher and W. D. Wallace, *Solid State Commun.* **8**, 1535-38 (1970).

1984. The Effect of Sr Ions on Magnetic Properties of $Co_2Ba_3Fe_{24}O_{41}$-$Co_2Sr_3Fe_{24}O_{41}$ Solid Solutions, I. I. Petrova, *Soviet Phys.-Solid State* **12**, 1135-37 (1970).

1985. Coexistence of Superconductivity and Short-Range Magnetic Order, L. J. Williams, W. R. Decker, and D. K. Finnemore, *Phys. Rev. B* **2**, 1287-93 (1970).

1986. Inelastic Neutron Scattering Investigation of Spin Waves and Magnetic Interactions in Cr_2O_3, E. J. Samuelsen, M. T. Hutchings, and G. Shirane, *Physica* **48**, 13-42 (1970).

1987. Temperature and Field Dependence of Magnetization of Amorphous (Fe,Mn)-P-C Alloys, Ashok K. Sinha, CALT-822-7 (July 1970).

1988. Electron Paramagnetic Resonance Linewidth of $CuCl_2 \cdot 2H_2O$ near the Néel Temperature, N. J. Zimmerman, F. P. van der Mark, and J. van den Handel, *Physica* **46**, 204-12 (1970).

1989. Mössbauer Effect Study of $KFeO_2$, Toshio Ichida, Teruya Shinjo, Yoshichika Bando, and Toshio Takada, *J. Phys. Soc. Japan.* **29**, 1109-10 (1970).

1990. Calorimetric Investigation of the Antiferromagnetic Transition in $NiCl_2 \cdot 6H_2O$, W. L. Johnson and W. Reese, *Phys. Rev. B* **2**, 1355-62 (1970).

1991. Neutron-Diffraction Study of $CoCl_2 \cdot 6H_2O$, Robert Kleinberg, *J. Chem. Phys.* **53**, 2660-62 (1970).

1992. Nuclear Antiferromagnetism in Solid Helium Three, Wiley Price Kirk, BNL-14363 (January 1970).

1993. Specific Heat of Alpha-Uranium, J. A. Lee, P. W. Sutcliffe, and K. Mendelssohn, *Phys. Letters* **30A**, 106-07 (1969).

1994. Effects of the Néel Transition on the Thermal and Electrical Resistivities of Cr and Cr:Mo Alloys, G. T. Meaden, K. V. Rao, and K. T. Tee, *Phys. Rev. Letters* **25**, 359-62 (1970).

REFERENCES

1995. Magnetic Form Factor of Gold-Chromium Alloys, Yutaka Nakai, Nobuhiko Kunitomi, and Yasusada Yamada, *J. Phys. Soc. Japan.* **29,** 978-86 (1970).

1996. Magnetic Properties of $TbIn_3$, $TbPt_3$, and $HoIn_3$, Norris Nereson and George Arnold, *J. Chem. Phys.* **53,** 2818-23 (1970).

1997. Magnetic Phase Transition in $Ni_xCo_{1-x}Cl_2 \cdot 6H_2O$, Kazuyoshi Takeda, Motohiro Matsuura, and Taiichiro Haseda, *J. Phys. Soc. Japan.* **29,** 885-89 (1970).

1998. The Magnetic Structure of the Spinel Fe_3S_4, J.M.D. Coey, M. R. Spender, and A. H. Morrish, *Solid State Commun.* **8,** 1605-08 (1970).

1999. Electrical and Magnetic Properties of $Cu_2FeSnSe_4$, Saburo Endo and Taizo Irie, *J. Phys. Soc. Japan.* **29,** 1393 (1970).

2000. Magnetic Structure and Two Dimensional Behavior of Rb_2MnCl_4 and Cs_2MnCl_4, A. Epstein, E. Gurewitz, J. Makovsky, and H. Shaked, *Phys. Rev. B* **2,** 3703 (1970).

2001. Magnetic Resonance of $CoCr_2O_4$ and $MnCr_2O_4$, Satoru Funahashi, Kiiti Siratori, and Yuzo Tomono, *J. Phys. Soc. Japan.* **29,** 1179-93 (1970).

2002. Magnetic Polytypes in Rb_2MnCl_4, E. Gurewitz, A. Epstein, J. Makovsky, and H. Shaked, *Phys. Rev. Letters* **25,** 1713 (1970).

2003. Exchange Coupling between Pairs of Fe(III) Ions in $FeF_3 \cdot 3H_2O$ at Low Temperatures, E. R. Jones, Jr., M. E. Hendricks, L. Cathey, S. L. Finklea, T. Auel, and E. L. Amma, *Solid State Comm.* **8,** 1657-59 (1970).

2004. Neutron Diffraction and Magnetic Structure of $CsMnCl_3$, M. Melamud, J. Makovsky, and H. Shaked (Nuclear Research Centre-Negev, P.O.B. 9001, Beer-Sheva, Israel) preprint received at RMIC Jan. 4, 1971.

2006. Magnetic Structure of $RbMnCl_3$, M. Melamud, J. Makovsky, H. Shaked, and S. Shtrikman, *Phys. Rev. B* **2,** 821 (1971).

2007. Neutron Diffraction and the Magnetic Structure of $TlMnCl_3$, M. Melamud, H. Pinto, G. Shachar, J. Makovsky, and H. Shaked, *Phys. Rev. B* **2,** 2344-48 (1971).

2008. Structure Magnetique du Metaborate de Fer $FeBO_3$, Michel Pernet, David Elmaleh, and Jean-Claude Joubert, *Solid State Commun.* **8,** 1583-87 (1970).

2009. First Order Phase Transition in EuSe, G. Petrich and T. Kasuya, *Solid State Commun.* **8,** 1625-29 (1970).

2010. Neutron Diffraction and Magnetic Measurements of Polycrystalline NH_4MnCl_3, G. Shachar, J. Makovsky, and H. Shaked (Dept. of Physics, Nuclear Research Centre-Negev, P.O.B. 9001, Beer-Sheva, Israel), preprint received at RMIC Jan. 4, 1971 (to be published in *Solid State Commun.*).

2011. Magnetic Structure of $RbNiCl_3$, A. Epstein, J. Makovsky, and H. Shaked, *Solid State Commun.* **9,** 249 (1971).

2012. Bromine Magnetic Resonance in Para- and Ferromagnetic $Cu(NH_4)_2Br_4 ,2H_2O$, J.-P. Renard and E. Velu, *Solid State Commun.* **8,** 1945-49 (1970).

2013. Magnetic Measurements in the Copper-Manganese System, Erich Scheil and Ernst Wachtel, ORNL-tr-2398, translated from *Z. Metallk.* **48,** 571-82 (1957) (Nov. 1970).

2014. Magnetic Phase Diagram of Chromium-Rich Chromium-Iron Alloys, S. Arajs, *J. Less-Common Metals* **22,** 519-22 (1970).

2015. Garnets Having A Single Magnetic Sublattice, K. P. Belov, B. V. Mill', G. Ronninger, V. I. Sokolov, and T. D. Hien, *Soviet Phys.-Solid State* **12,** 1393-96 (1970).

2016. Anomalies de Dilatation et Deplacement des Atomes à la Transition Magnetique dans $FeLiO_2$, Michel Brunel and Francois de Bergevin, *Compt. Rend. Ser. B.* **270,** 1050-53 (1970).

2017. Faraday Effect of EuSe and EuTe and Experimental Evidence of the Magnetic Brillouin Zone, G. Busch, J. Schoenes, and P. Wachter, *Solid State Commun.* **8,** 1841-44 (1970).

2018. Electrical Properties of Alpha-$TiCl_3$ Crystals, F. Cavallone, I. Pollini, and G. Spinolo, *Lettre al Nuovo Cimento Serie I,* **4,** 764-66 (Oct. 24, 1970).

2019. Low Temperature Magnetic Phase Transition in $ErCrO_3$, M. Eibschutz, L. Holmes, J. P. Maita, and L. G. van Uitert, *Solid State Commun.* **8,** 1815-17 (1970).

2020. The Magnetic Behavior of Fe-Al Alloys of B2-Structure Investigated by Mössbauer Spectroscopy, R. Hergt, E. Wieser, H. Gengnagel, and A. Gladun, *Phys. Stat. Sol.* **41,** 255 (1970).

2021. Heat Capacity of Gadolinium-Lutetium Alloys at Superlow Temperatures, A. V. Kogan, E. I. Nikulin, and Yu. B. Patrikeev, *Soviet Phys.-Solid State* **12,** 1565 (1971).

2022. Nqr of ^{35}Cl in Paramagnetic and Antiferromagnetic K_2IrCl_6, A. J. Lindop, *J. Phys. C., Solid State Phys.* **3**, 1984-95 (1970).

2023. Pulsed Field Transition in Hematite, P. Novotny and J. Kaczer, *Czech. J. Phys. B* **20**, 979-84 (1970).

2024. On Magnetic Properties of Gadolinium-Cobalt Compounds, O.A.W. Strydom and L. Alberts, *J. Less-Common Metals* **22**, 503-09 (1970).

2025. Features of the Electrical and Magnetic Properties of the Alloy Ni_3Mn in the Temperature Range 4.2-345°K, N. V. Volkenshtein, T. D. Zotov, S. F. Savchenkova, and Yu. N. Tsiovkin, *Soviet Phys.-Solid State* **12**, 1466-67 (1970).

2026. Etude de la Susceptibilité Magnétique de l'Oxyde de Manganese et du Sulfure de Manganese en Poudre dans l'Etat Ordonné, D. Bloch, R. Georges, and I. S. Jacobs, *J. Phys.* **31**, 589-95 (1970).

2027. Temperature Dependence of the Initial Permeability of a Ferromagnetic Amorphous Co-P Alloy, J.G.M. de Lau, *J. Appl. Phys.* **41**, 5355-56 (1970).

2028. Distant-Neighbor B-B Interactions in Cobalt Chromite, K. Dwight and N. Menyuk, *J. Appl. Phys.* **40**, 1156-57 (1969).

2029. Magnetic Structure of $NaCoF_3$, Z. Friedman, M. Melamud, J. Makovsky, and H. Shaked, *Phys. Rev. B* **2**, 179-81 (1970).

2030. Evidence of Magnon Conductivity and Critical Phonon Scattering in the Thermal Conductivity of $Cu(NH_4)_2Cl_4 \cdot 2H_2O$, J. N. Haasbroek and W. J. Huiskamp, *Phys. Letters* **33A**, 173-74 (1970).

2031. Exchange Strictions of Ferromagnetic Compound Mn_5Ge_3, Hideaki Ido, Takanobu Suzuki, and Takejiro Kaneko, *J. Phys. Soc. Japan.* **29**, 1490-94 (1970).

2032. Magnetic Properties of $(Fe_{1-x}Cr_x)_3Se_4$, Kiyoshi Kojima, Masaaki Matsui, Kiyoo Sato, and Kengo Adachi, *J. Phys. Soc. Japan.* **29**, 1643-44 (1970).

2033. An X-Ray Study of the Structures of UN, UP, US and USe at Cryogenic Temperatures, J.A.C. Marples, *J. Phys. Chem. Solids* **31**, 2431-39 (1970).

2034. Neutron Scattering in the Linear Chain Antiferromagnet $CsMnCl_3 \cdot 2H_2O$, J. Skalyo, Jr., G. Shirane, S. A. Friedberg, and H. Kobayashi, *Phys. Rev. B* **2**, 1310-17 (1970).

2035. Propriétés Metamagnetiques d'un Monocristal de $ErCrO_3$, C. Veyret, J. B. Ayasse, J. Chaussy, J. Mareschal, and J. Sivardiere, *J. Phys.* **31**, 607-611 (1970).

2036. Preparation et Propriétés Magnetiques de Deux Nouveaux Ferrites Oxyfluores a Structure Grenat: $Y_3Fe_5O_{12-x}F_x$ ($0 \leq x \leq 1$) et $Y_3Fe_{5-x}O_{12-x}F_x$ ($0 \leq x \leq 0.7$), B. Tanguy, J. Portier, A. Morell, R. Olazcuaga, M. Francillon, R. Pauthenet, and P. Hagenmuller, *Mat. Res. Bull.* **6**, 63-68 (1971).

2037. Crystal Structure and Magnetic Properties of Cerium-Iron Compounds, K.H.J. Buschow and J. S. van Wieringen, *Phys. Stat. Sol.* **42**, 231 (1970).

2038. On the Double-Exchange Interactions in $La_{1-x}Ba_xMnO_3$, F. K. Lotgering, *Philips Res. Repts.* **25**, 8-16 (1970).

2039. Structural and Magnetic Characteristics of the Pseudobinary System $PrMn_2-ErMn_2$, H. Oesterreicher, *J. Less-Common Metals* **23**, 7-19 (1971).

2040. Anomalous Behavior of the Electrical Resistivity in the Compound Ce_3Al_{11}, H. J. van Daal and K.H.J. Buschow, *Phys. Letters* **31A**, 103-04 (1970).

2041. Neutron Diffraction Study of TmC_2 at 300–1.7°K, Masao Atoji, *J. Chem. Phys.* **52**, 6431-33 (1970).

2042. Neutron Diffraction Study of YbC_2 at 300–2°K, Masao Atoji and R. H. Flowers, *J. Chem. Phys.* **52**, 6430-31 (1970).

2043. Etude par Diffraction de Neutrons de Ho_2C, G. L. Bacchella, P. Meriel, M. Pinot, and R. Lallement, *Bull. Soc. Franc. Miner. Crist.* **89**, 226-28 (1966).

2044. Magnetic Susceptibilities of Uranium (V) and Uranium (III) Chlorides, Paul Handler and Clyde A. Hutchison, Jr., *J. Chem. Phys.* **25**, 1210-13 (1956).

2045. Magnetic Resonance Study of Polycrystalline $MnAu_2$ at 100 GHz, L. W. Hart and J. L. Stanford, *J. Appl. Phys.* **41**, 2523-25 (1970).

2046. Magnetic Structures and Exchange Interactions in the Mn-Pt System, E. Kren, G. Kadar, L. Pal, J. Solyom, P. Szabo, and T. Tarnoczi, *KFKI* 2/1968 (1968).

2047. Magnetic Susceptibilities of bcc Cr-Rh, Cr-Ru, Cr-Re, and Cr-Os Alloys, J. G. Booth, *Phys. Stat. Sol.* **7**, K157-60 (1964).

REFERENCES

2048. Magnetic Properties of Cr Alloys Containing Dilute Concentrations of Co, Ni, and Fe, J. G. Booth, *J. Phys. Chem. Solids* **27**, 1639-45 (1966).

2049. Crystal Chemistry and Magnetism of Oxide Materials. II. Magnetic Compounds with Spinel Structure, P. F. Bongers, *Philips Tech. Rev.* **28**, 23-30 (1967).

2050. Critical Magnetic Behavior of Nickel near the Curie Point, S. Arajs, B. L. Tehan, E. E. Anderson, and A. A. Stelmach, *Phys. Stat. Sol.* **41**, 639-48 (1970).

2051. Magnetic Properties of Gadolinium Ortho-Aluminate, J. D. Cashion, A. H. Cooke, T. L. Thorp, and M. R. Wells, *Proc. Roy. Soc. London* **318A**, 473-495 (1970).

2052. Spin Fluctuation Effects in Nearly Antiferromagnetic Vanadium and Chromium Diboride Alloys, Jacques Castaing, Paul Costa, Michel Heritier, and Pascal Lederer (Office National d'Etudes et de Recherches Aerospatiales, 92 Chatillon, France), T.P. No. 859 (1970) (to be published in *J. Phys. Chem. Solids*).

2053. Magnetic and Crystallographic Properties of Pr-Gd Alloys, L. Tissot and A. Blaise, in CONF-691103-3; CEA-CONF-1476 (*Fifteenth Annual Conf. on Magnetism and Magnetic Materials*. Philadelphia, Nov. 18-21, 1969).

2054. On the Compound Eu_pNbS_2, F. Jellinek, *Mat. Res. Bull.* **6**, 169-172 (1971).

2055. Magnetic Ordering in PrB_6, K. N. Lee, R. Bachmann, T. H. Geballe, and J. P. Maita, *Phys. Rev. B* **2**, 4580-85 (1970).

2056. High-Pressure $RbFeCl_3$ — A Transparent Ferrimagnet, J. M. Longo, J. A. Kafalas, N. Menyuk, K. Dwight, D. A. Batson, and D. M. Tracy (Lincoln Laboratory, M.I.T., Mass.), pp. 20-23 in *Solid State Research, Quarterly Tech. Summary for Aug. 1 through Oct. 31, 1970*, ESD-TR-70-350 (Nov. 15, 1970).

2057. Magnetic Properties of $(V_{1-x}Cr_x)_2O_3$, A. Menth and J. P. Remeika, *Phys. Rev. B* **2**, 3756-62 (1970).

2058. Magnetic Susceptibility of $Al_2O_3:Ti^{3+}$, Alfred R. Smith, D. J. Arnold, and R. W. Mires, *Phys. Rev. B* **2**, 2323-34 (1970).

2059. Phonons and Magnetic Order in Ferromagnetic $CdCr_2Se_4$ and $CdCr_2S_4$, E. F. Steigmeier and Gunther Harbeke, *Phys. Kondens. Materie* **12**, 1-15 (1970).

2060. Neutron Diffraction Study of $Ni(OH)_2$, A. Szytula, A. Murasik, and M. Balanda, *Phys. Stat. Sol. B* **43**, 125 (1971).

2061. Crystal-Field Effects and Anomalous Susceptibility of Antiferromagnets: Application to Ce-Group-V Compounds, Yung-Li Wang and Bernard R. Cooper, *Phys. Rev. B* **2**, 2607-17 (1970).

2062. Measurement of the Magnetic Susceptibility of $Cu(NH_4)_2Br_4 \cdot 2H_2O$ near its Curie Temperature. Comparison with Heisenberg Model Predictions, L. J. de Jongh, A. R. Miedema and R. F. Wielinga, *Physica* **46**, 44-58 (1970).

2063. Effect of Alloying on the Thermodynamic Characteristics of the Antiferromagnetic Phase Transformation in MnTe, E. D. Devyatkova and V. V. Tikhonov, *Soviet Phys.-Solid State* **12**, 2168-70 (1971).

2064. Critical Behavior of the Resistivity of a Dilute Pd-Mn Alloy, G. J. Nieuwenhuys and B. M. Boerstoel, *Phys. Letters* **33A**, 281-82 (1970).

2065. Magnetic Properties of Dysprosium Thallium Three, Clayton E. Olson, G. P. Arnold, and N. G. Nereson (Los Alamos Scientific Lab., Los Alamos, New Mex.), pp. 63-73 in *Proc. 8th Rare Earth Research Conf.*, April 19-22, 1970, Reno, Nevada, T. A. Henrie and R. E. Lindstrom, eds., Vol. 1, CONF-700403.

2066. Magnetic Properties of Mn_2Ge_3 Thin Films, H. Ratajczak and Z. Sczaniecki, *Phys. Stat. Sol. A* **1**, 171-76 (1970).

2067. Heat Capacity and Magnetic Phase Transitions of fcc Manganese Hexammine Halides, H. Suga, E. Lagendik, and W. J. Huiskamp, *Phys. Letters* **32A**, 297-98 (1970).

2068. Ordre Magnetique dans Li_2CoGeO_4, Na_2CoGeO_4, et Na_2CoSiO_4, Micheline Wintenberger, *Compt. Rend.* **271B**, 669-72 (1970).

2069. Magnetic Susceptibility of Single-Crystal Hexagonal Nickel Sulphide, J. L. Horwood and M. G. Townsend, *Solid State Commun.* **9**, 41-43 (1971).

2070. Acoustic Study of the Critical Phenomena in FeF_2 near the Néel Temperature, A. Ikushima and R. Feigelson, *J. Phys. Chem. Solids* **32**, 417-25 (1971).

2071. Magnetic Properties of the Mn_3PtN_x System, E. Kren, E. Zsoldos, M. Barberon, and R. Fruchart, *Solid State Commun.* **9**, 27-31 (1971).

2072. A First Order Phase Transition in Europium Metal, R. L. Cohen, S. Hufner, and K. W. West, *Phys. Letters* **28A**, 582-83 (1969).

2073. Propriétés Thermomagnetiques de l'Uranium Alpha Polycristallin, Jean-Marc Fournier, *Compt. Rend.* **271B**, 1156-58 (1970).

2074. Magnetic Susceptibilities of Several Rare Earth Selenides and Tellurides, D. E. Hegland and D. R. Behrendt (National Aeronautics and Space Administration, Lewis Research Center, Cleveland, Ohio), NASA-TM-X-52921 (Nov. 1970), 17 pp.

2075. Mössbauer-Effect Studies of Iron-Tin Alloys, G. Trumpy, E. Both, C. Djega-Mariadassou, and P. Lecocq, *Phys. Rev. B* **2**, 3477-90 (1970).

2076. Influence of a Magnetic Transition on the Faraday Effect in EuO near the Absorption Band Edge, Yu. V. Knyazev and M. M. Noskov, *Soviet Phys.-Solid State* **12**, 2235-36 (1971).

2077. Influence de la Distribution des Cations sur la Temperature de Curie dans les Ferrites Cuivre-Zinc, Determinée par des Mesures d'Aimantation à Saturation, Nicolae Rezlescu, *Compt. Rend.* **270B**, 1134-37 (1970).

2078. Inelastic Neutron Scattering Investigation of Spin Waves and Magnetic Interactions in Alpha Fe_2O_3, E. J. Samuelsen and G. Shirane, *Phys. Stat. Sol.* **42**, 241-56 (1970).

2079. Preparation and Structure of Fe-Doped EuO Films, K. Y. Ahn, K. N. Tu, and W. Reuter, *J. Appl. Phys.* **42**, 1769-78 (1971).

2080. The Magnetic Structure of Cr_2Te_3, Cr_3Te_4, and Cr_5Te_6, Arne F. Andresen, *Acta Chem. Scand.* **24**, 3495-3509 (1970).

2081. Antiferromagnetism and Electrical Transport Properties of Chromium-Aluminum Alloys, Sigurds Arajs, N. L. Reeves, and E. E. Anderson, *J. Appl. Phys.* **42**, 1691-92 (1971).

2082. Installation d'un Appareillage de Mesures d'Aimantation. Propriétés Magnetiques des Composes T_3Al_2 ou T Represente un Element Terre Rare, B. Barbara, These Doct. 3e. Cycle Magnet. Phys. Solide, Grenoble, 1968.

2083. Anomalous Exchangestriction in Ferromagnetic Pyrite and Chromium Chalcogenide Spinel Compounds, W. Bindloss, *J. Appl. Phys.* **42**, 1474-75 (1971).

2084. Mössbauer Measurements of $BiFeO_3$ and $BiFeO_3$-$PbZrO_3$ Systems, A. Biran, P. A. Montano, and U. Shimony, *J. Phys. Chem. Solids* **32**, 327 (1971).

2085. Magnetic Structure of Mn_3O_4 by Neutron Diffraction, B. Boucher, R. Buhl, and M. Perrin, *J. Appl. Phys.* **42**, 1615-17 (1971).

2086. Magnetization of Compounds of Rare Earths with Platinum Metals, R. M. Bozorth, B. T. Matthias, H. Suhl, E. Corenzwit, and D. D. Davis, *Phys. Rev.* **115**, 1595-96 (1959).

2087. An Investigation of the Antiferromagnetic Transition of CrN, J. D. Browne, P. R. Liddell, R. Street, and T. Mills, *Phys. Stat. Sol. A* **1**, 715 (1970).

2088. Magnetism, Metal-Insulator Transition, and Optical Properties in Sm- and Some Other Divalent Rare-Earth Monochalcogenides, E. Bucher, V. Narayanamurti, and A. Jayaraman, *J. Appl. Phys.* **42**, 1741 (1971).

2089. Crystal-Field Splitting and Spin Waves in Antiferromagnetic CeSb, G. Busch, W. Stutius, and O. Vogt, *J. Appl. Phys.* **42**, 1493-94 (1971).

2090. Magnetic Susceptibility of Au-Fe Alloys, V. Cannella, J. A. Mydosh, and J. I. Budnick, *J. Appl. Phys.* **42**, 1689-90 (1971).

2091. Chaleur Specifique à Basse Temperature et Susceptibilité Magnetique des Borures Mixtes de Molybdène et de Chrome, J. Castaing, R. Caplain, and P. Costa, *Solid State Commun.* **9**, 297-300 (1971).

2092. A Study on the $NiVO_3$ System, B. L. Chamberland, *J. Solid State Chem.* **2**, 521-24 (1970).

2093. Alkaline-Earth Vanadium (IV) Oxides Having the AVO_3 Composition, B. L. Chamberland and P. S. Danielson (Dept. of Chemistry and Institute of Materials Science, Univ. of Conn., Storrs, Conn.), preprint received RMIC April 21, 1971.

2094. Specific Heat of Nickel Near the Curie Temperature, D. L. Connelly, J. S. Loomis, and D. E. Mapother, *Phys. Rev. B* **3**, 924-34 (1971).

2095. Structure Electronique des Carbonitrures d'Uranium $UC_{1-x}N_x$, Charles Henri de Novion and Paul Costa, *Compt. Rend., Ser. B*, **270**, 1415-18 (1970).

2096. Magnetic and Electrical Properties of Ternary Pyrite-Type $Cr_xCo_{1-x}S_2$ Phases Prepared at High Pressure, P. C. Donohue, T. A. Bither, W. H. Cloud, and C. G. Frederick, *Mat. Res. Bull.* **6**, 231-38 (1971).

2097. Investigation of the Magnetic Behavior of Some Conducting Am Systems, B. D. Dunlap, D. J. Lam, G. M. Kalvius, and G. K. Shenoy, *J. Appl. Phys.* **42**, 1719-20 (1971).

REFERENCES

2098. Magnetic Study of a Reversible Phase Transition in $CoZrF_6 \cdot 6H_2O$, $Co_{1-x}Zn_xSiF_6 \cdot 6H_2O$ and $Co_{1-x}Zn_xF_2 5HF \cdot 6H_2O$, S. K. Dutta Roy, B. Ghosh, and Supratik Kar, *J. Phys. Chem. Solids* **32**, 857 (1971).

2099. Absence of Antiferromagnetism in $NbSe_2$ and $TaSe_2$, E. Ehrenfreund, A. C. Gossard, F. R. Gamble, and T. H. Geballe, *J. Appl. Phys.* **42**, 1491-93 (1971).

2100. Magnetic Behavior of a Layer-Type Antiferromagnet $RbFeF_4$, M. Eibschutz, H. J. Guggenheim, and L. Holmes, *J. Appl. Phys.* **42**, 1485 (1971).

2101. Spin Flopping in $LiCuCl_3 \cdot 2H_2O$, H. Forstat, P. T. Bailey, and J. R. Ricks, *J. Appl. Phys.* **42**, 1559-60 (1971).

2102. Are There Magnetic Moments at the Transition Metal Sites in FeAl, CoAl and NiAl?, R. B. Frankel, D. J. Sellmeyer, and N. A. Blum, *Phys. Letters A* **33**, 13-14 (1970).

2103. Garnet Compositions for Bubble Domain Systems Utilizing Stress-Induced Uniaxial Anisotropy, E. A. Giess, B. A. Calhoun, E. Klokholm, T. R. McGuire, and L. L. Rosier, *Mat. Res. Bull.* **6**, 317-28 (1971).

2104. Propriétés Cristallographiques et Magnétiques des Composés Entre le Cobalt et le Lutecium, Francoise Givord and Remy Lemaire, *Solid State Commun.* **9**, 341-46 (1971).

2105. Crystallographic and Magnetic Properties of the Intermetallic Compounds of Cobalt with Lutetium, Francoise Givord and Remy Lemaire, *Cobalt* **50**, 16 (1971).

2106. Thermal Expansivity of $RbMnF_3$ Near T_N, B. Golding, *J. Appl. Phys.* **42**, 1381 (1971).

2107. Problem of the Magnetic Structure of Cu_2Sb, V. S. Golovkin, V. N. Bykov, and V. A. Levdik, *Soviet Phys. Cryst.* **15**, 935 (1971).

2108. Magnetic Structure of FeOCl, R. W. Grant, *J. Appl. Phys.* **42**, 1619 (1971).

2109. Dependence of the E_1 Reflectivity Structure in EuO on Temperature and Doping, Paul M. Grant, *J. Appl. Phys.* **42**, 1771 (1971).

2110. The Magnetic Properties of $EuLiH_3$, J. E. Greedan, *J. Phys. Chem. Solids* **32**, 819 (1971).

2111. Magnetic Anisotropy of V_2O_3 Single Crystals, Morris Greenwood, Raymond W. Mires, and Alfred R. Smith, *J. Chem. Phys.* **54**, 1417 (1971).

2112. Ferromagnetism in Heusler-Type Alloys Based on Platinum-Group or Palladium-Group Metals, F. A. Hames and J. Crangle, *J. Appl. Phys.* **42**, 1336-38 (1971).

2113. Magnetic Properties of Au-Mn Alloys with 10-20 at.% Mn, L. E. Hedman, *J. Phys. Chem. Solids* **32**, 825 (1971).

2114. Crystal Structure and Magnetic Properties of the Two-Dimensional Antiferromagnet $KFeF_4$, G. Heger, R. Geller, and D. Babel, *Solid State Commun.* **9**, 335 (1971).

2115. Low-Temperature Phase Transition in $PrCl_3$, Jan P. Hessler and E. H. Carlson, *J. Appl Phys.* **42**, 1316 (1971).

2116. Compounds with the Marcasite Type Crystal Structure. VI. Neutron Diffraction Studies of $CrSb_2$ and $FeSb_2$, Hans Holseth, Arne Kjekshus, and Arne F. Andresen, *Acta Chem. Scand.* **24**, 3309-16 (1970).

2117. Electrical and Magnetic Properties of Single-Crystal Hexagonal Nickel Sulfide, J. L. Horwood, L. G. Ripley, M. G. Townsend, and R. J. Tremblay, *J. Appl. Phys.* **42**, 1476-77 (1971).

2118. Mössbauer Study and Molecular Field Theory of the Magnetic Properties of Fe-Al Alloys, G. P. Huffman, *J. Appl. Phys.* **42**, 1606-07 (1971).

2119. Antiferromagnetic Resonance in MnO: Co: A Measurement of the Magnetoelastic Properties of the Co^{2+} Ion, A. E. Hughes, *Phys. Rev. B* **3**, 877-94 (1971).

2120. Clustering and Giant Moments in Cu-Ni Alloys, J. Jach, R. J. Borg, and D. Y. F. Lai, *J. Appl Phys.* **42**, 1611-12 (1971).

2121. Synthesis and Characterization of Single Crystal $(RE, Pb)MnO_3$ Perovskites, D. L. Janes and R. E. Bodnar, *J. Appl. Phys.* **42**, 1500-01 (1971).

2122. Effect of Pressure on the Magnetic Properties of $Ca_{1-x}Sr_xMnO_3$, J. A. Kafalas, N. Menyuk, K. Dwight, and J. M. Longo, *J. Appl. Phys.* **42**, 1497-98 (1971).

2123. Propriétés Magnétiques des Solutions Solides $TbFe_{1-x}Co_xO_3$ à Structure Perovskite Déformée (x = 0.1 et 0.2), A. Kappatsch, P. Burlet, and J. Sivardiere, *Solid State Commun.* **9**, 435-39 (1971).

2124. Ferromagnetism of $ZrZn_2$, G. S. Knapp, F. Y. Fradin, and H. V. Culbert, *J. Appl. Phys.* **42**, 1341-46 (1971).

2125. Etude des Monocarbures d'Europium et d'Ytterbium–Structure Cristalline et Valence du Metal, Alain Laplace and Robert Lorenzelli, pp. 385-95 in Les Elements des Terres Rares, Paris-Grenoble, 5-10 Mai 1969, Tome I, Colloques Internationaux du Centre National de la Recherche Scientifique, No. 180 (Editions du Centre National de la Recherche Scientifique, Paris, 1970).

2126. Crystal Growth and Magnetic Behavior of $FeVO_4$, Lionel M. Levinson and B. M. Wanklyn, *J. Solid State Chem.* **3**, 131-33 (1971).

2127. High-Pressure $RbFeCl_3$–A Transparent Ferrimagnet, J. M. Longo, J. A. Kafalas, N. Menyuk, and K. Dwight, *J. Appl. Phys.* **42**, 1561-62 (1971).

2128. On the 16K Anomaly in Europium Metal, O. V. Lounasmaa and G. M. Kalvius, *Phys. Letters A* **26**, 21-22 (1967).

2129. Electrical Resistivity and Magnetization of Gd-Th and Gd-Mg Alloys, D. W. Mellon and S. Legvold, *J. Appl. Phys.* **42**, 1295-97 (1971).

2130. Magnetic Properties of $CoAs_xS_{2-x}$, J. C. Mikkelsen and A. Wold, *J. Solid State Chem.* **3**, 39 (1971).

2131. Magnetic Properties of CeBi, NdBi, TbBi, and DyBi, N. Nereson and G. Arnold, *J. Appl. Phys.* **42**, 1625-27 (1971).

2132. Neutron Diffraction Study of the $ZnFe_{2-x}Cr_xO_4$ Series, A. Oles, *Phys. Stat. Sol. A* **3**, 569 (1970).

2133. Ordre Magnetique a Deux Dimensions dans Ca_2MnO_4, G. Ollivier and G. Buisson, *Solid State Commun.* **9**, 235 (1971).

2134. Sur Quelques Nouvelles Phases de Formule Na_xMnO_2 ($x < 1$), Jean-Paul Parant, Roger Olazcuaga, Michel Devalette, Claude Fouassier, and Paul Hagenmuller, *J. Solid State Chem.* **3**, 1-11 (1971).

2135. Spin Structure and Magnetic Anisotropy of Cr_5S_6 and Rhombohedral Cr_2S_3, T.J.A. Popma, C. Haas, and B. van Laar, *J. Phys. Chem. Solids* **32**, 581 (1971).

2136. Semiconductor-Metal Transition in Ti_3O_5, C.N.R. Rao, S. Ramdas, R. E. Loehman, and J. M. Honig, *J. Solid State Chem.* **3**, 83 (1971).

2137. Cation Distribution and Curie Temperature in Some Ferrites Containing Copper and Manganese, N. Rezlescu and E. Cuciureanu, *Phys. Stat. Sol. A* **3**, 873 (1970).

2138. Spin Waves in Ferromagnetic $CrBr_3$ Studied by Inelastic Neutron Scattering, E. J. Samuelsen, R. Silberglitt, G. Shirane, and J. P. Remeika, *Phys. Rev. B* **3**, 157-66 (1971).

2139. Magneto-Optical Properties of Sputtered Mn_5Ge_3 Films, E. Sawatzky, *J. Appl. Phys.* **42**, 1706-07 (1971).

2140. Magnetization and Lattice Parameters of Ti Substituted Fe-Cr Spinels, E. Schmidbauer, *J. Phys. Chem. Solids* **32**, 71-76 (1971).

2141. Measurement of Dynamical Scaling in FeF_2, M. P. Schulhof, M. T. Hutchings, and H. J. Guggenheim, *J. Appl. Phys.* **42**, 1376-77 (1971).

2142. Crystal Preparation, Structures, and Magnetic Properties of Layer-Type $Cs_4B_3F_{10}$ Compounds (B = Fe, Co, Ni), M. W. Shafer, T. R. McGuire, and J. E. Weidenborner, *J. Appl. Phys.* **42**, 1489-90 (1971).

2143. Paramagnetic-to-Antiferromagnetic Phase Boundaries of FeF_2 from Ultrasonic Measurements, Y. Shapira, *Phys. Rev. B* **2**, 2725-34 (1970).

2144. Raman Scattering in $FeBO_3$, I. W. Shepherd, *J. Appl. Phys.* **42**, 1482-83 (1971).

2145. MnAlGe Films for Magneto-Optic Applications, R. C. Sherwood, E. A. Nesbitt, J. H. Wernick, D. D. Bacon, A. J. Kurtzig, and R. Wolfe, *J. Appl. Phys.* **42**, 1704-05 (1971).

2146. Magnetic Properties of the System Au_4R(R = Er, Tm, and Yb), L. R. Sill, W. J. Mass, A. J. Fedro, J. C. Shaffer, and C. W. Kimball, *J. Appl. Phys.* **42**, 1297-98 (1971).

2147. Mössbauer Investigations of Magnetic Phases of Anhydrous $FeCl_3$ in High Fields, J. P. Stampfel, W. T. Oosterhuis, and F. de S. Barros, *J. Appl. Phys.* **42**, 1721-22 (1971).

2148. High Pressure Synthesis of Ilmenite and Perovskite Type $MnVO_3$ and Their Magnetic Properties, Yasuhiko Syono, Syun-iti Akimoto, and Endoh, *J. Phys. Chem. Solids* **32**, 243-49 (1971).

2149. Magnetic Properties of Cobalt-Nickel Ferrite, H. Szydlowski, *Phys. Stat. Sol. A* **2**, K37-39 (1970).

2150. Magnetic Properties of a Unique 18-Layer Hexagonal Ferrite, A. Tauber, R. O. Savage, Jr., and M. D. Grebenau, *J. Appl. Phys.* **42**, 1738-40 (1971).

2151. Low-Temperature Magnetic Susceptibilities of $NbSe_2$ Containing the First-Row Transition Metals, J. M. Voorhoeve-van den Berg and R. C. Sherwood, *J. Phys. Chem. Solids* **32**, 167-73 (1971).

2152. Antiferromagnetic Solid Oxygen, E. J. Wachtel and R. G. Wheeler, *J. Appl. Phys.* **42**, 1581-87 (1971).

2153. Antiferromagnetic-Ferromagnetic Transition in the Alloy System Au_4-$(Mn_{1-x}Cr_x)$, M. Yessik, J. Noakes, and H. Sato, *J. Appl. Phys.* **42**, 1349-50 (1971).

2154. Ferromagnetism and Superconductivity in Uranium Compounds, B. T. Matthias, C. W. Chu, E. Corenzwit, and D. Wohlleben, *Proc. Natl. Acad. Sci.* **64**, 459-61 (1969).

2155. An Anomaly of the Magnetic Susceptibility of Dysprosium Near the Néel Point, L. A. Boyarskii and M. A. Starikov, *Soviet Phys.-Solid State* **12**, 2569-72 (1971).

2156. A Neutron Diffraction Study of $RbFeF_3$, F.F.Y. Wang (Dept. of Materials Science, State Univ. of New York, Stony Brook, N. Y.), D. E. Cox (Phys. Dept., Brookhaven National Laboratory, Upton, N. Y.), and M. Kestigian (Sperry Rand Research Center, Sudbury, Mass.), preprint received RMIC June 1971.

2157. Structure aux Rayons X, Neutrons et Propriétés Magnétiques des Orthovanadates de Nickel et de Cobalt, H. Fuess, E. F. Bertaut, R. Pauthenet, and A. Durif, *Acta Cryst. B* **26**, 2036 (1970).

2158. Etude Cristallographique et Magnétique de $MnGeO_3$, P. Herpin, A. Whuler, B. Boucher, and M. Sougi, *Phys. Stat. Sol. B* **44**, 71 (1971).

2159. Investigation of Transport Phenomena in Neodymium Sulfide in the Magnetic Ordering Region, V. I. Novikov and S. S. Shalyt, *Soviet Phys.-Solid State* **12**, 2628 (1971).

2160. Ferrimagnetism of Calcium-Vanadium Garnets Which Do Not Contain Yttrium, A. T. Starovoitov and V. I. Ozhogin, *Soviet Phys.-Solid State* **12**, 2573 (1971).

2161. Mössbauer Spectra and Magnetic Susceptibility of Neptunium Tetra-Chloride Below 85°K, J. A. Stone and E. R. Jones, Jr., *J. Chem. Phys.* **54**, 1713-18 (1971).

2162. The Specific Heat of Lanthanum Nickel Double Nitrate, George R. Unrine, Jr., thesis, University of Cincinnati, 1970, 91 p.

2163. Rare-Earth Intermediate Phases. VI. Pseudo-Binary Systems Between Cubic Laves Phases Formed by Rare-Earth Metals with Iron, Cobalt, Nickel, Aluminium and Rhodium, R. C. Mansey, G. V. Raynor, and I. R. Harris, *J. Less-Common Metals* **14**, 337-47 (1968).

2164. Thermal and Electronic Behavior of the Rare Earth Hexaborides from Cryogenic Calorimetry, Edgar F. Westrum, Jr., pp. 443-50 in Les Elements des Terres Rares, Paris-Grenoble, 5-10 Mai 1969, Tome II, Colloques Internationaux du Centre National de la Recherche Scientifique, No. 180 (Editions du Centre National de la Recherche Scientifique, Paris, 1970).

2165. Magnetic Susceptibilities of Cl_b Compounds CoVSb and CoTiSb, Michiko Terada, K. Endo, Y. Fujita, T. Ohoyama, and Ren'iti Kimura, *J. Phys. Soc. Japan.* **29**, 1091 (1970).

2166. Michiko Terada, Faculty of Engineering, Keio University, Kogamei, Tokyo, Japan, private communication.

2167. The Specific Heat of Dysprosium Vanadate, A. H. Cooke, D. M. Martin, and M. R. Wells, *Solid State Commun.* **9**, 519 (1971).

2168. A. H. Cooke, D. M. Martin, and M. R. Wells, Clarendon Laboratory, Parks Road, Oxford, England, private communication.

2169. Contribution a l' Etude de CrN a Bases Temperature par Diffraction de Rayons X et de Neutrons, Mahmoud Nasr Eddine, Francoise Sayetat, et Erwin Felix Bertaut, *Compt. Rend.* **269**, 574-577 (1969).

2170. Effect of Pressure on the Curie Point of Cr_7Te_8 with the NiAs-Type Structure, K. Ozawa, T. Yoshimi, and S. Yanagisawa, *Phys. Stat. Sol.* **681** (1971).

2171. The First Order Magnetic Transition of Equiatomic FeRh, F. J. Bartis, *Scripta Met.* **4**, 783-788 (1970).

2172. Das kritische magnetische Verhalten von EuO, Gunter Groll, *Z. Physik.* **243**, 60-78 (1971).

2173. Magnetic Structure of Iron Monophosphide, G. P. Felcher and F. A. Smith, *Phys. Rev. B* **3**, 3046-52 (1971).

2174. Preparation and Properties of Fe_2P, David Bellavance, James Mikkelsen, and Aaron Wold, *J. Solid State Chem.* **2**, 285-290 (1970).

2175. Magnetic Properties of $DyAsO_4$ at Low Temperatures, H. G. Kahle, L. Klein, G. Muller-Vogt, and H. C. Schopper, *Phys. Stat. Sol. B* **44**, 619 (1971).

2176. Magnetic Phase Diagram of Hexagonal FeGe, K. Carrander, H. Schwartz and O. Becjman, *Physica Scripta* **2**, 313-315 (1970).

2177. Magnetic Structure of Manganese Pyrophosphate, $Mn_2P_2O_7$, M. F. Collins, G. S. Gill, and C. V. Stager, *Can. J. Phys.* **49**, 979-982 (1971).

2178. The Hall Effect in Oriented α-Pu, T. R. Loree and H. T. Pinnick, *J. Nucl. Mater.* **38**, 143-149 (1971).

2179. A note on the Phase Transition in Cerium Magnesium Nitrate, G. O. Zimmerman, D. J. Abeshouse, E. Maxwill, and D. R. Kelland, *Physica* **51**, 623-26 (1971).

2180. The Magnetic Transition, Heat Capacity, and Thermodynamic Properties of Uranium Dioxide from 5 to 350°K, James J. Huntsicker and Edgar F. Westrum, Jr., *J. Chem. Thermodynamics* **3**, 61-76 (1971).

2181. The Low Temperature Magnetic Thermal Expansion Coefficients of $NiCl_2 \cdot 6H_2O$, P. C. Lanchester, *J. Phys. Chem. Sol.* **3**, 2411 (1970).

2182. The Resistivity Anomaly in Chromium Near the Néel Temperature, Bengt Stebler, *Physica Scripta* **2**, 53-56 (1970).

2183. The Orthorhombic Distortions and Magnetic Properties of Manganese Arsenide, N. N. Sirota, E. A. Vasilev, and G. A. Gover, *Phys. Stat. Sol. A* **5**, K43 (1971).

2184. Magnetothermodynamics of $CuK_2(SO_4)_2 \cdot 6H_2O$. III. Fields Along the γ Axis, Thermodynamic Temperature without Heat Introduction below 0.4°K, W. F. Giauque, R. A. Fisher, G. E. Brodale, and E. W. Hornung, *J. Chem. Phys.* **54**, 4908 (1971).

2185. Neutron Diffraction and Magnetic Measurements of Polycrystalline NH_4MnCl_3, G. Shachar, J. Makovosky, and H. Shaked, *Solid State Commun.* **9**, 493-495 (1971).

2186. Anomalous Thermal Expansion of Chromium near TN, Gengt Stebler, Claes-Goran Andersson, and Olof Kristensson, *Physica Scripta* **1**, 281-285 (1970).

2187. Magnetic Interactions in Compounds of Graphite with Molybdenum Pentachloride, A. V. Zvarykina, Y. S. Karimov, M. E. Vol'pin, and Y. N. Novikov, *Soviet Phys.-Solid State* **13**, 21 (1971).

2188. Observation of Magnetically Controllable Distortion in $TbVO_4$ by Optical Spectroscopy, K. A. Gehring, A. P. Malozemoff, W. Staude, and Y. N. Tyte, *Solid State Commun.* **9**, 511-515 (1971).

2189. Thermal Expansion Coefficient of Fe-Ni (fcc) Alloys, Yasunori Tanji and Yuki Shirakawa, *Sci. Rept. Res. Inst. Tohoku Univ. Ser. A* **22**(4), 135-45 (1971).

2190. Optical Properties of a-$TiCl_3$ at the Magnetic Transition Temperature, F. Cavallone, I. Pollini, and G. Spinolo, *Phys. Stat. Sol. B* **45**, 405 (1971).

2191. Study of the Spin Reordering Transition in Cr_5S_6, K. Dwight, N. Menyuk, and J. A. Kafalas, *Phys. Rev. B* **2**, 3630-8 (1970).

2192. Heat Capacity Measurements on Maganese Dibromide Tetrahydrate Near its Néel Temperature, R. D. Hempstead, Ph.D. thesis, University of Illinois, Urbana, Illinois (1970).

2193. Antiferromagnetism in $CsCoCl_3 \cdot 2H_2O$, W. J. M. deJonge, K. V. S. Rama Rao, C. H. W. Swuste, and A. C. Botterman, *Physica* **51**, 620-22 (1971).

2194. Magnetic Properties of Ho_6Re_{23} and $HoFe_2$, G. J. Roe, T. J. O'Keefe, J. M. Moreau, C. Michel, and W. J. James, in *Les Elements des Terres Rares*, Vol. II, pp. 251-9, Centre National de la Recherche Scientifique, Paris (1970).

2195. Magnetic Structure of ErSi, N. V. Nhung, J. Sivardiere, and A. Apostolov, in *Les Elements des Terres Rares*, Vol. II, pp. 261-274, Centre National de la Recherche Scientifique, Paris (1970).

2196. Structure Cristalline de Ho et Er de 77 at 300°, V. A. Finkel and M. I. Palatnik, *Zh. Eksperim. i Teor. Fiz., S.S.S.R.* **59**(5), 15-18-23 (1970).

2197. Contribution à l'Etude des Transitions Magnetiques de Premier Ordre des Nitrures Metalliques du Manganese de Type Perowskite, R. Madar, Ph.D. thesis, Univ. of Paris, CNRS No. 4768 (1970), 95 pp.

2198. Magnetic and Elastic Properties of Zinc-Chromite, Y. Kino and B. Luthi, *Solid State Commun.* **9**, 805-8 (1971).

2199. A Comment on A. Bachellerie's Paper "Critical Attenuation of 1 GHz Ultrasonic Waves in MnF_2," Y. Shapira, *Solid State Commun.* **9**, 809-11 (1971).

2200. Study of Pure and Doped Cobaltous and Nickelous Oxide, William R. Helms and J. G. Mullen, *Phys. Rev. B* **4**, 750 (1971).

2201. Mössbauer Effect Study of Ferromagnetism and Ferromagnetic Relaxation in $ErAl_2$, W. Wiedemann and W. Zinn, *Phys. Letters A* **24**, 506-8 (1967).

2202. Structural and Magnetic Characteristics of Th-Co and Th-Fe Compounds, K. H. J. Buschow, *J. Appl. Phys.* **42**, 3433-37 (1971).

2203. Specific Heat of Chromium Hydride CrH_x from 11 to 300°K, G. Wolf, *Phys. Stat. Sol. A* **5**, 627 (1971).

2204. Slope Change in the Temperature Dependence of the Electrical Resistivity of Thulium near 22°K, S. Janos and R. Mlynek, *Phys. Stat. Sol. A* **5**, K211 (1971).

2205. Structure and High Temperature Magnetic Studies of the Systems $PrMn_2$-$GdMn_2$ and $PrMn_2$-$TbMn_2$, C. Nair and H. Oestereicher, *J. Less-Common Metals* **24**, 237-242 (1971).

REFERENCES

2206. Lattice Constants and Curie Temperatures of $CaCu_5$-Type Thorium-Cobalt Compounds, A. S. van der Goot and K. H. J. Buschow, *Phys. Stat. Sol. A* **5**, 665 (1971).

2207. Magnetic Characteristics of the Aluminum-Rich Ternary Laves Phases Containing Rare Earths, Nickel, and Aluminum, B. Leon, V. U. S. Rao, and W. E. Wallace, *J. Less-Common Metals* **24**, 247-51 (1971).

2208. Alkaline-Earth Vanadium. IV. Oxides Having the AVO_3 Composition, B. L. Chamberland and P. S. Danielson, *J. Solid State Chem.* **3**, 243-247 (1971).

2209. Magnetic and Caloric Properties of $TiFe_{0.5}Co_{0.5}$, Yuji Asada and Hiroshi Nose, *J. Phys. Soc.* **297** (1971).

2210. Structural and Magnetic Studies of the Pseudo-Binary Compounds $(Gd_{1-x}Nd_x)Co_2$, G. J. Primavesi, K. N. R. Taylor, and I. R. Harris, *J. Phys.* (France) 32 Suppl., 661-2 (1971). (From 7th Internatl. Conf. on Magnetism, Grenoble, Sept. 14-19, 1970.)

2211. Propriétés Ferrimagnetiques des Pervoskites de Type $Sr_3U_{1-x}W_xFeO_9$, J. C. Bernier, F. Seveque, P. Delamoye, and P. Poix, *Solid State Commun.* **9**, 1225-30 (1971).

2212. Magnetic and Chemical Order in Heusler Alloys Containing Cobalt and Maganese, P. J. Webster, *J. Phys. Chem. Solids* **32**, 1221-31 (1971).

2213. A Model for the Invar Alloys and the Fe-Ni System, W. F. Schlosser, *J. Phys. Chem. Solids* **32**, 939-49 (1971).

2214. Magnetic Properties of Flouride Substituted Orthoferrites, M. Robbins, R. D. Pierce, and R. Wolfe, *J. Phys. Chem. Solids* **32**, 1789-96 (1971).

2215. Crystal Field and Magnetic Heat Capacity in $PrIn_3$ and $CeIn_3$, A. M. van Dieper, R. S. Craig, and W. E. Wallace, *J. Phys. Chem. Solids* **32**, 1867-72 (1971).

2216. Reversal of Coupling in Intermetallic Compounds Containing Two Rare Earth Elements, W. M. Swift and W. E. Wallace, *J. Solid State Chem.* **3**, 180-3 (1971).

2217. Magnetic Properties of Neodymium Sulfides $xNd_3S_4 \cdot (1-x)Nd_2S_3$ in the Temperature Range 4.2-300°K, G. M. Loginov, V. M. Sergeeva, and M. F. Bryzhina, *Soviet Phys.-Solid State* **12**, 2942 (1971).

2218. Structure et Propriétés Magnetiques du Manganite de Chrome, B. Boucher, R. Buhl, and M. Perrin, *J. Phys. Chem. Solids* **32**, 1471-88 (1971).

2219. Low Temperature Heat Capacities of Laves Phase Lanthanide-Aluminum Compounds, C. Deenadas, A. W. Thompson, R. S. Craig, and W. E. Wallace, *J. Phys. Chem. Solids* **32**, 1853-66 (1971).

2220. Neutron Diffraction Study of Ho_{25}-Dy_{75} and Ho_{50}-Dy_{50} in an External Magnetic Field, Q. H. Khan and W. C. Koehler, *J. Physique* **32**, C1-362 (1971).

2221. Multiband Model for the Electronic Heat Capacity of Chromium, J. F. Goff, *Phys. Rev. B* **4**, 1121 (1971).

2222. Magnetoelectric Effect in $BaCoF_4$, B. I. Al'shin, D. N. Astrov, A. V. Tishchenko, and S. V. Petrov, *JETP Letters* **12**(5), 142-144 (1970).

2223. Magnetic Properties of $TbPO_4$, A Canted Antiferromagnet, J. N. Lee, H. W. Moos, and B. W. Mangum, *Solid State Commun.* **9**, 1139-41 (1971).

2224. An Initial Comparison of the Elastic Properties of Dy and Th-50% Ho, R. G. Jordan, *Solid State Commun.* **9**, 857-859 (1971).

2225. Magnetic Susceptibility of $NiAs_2O_4$ and $NiSb_2O_4$, H. T. Witteveen, *Solid State Commun.* **9**, 1313-15 (1971).

2226. Diffuse Scattering of Neutrons Near the Critical Composition in Au-Cr Alloys, Y. Nakai, N. Kunitomi, Y. Endoh, and Y. Ishikawa, *Solid State Commun.* **9**, 921-23 (1971).

2227. Fluctuations and Critical Indices Near the Néel Temperature of Europium, G. T. Meaden and N. H. Sze, in *Les Elements Des Terres Rares*, Vol. II, pp. 109-117, Colloques Internationaux du Centre National de la Recherche Scientifique No. 180, Paris-Grenoble, May 5-10, 1969 (Editions du CNRS, 15, quai Anatole-France, Paris, 1970).

2228. Ferromagnetism in Granular Nickel Films, Y. Goldstein and J. I. Gittleman, *Solid State Commun.* **9**, 1197-1200 (1971).

2229. Influence de la Preparation Chimique sur les Structures Magnetiques de Ca_2MnO_4, G. Ollivier and G. Buisson, *J. Phys. Chem. Solids* **32**, 1189-94 (1971).

2230. Observation of an Anomaly in the Thermal Resistivity and the Lorentz Number of the Metallic Antiferromagnet CeAl, A. Fote and T. Mihalisin, *Phys. Rev. Letters* **27**(3), 148-50 (1971).

2231. The Gadolinium-Cadmium System, G. Bruzzone, M. L. Fornasini, and F. Merlo, *J. Less-Common Metals* **25**, 295-301 (1971).

2232. Magnetic Properties of the Rare Earth Hydroxides, Carmen A. Catanese, thesis, Yale University, New Haven, Conn., Yale-3227-65 (1970), 151 pp.
2233. Etude des Propriétés Cristallographiques Magnetiques et Elastiques de $RbCoF_3$, Y. Allain, J. Denis, A. Herpin, J. Lecomte, P. Meriel, I. Nouet, F. Plicque, and A. Zarembovitch, *J. Phys.* (France) **32**(2/3) Suppl. 611-13 (1971).
2234. Observation d'une Transition du Premier Ordre Antiferroparamagnetique dans $ZnCr_2Se_4$ and $MgCr_2O_4$, F. Varret, A. Gerard, F. Hartmann-Boutron, P. Inbert, and R. Kleinberger, *Proceeding of the Conference on the Application of the Mössbauer Effect*, held June 17-21, 1969, pp. 561-92, 1. Dezsi, Ed. (Akademiai Kiado, Budapest, 1971).
2235. Photoconductivity and Electronic Structure of Ferromagnetic Semiconductor $CdCr_2Se_4$, K. Sato, T. Teranishi, and S. Imamura, in *Proceedings of the Tenth International Conference on the Physics of Semiconductors*, Cambridge, Massachusetts, August 17-21, 1970, p. 247 CONF-700801 (1970).
2236. Magnetic and Electrical Properties of $CuCr_2Se_{4-x}Br_x$ and $CuCr_2Se_{4-x}Cl_x$, K. Miyatani, K. Minematsu, Y. Wada, F. Okamoto, K. Kato, and P. K. Baltzer, *J. Phys. Chem. Solids* **32**, 1429-34 (1971).
2237. Magnetic Characteristics of $TbCo_{0.3}Al_{1.7}$, H. Oestereicher, *J. Phys. Chem. Solids* **32**, 1402-8 (1971).
2238. Single Crystal Growth and Some Properties of $HgCr_{2-x}In_xSe_4$ Solid Solutions, T. Takahashi, K. Minematsu, and K. Miyatain, *J. Phys. Chem. Solids* **32**, 1007-13 (1971).
2239. Electronic Energy Levels in Ferromagnetic Eu_3P_2 and Eu_3As_2, G. Busch, M. Campagna, F. Hulliger, and H. C. Siegmann, *J. Phys. Chem. Solids* **32**, 2173-9 (1971).
2240. The Magnetic Phase Diagram of the UAs-UP Solid Solutions, J. Leciejewicz, A. Murasik, and T. Palewski, *Phys. Stat. Sol. B* **46**, K67 (1971).
2241. Temperature Dependence of Infrared Active Phono Frequencies of $CdCr_2Se_4$, P. Bruesch, H. Kalbfleisch, and F. Lehmann, *Phys. Stat. Sol. B* **46**, K99 (1971).
2242. Investigations of the Magnetic Properties of $TbPO_4$ at Low Temperatures, H. C. Schopper, P. J. Becker, W. Bohm, G. Dummer, H. G. Kahle, L. Klein, and G. Muller-Vogt, *Phys. Stat. Sol. B* **46**, K115 (1971).
2243. Thermal Expansion of Ordered FePt Alloy, M. Hayase, M. Shiga, and Y. Nakamura, *Phys. Stat. Sol. B* **46**, K117 (1971).
2244. Magnons in the Linear-Chain Antiferromagnet $CsMnCl_3 \cdot 2H_2O$, J. Akalyo, Jr., G. Shirane, S. A. Friedberg, and H. Kobayashi, *Phys. Rev. B* **2**(11), 4632 (1970).
2245. Propriétés Magnetiques du Grenat de Gallium et de Dysprosium a Tres Basses Temperatures, Jean Filippi, *Compt. Rend. B* **273**, 72 (1971).
2246. Magnetic First-Order Phase Transition in MnAs Films, Axel M. Stoffel, Jochen W. Schneider, and Fritz G. Hartmann, *IEEE Trans. Magnetics,* **MAG-6**(3), 545-8 (1970).
2247. Low-Temperature Magnetic Behavior of Several Oxides of Gadolinium, A. E. Miller, F. J. Jelinek, K. A. Gschneidner, Jr., and B. C. Gerstein, *J. Chem. Phys.* **55**(6), 2647-48 (1971).
2248. Magnetic Parameters of Cobalt-Bearing Yttrium Garnet Ferrites, K. P. Belov, B. V. Mill', V. I. Sokolov, and T. D. Khien, *Soviet Phys.-Solid State* **13**(3), 721 (1971).
2249. Conductivity Type of 3d Transition Metal Silicides, E. N. Nikitin and V. I. Tarasov, *Soviet Phys.-Cryst.* **16**(2), 305 (1971).
2250. Specific Heat of Mn_5Ge_3, G. I. Kalishevich, M. E. Sheinker, V. L. Zagryazhskii, and P. V. Gel'd, *Soviet Phys.-Solid State* **13**(2), 524 (1971).
2251. Inelastic Neutron Scattering from MnF_2 in the Critical Region, Michael P. Schulhof, Robert Nathans, Peter Heller, and Arthur Linz, *Phys. Rev.* **4**(7), 2254 (1971).
2252. Magnetic Properties of Ferrous Chloride Complexes with Two-Dimensional Structures, Mohgah F. Mostafa and Roger D. Willett, *Phys. Rev. B* **4**(7), 2213 (1971).
2253. Magnetic Ordering and Low Ni^{2+} Moment in $CsNiCl_3$, D. E. Cox and V. J. Minkiewicz, *Phys. Rev. B* **4**(7), 2209 (1971).
2254. Magneto-Elastic Effects and 3-mm Microwave Resonance in Single-Crystal $Tb_{85}Y_{15}$, J. L. Stanford and T. K. Wagner, *Phys. Rev. B* **4**(7), 2085 (1971).
2255. Temperature Dependence of the First- and Second-Order Cubic Anisotropy Constants in EuSe, Glen E. Everett and Ray C. Jones, *Phys. Rev. B* **4**(5), 1561 (1971).
2256. The Elastic Constants of Chromium, S. B. Palmer and E. W. Lee, *Phil. Mag.* **24**, 311-18 (1971).

REFERENCES

2257. Magnetic Susceptibility of $ErCl_3$, C. W. Fairall, J. A. Cowen, and C. Grabowski, *Phys. Letters A* **35**(6), 405-6 (1971).
2258. Magnetostriction, Thermal Expansion, and Phase Transitions in $CsMnCl_3 \cdot 2H_2O$, P. R. Aron, J. A. Woollam, and G. J. Butterworth. *Phys. Letters A* **35**(6), 422-23 (1971).
2259. Mössbauer Effect Study of FeOCl, E. Kostiner and J. Steger, *J. Solid State Chem.* **3**, 273-75 (1971).
2260. Metamagnetism of $Ni(NO_3)_2 \cdot 2H_2O$, V. A. Schmidt and S. A. Friedberg, *Phys. Rev. B* **1**(5), 2250-56 (1970).
2261. Etude de Quelques Composés Fluores du Chrome, Ariel de Kozak, Ph.D. thesis, Univ. Paris, CNRS A.O. 4714 (Gauthier-Villars Editeurs, Paris, 1971).
2262. Magnetic Phase Transitions in Four Manganese Double Chlorides, H. W. J. Blote and W. J. Huiskamp, *Physica* **53**, 445-70 (1971).
2263. Magnetic Behavior of Europium Silicates Using Nuclear Gamma Resonance Spectroscopy, G. M. Kalvius and G. K. Shenoy, *Z. Naturforsch.* **26a**, 353-56 (1971).
2264. Low-Temperature Heat Capacity of the Metamagnet $Ni(NO_3)_2 \cdot 2H_2O$, L. G. Polgar and S. A. Friedberg, *Phys. Rev. B* **4**(9), 3110 (1971).
2265. Influence of Magnetic Ordering on a Phonon Resonant Mode in Chlorine-Doped MnF_2, D. Bauerle, *Phys. Rev. B* **4**(8), 2347 (1971).
2266. Electrical and Magnetic Properties of $Fe_{1-x}Cu_xCr_2S_4$ Alloys, M. D. Banus and A. J. Strauss, (Lincoln Laboratory, M.I.T.), *Solid State Research* (1971:3); ESD-TR-71-247 pp. 27-9 (1971).
2267. The Mott Transition in NiS, Richard Koehler, (Stanford University, Center for Materials Research) p. 252, in Rept. CMR-71-17 (1971).
2268. Magnetic Structures of $Cr_{2-x}Ti_xO_3$ and $Cr_{2-x}V_xO_3$ Solid Solutions, V. G. Subkov, I. I. Matveenko, M. E. Sheinker, and P. V. Gel'd, *Soviet Phys.-Solid State* **13**(3), 644 (1971).
2269. Magnetic and Electrical Transport Properties of $CaCrO_3$, J. F. Weiher, B. L. Chamberland, and J. L. Gillson, *J. Solid State Commun.* **3**, 529-32 (1971).
2270. Two-dimensional Antiferromagnetism in $Mn(HCOO)_2 \cdot 2H_2O$, J. Skalyo, Jr., and G. Shirane, *Phys. Rev.* **188**(2), 680-87 (1968).
2271. Heat Capacity of $Mn(HCOO)_2 \cdot 2H_2O$ between 1.4 and 20°K, R. D. Pierce and S. A. Friedberg, *Phys. Rev.* **165**(2), 1037-41 (1969).
2272. Heat Capacity of the Antiferromagnet $CoCl_2 \cdot 6H_2O$ Near its Néel Point, J. Skalyo, Jr., and S. A. Friedberg, *Phys. Rev. Letters* **13**(4), 133 (1964).
2273. Low-Temperature Magnetic Phase Transitions in $CoCl_2 \cdot 6H_2O$ and $MnBr_2 \cdot 4H_2O$, V. A. Schmidt and S. A. Friedberg, *J. Appl. Phys.* **38**(13), 5319-26 (1967).
2274. Halide Doping of Ferromagnetic Europium Chalcogenides, M. W. Shafer, *J. Solid State Chem.* **3**, 505-9 (1971).
2275. Low-Temperature Magnetic Susceptibility of Ferrous Oxalate Dihydrate, S. de Barros and S. A. Friedberg, *Phys. Rev.* **141**(2), 637-40 (1966).
2276. Neutron Diffraction Study of the Magnetic Structure of Mn_3B_4, S. Neov and E. Legrand, *Phys. Stat. Sol.* **49B**, 589-96 (1972).
2277. The Samarium-Iron System, K. H. J. Buschow, *J. Less-Common Metals* **26**, 131-34 (1971).
2278. Magnetic Ordering in $CsMnCl_3 \cdot 2H_2O$, J. Skalyo, Jr., G. Shirane, S. A. Friedberg, H. Kobayashi, and I. Tsujikawa, in *Proceedings of the 12th International Conference on Low Temperature Physics*, Kyoto, Japan, Sept. 4-10, 1970, p. 799 (1970).
2279. Magnetic Properties of Rare Earth-Lead Intermetallics, R. D. Hutchens and W. E. Wallace, *J. Solid State Chem.* **3**, 564-66 (1971).
2280. 5f Magnetism in Palladium-Actinide Solid Solutions, W. J. Nellis and M. B. Brodsky, *Phys. Rev. B* **4**(5), 1594 (1971).
2281. Néel Temperature of Europium Telluride, Fred Massei, *Phys. Rev. B* **3**(7), 2364-5 (1971).
2282. Magnetic Properties of Ni-Cr Alloys, M. J. Besnus, Y. Gottehrer, and G. Munschy, *Phys. Stat. Sol. B* **49**, 597-607 (1972).
2283. Paramagnetic Resonance and Static Susceptibility Measurement in $GdZn_2$, D. Debray and E. Ryba, *J. Phys.* (France), Colloq. 32(2-3) Suppl., C1.1130-2 (1971). (From International Conf. on Magnetism, Grenoble, France, Sept. 1970; CONF-700940.)

2284. Magnetic Order of the Compound Series $RE_6(Mn_xFe_{1-x})_{23}$; (RE=Y, Gd), H. R. Kirchamyr and W. Steiner, *J. Phys.* (France), Colloq. 32(2-3) Suppl., C1.665-7 (1971).

2285. Ferromagnetism in TbF_3, L. Holmes and H. J. Guggenheim, *J. Phys.* (France), Colloq. 32(2-3) Suppl., C1.375-7 (1971).

2286. Néel and Curie Temperatures for Six of the Crystal Structures of Cerium and Praseodymium, G. T. Meaden, N. H. Sze, G. Krithivas, and M. J. Suckermann, *J. Phys.* (France), Colloq. 32(2-3) Suppl., Cl. 375-7 (1971).

2287. Die magnetische Struktur und das magnetische Verhalten von $ErFe_2$: Eine Neutronenbeugungsuntersuchung, G. Will and M. O. Bargouth, *Phys. Kondens. Materie* **13**, 137-48 (1971).

2288. Neutron Diffraction Study of Short Range Order in γ-MnNi, P. Wells and J. H. Smith, *J. Phys.* (France), Colloq. 32(2-3) Suppl., C1-70 (1971).

2289. The Magnetic Structures of CeBi, J. W. Cable and W. C. Koehler, AIP Conf. Series, in *Proceedings of the 17th Annual Conf. on Magnetism and Magnetic Materials,* Chicago, Illinois, Nov. 16-19, 1971.

2290. Research and Development of Rare Earth-Transition Metal Alloys as Permanent-Magnet Materials, Alden E. Ray and Karl J. Strnat, (University of Dayton Research Institute, Ohio), UDRI-TR-71-32; AFML-TR-71-210, (October, 1971), 120 pp.

2291. Mössbauer Effect and Resistivity Studies in $Fe_{11.1}Te$, E. Hermon, R. D. Nolan, and S. Shtrikman, *Israel J. Chem.* **9**(1), 1-5 (1971).

2292. Strong Anisotropy in the Cubic Ferrimagnet $FeCr_2S_4$, R. P. Van Stapele, J. S. Van Wieringen, and P. F. Bongers, *J. Phys.* (France), Colloq. 32(2-3), Pt. 1, 53-4 (1971).

2293. Observation of Induced Antiferromagnetism above the Néel Point in FeF_3, V. Ya. Gamlitskii, S. S. Yakimov, V. I. Nikolaev, and N. F. Simonov, *Zh. Eksper. Teor. Fiz. Pis'ma v Redaktsiyu* **13**(5), 129-30 (1971). (1971).

2294. New Compounds in the Systems $Bi_2O_3-Me_2O_3$ (Fe^{3+}, Al^{3+}, Ga^{3+}, Mn^{3+}), A. G. Tutov, I. E. Myl'nikova, N. N. Parfenova, V. A. Bokov, and S. A. Kizhaev, *Soviet Phys.-Solid State* **6**(4), 963-64 (1964).

2295. On Structural Effects at the Curie Point of Gadolinium, V. V. Vorobev, M. I. Palatnik, and V. A. Finkel, *Phys. Stat. Sol. B* **47**, K53 (1971).

2296. Intrinsic Coercivity of $BaFe_{12-x}Cr_xO_{19}$, K. Haneda and H. Kojima, *Phys. Stat. Sol. A* **6**, 259 (1971).

2297. Magnetic Properties of Uranium Silicides, A. Misiuk and W. Trzebiatowski, *Bull. Acad. Polon. Sev. Sci. Chem.* **18**(11-12), 633-37 (1970).

2298. Mössbauer Study of Cobalt-Zinc Ferrites, G. A. Petitt and D. W. Forester, *Phys. Rev. B* **4**(11), 3912-24 (1971).

2299. de Haas-van Alphen Effect in Antiferromagnetic Cr-Rich Allows, E. J. Gutman and J. L. Stanford, *Phys. Rev. B* **4**(11), 4026-33 (1971).

2300. Magnetic Properties of the Antiferromagnet $DyPO_4$ in Applied Fields, C. S. Koonce, B. W. Mangum, and D. D. Thornton, *Phys. Rev. B* **4**(11), 4094-99 (1971).

2301. Magneto-Elastic Dependence of the Propagation of Sound in Gadolinium at the Critical Point, M. Long, Jr. and R. Stern, *Phys. Rev. B* **4**(11), 4054-69 (1971).

2302. Structural and Magnetic Studies on $ErFe_2-ErAl_2$, Hans Oesterreicher, *J. Appl. Phys.* **42**(12), 5137-43 (1971).

2303. First-Order Crystallographic and Magnetic Phase Transition in CrAs, H. Boller and A. Dallel, *Solid State Commun.* **9**, 1699-1706 (1971).

2304. Electronic Specific Heat of Metallic Ti-Doped V_2O_3, D. B. McWhan, J. P. Remeika, T. M. Rice, W. F. Brinkman, J. P. Maita, and A. Menth, *Phys. Rev. Letters* **27**(14), 941 (1971).

2305. Elaboration et Etude de Quelques Propriétés de l'Alliage d'Heusler Cu_2MnAl, J. P. Lisse and B. DuBois, *Mem. Sci. Rev. Met.* **68**(7-8), 521-34 (1971).

2306. Substitution von Chrom durch Vanadium und Mangan im Ferromagneten CrTe, J. Nosselt, U. Sondermann, and W. Treutmann, *Z. Angew. Physik* **32**, 66 (1971).

2307. Magnetische Eigenschaften des zweidimensionalen Antiferromagneten $KFeF_4$, G. Heger and R. Geller, *Z. Angew. Physik* **32**, 63 (1971).

2308. Elastic Moduli of Thulium and Ytterbium from 4.2 to 300°K, M. Rosen, *J. Phys. Chem. Solids* **32**, 2351-56 (1971).

2309. Propriétés et Structure Magnetique de Mn_3O_4, B. Boucher, R. Buhl, and M. Perrin, *J. Phys. Chem. Solids* **32**, 2429-37 (1971).

REFERENCES

2310. Surface Magnetization Near the Critical Temperature and the Temperature Dependence of Magnetic-Electron Scattering from NiO, T. Wolfram, R. E. DeWames, W. F. Hall, and P. W. Palmberg, *Surface Sci.* **28**, 45-60 (1971).
2311. Recent Studies on the Optical Properties of Rare-Earth Metals, C. Chr. Schuler, *Proc. Conf. on Optical and Electronic Structures of Metal Alloys,* Paris, 1965, pp. 221-235 (North Holland Publ. Co., Amsterdam, 1966).
2312. Seebeck Coefficient at the Curie Temperature: Specific Heat of Charge Carriers in Ferromagnets, S. H. Tang, P. P. Craig, and T. A. Kitchens, *Phys. Rev. Letters* **27**(9), 593-95 (1971).
2313. Short-Range Magnetic Order in Gamma-Manganese Alloys, E. R. Vance, J. H. Smith, and T. M. Sabine, *J. Phys. C: Metal Phys. Suppl.,* No. 1, 534-45 (1970).
2314. Antiferromagnetism in γ-Phase Manganese-Palladium and Manganese-Nickel Alloys, T. J. Hicks, A. R. Pepper, and J. H. Smith, *J. Phys. C. (Proc. Phys. Soc.), Series 2,* **1**, 1683 (1968).
2315. Electrical Resistance and Specific Heat of a Metallic Metamagnet in the Presence of a Magnetic Field, A. Fote and T. Mihalisin, *Solid State Commun.* **9**, 1391-2 (1971).
2316. Magnetoelectric Effect and Critical Behavior in the Ising-Like Antiferromagnet $DyAlO_3$, L. M. Holmes, L. G. van Uitert, and G. W. Hull, *Solid State Commun.* **9**, 1373-76 (1971).
2317. Mössbauer Study of Dysprosium Equiatomic Compounds with Copper, Silver, Zinc, and Rhodium, M. Belakhovsky and J. Pierre, *Solid State Commun.* **9**, 1409-15 (1971).
2318. Optical Properties of α-MnSe, Donald L. Decker and R. L. Wild, *Phys. Rev. B* **4**(10), 3425-37 (1971).
2319. The Metal-Nonmetal Transition in Nickel Sulphide (NiS), R. M. White and N. F. Mote, *Phil. Mag.* **24**, 845-56 (1971).
2320. Elasticity and Antiferromagnetism of Metallic Antiferromagnetics, R. Street and J. H. Smith, *J. Phys. Radium* **20**, 82 (1959).
2321. Antiferromagnetism in Au_5Mn_2, J. H. Smith and P. Wells, *J. Phys. C. (Solid State Phys.)* Sev., **2**, 2, 356 (1969).
2322. Resistivity and Magnetoresistance of Au_2Mn, J. H. Smith and R. Street, *Proc. Phys. Soc.* **B70**, 1089 (1957).
2323. Critical Magnetic Thermal Expansivity of $RbMnF_3$, Brage Golding, *Phys. Rev. Letters* **27**(17), 1142 (1971).
2324. Neutron-Diffraction Determination of the Oxygen Parameter and the Cation Distribution for the Ferrite Spinels $NiSc_{0.15}Fe_{1.85}O_4$, and $NiSc_{0.4}Fe_{1.6}O_4$, A. G. Kocharov, I. I. Yamzin, and M. K. Faek, *Soviet Phys.-Cryst.* **16**(3), 541 (1971).
2325. Low-Temperature Magnetic Transition in Mn_3Ge_2, I. G. Fakidov, N. P. Grazhdankina, R. I. Zainullina, A. M. Burkhanov, and Y. S. Bersenev, *Soviet Phys.-Solid State* **13**(5), 1181-87 (1971).
2326. Study of the Structure and Magnetic Properties of β-FeOOH in the Process of Dehydration to Hematite, M. N. Pukosuev, I. I. Yamzin, V. A. Chikhachev, and V. A. Lyubimtsev, *Soviet Phys.-Cryst.* **16**(3), 450-54 (1971).
2327. Etude par l'Effet Mössbauer de $MnFe_{2-x}Sc_xO_4$, K. V. Pocholok, B. I. Pocrovskii, A. M. Badeshkin, A. K. Gapeev, and L. N. Komissarova, *Solid State Commun.* **9**, 1727-8 (1971).
2328. Antiferromagnetic Transition in $CoCl_2 \cdot 6H_2O$ and Fisher's Relation, J. Skalyo, Jr., A. F. Cohen, S. A. Friedberg, and Robert B. Griffiths, *Phys. Rev.* **164**(2), 705-9 (1967).
2329. The Antiferromagnetism of Anhydrous Sulfates of Ni^{++}, Fe^{++}, Co^{++}, and Cu^{++}, A. S. Borovik-Romanov, V. R. Karasik, and N. M. Kreins, *Soviet Phys. JETP* **4**(1), 109-14 (1957).
2330. The Antiferromagnetism of Manganese Copper Alloys, G. E. Bacon, I. W. Dunmur, J. H. Smith, and R. Street, *Proc. Roy. Soc. A* **241**, 223-38 (1957).
2331. Magnetic Studies of TbFeAl, H. Oesterreicher, *Phys. Stat. Sol. A.* **7**, K55-57 (1971).
2332. Elastic Constants of a NiO Single Crystal, I, P. de V. Du Plessis, S. J. Van Tonder, and L. Alberts, *J. Phys. C Solid State Phys.* **4**, 1983-87 (1971).
2333. Studies of the V_2O_3-MoO_3 System: Magnetic Properties, A. Bielanski, K. Dyrek, I. Kracik, and E. Wenda, *Bull. Acad. Polon. Ser. Sci. Chem.* **19**, 513-21 (1971).
2334. A Study of the Spontaneous Magnetization of $EuLiH_3$ Using the Mössbauer Effect, Chia-Ling Chien and J. E. Greedan, *Phys. Letters* **36A**(3), 197-8 (1971).
2335. Determination par Diffraction des Neutrons de la Structure Antiferromagnétique du Grenat $Ca_3Mn_2Ge_3O_{12}$, R. Plumier, *Solid State Commun.* **9**, 1723-25 (1971).

2336. Eutude par Diffraction Neutronique de Formiate de Fer Dihydrate, P. Burlet and E. F. Bertaut, *Solid State Commun.* **9**, 1633-37 (1971).

2337. The Temperature Variation of Magnetization of Bulk Amorphous $Ni_{85}P_{15}$: An Amorphous Ferrimagnet?, A. W. Simpson and D. R. Brambley, *Phys. Stat. Sol. B* **49**, 685 (1972).

2338. Radio Frequency Measurements of Interactions in Magnetic Insulators, A. T. Skjeltorp, Ph.D. thesis, Yale University, New Haven, Conn., YALE-3227-74 (1971), 308 pp.

2339. Transient Photoconductivity and Magnetoresistance of Ferromagnetic Semi-Conductor EuS, Koji Kajita and Taizo Masumi, *J. Phys. Soc. Japan* **31**, 964 (1971).

2340. First-Order Transition in Chromium at the Néel Temperature, P. R. Garnier and M. B. Salamon, *Phys. Rev. Letters* **27**(22), 1523-25 (1971).

2341. Mössbauer Effect in Ordered Fe_3Pt Alloy, M. Shiga, A. Miyoshi, and Y. Nakamura, *Phys. Stat. Sol. B* **49**, K195 (1972).

2342. Study of the Magnetic Transition of CrAs, N. Kazama and H. Watanase, *J. Phys. Soc. Japan* **31**, 943 (1971).

2343. Induced Magnetic Ordering in Diamagnets, Yu. A. Kudenko and A. P. Slivinskii, *Soviet Phys.-Solid State* **13**(4), 1046-47 (1971).

2344. Magnetic Properties of Single Crystals $Cr_{2-x}Te_3$, T. Hashimoto, K. Hoya, M. Yamaguchi, and I. Ichitsubo, *J. Phys. Soc. Japan* **31**(3), 679-82 (1971).

2345. Invar Behavior in Antiferromagnetic Fe-Mn-(Ni) Alloys, D. A. Colling and M. P. Mathur, *J. Appl. Phys.* **42**(13), 5699-703 (1971).

2346. Intermetallic Compounds of Rare Earth Elements and Ni, Co, or Fe, K. H. J. Buschow, *Phys. Stat. Sol. A* **7**, 199-210 (1971).

2347. Ferromagnetic $Cu_{1+x}Cr_2Te_4$ and $CuAg_yCr_2Te_4$ with Metal-Excessive Spinel Structure, F. K. Lotgering and G. H. A. M. van der Steen, *Solid State Commun.* **9**, 1741-44 (1971).

2348. Preparation et Propriétés Physiques de la Phase de Type Spinelle $Zn_xFe_{3-x}O_{4-x}F_x$, J. Claverie, H. Dexpert, J. Portier, R. Pauthenet, and P. Hagenmuller, *Mat. Res. Bull.* **6**, 1125-30 (1971).

2349. The Electronic Structure of Lithium Metachromite, C. Simo and S. L. Holt, *J. Solid State Chem.* **4**, 76-79 (1972).

2350. $La_xSr_{1-x}RuO_3$: A New Pervoskite Series, R. J. Bouchard and J. F. Weiher, *J. Solid State Chem.* **4**, 80-86 (1972).

2351. Magnetic Transition Points in U_3Bi_4 and UGa_2, J. Sternberk, V. Ansorge, J. Hrebik, A. Minvosky, and Z. Smetana, *Czech. J. Phys.* **21**(9), 969-78 (1971).

2352. Neutron Diffraction and Magnetic Structures of Rare Earth Elements and of 3d-4f Intermetallic Compounds, G. Will, *Z. Angew. Phys.* **32**(1), 1-8 (1971).

2353. Magnetic Properties of Some Actinide Halide Compounds, M. Hendricks, (DuPont de Nemours, and Co.), Savannah River Lab., Aiken, S. C. DP-MS-71-46 (1971), 195 pp.

2354. Heat Capacity of Cerium-Lanthanum and Cerium-Lutecium Alloys at Very Low Temperatures, E. I. Nikulin and Y. B. Patrikeev, *Soviet Phys.-Solid State* **13**(6), 1527.

2355. Crystal Structure and Mössbauer Effect Investigation of $FeVO_4$, B. Robertson and E. Kostiner, *J. Solid State Chem.* **4**, 29-37 (1972).

2356. Mössbauer and Magnetic Investigation of the System $Mn_{5-x}Fe_xGe_3$ (x = 0.5, 1.0, and 1.5), W. M. Reiff, K. S. V. L. Narasimhan, and H. Steinfink, *J. Solid State Chem.* **4**, 38-45 (1972).

2357. Hydrothermal Preparation and Low Temperature Magnetic Properties of TbOOH, DyOOH, HoOOH, ErOOH, A. Norlund Christensen, *J. Solid State Chem.* **4**, 46-51 (1972).

2358. Hydrothermal and High-Pressure Prpearation of Some $BaMnO_3$ Modifications and Low-Temperature Magnetic Properties of $BaMnO_3$(2H), A. Norlund Christensen and G. Ollivier, *J. Solid State Chem.* **4**, 131-37 (1972).

2359. Magnetic Ordering in $LiCr_{1-x}Fe_xO_2$, A. Tauber, W. M. Moller, and E. Banks, *J. Solid State Chem.* **4**, 137-52 (1972).

2360. Evolution Structurale des Phosphures, Arseniures et Arseniophosphures M_2P, M_2As, et $M_2(P_{1-x}As_x)$, A. Nylund, A. Roger, J. P. Senateur, and R. Fruchart, *J. Solid State Chem.* **4**, 115-22 (1972).

2361. The Effect of High Pressure on the Néel Temperature of γ-Manganese, A. Sawaoka, T. Soma, S. Saito, and Y. Endoh, *Phys. Stat. Sol. B* **47**, K99 (1971).

REFERENCES

2362. Static Magnetization Studies on the Family of Compounds $Cd_{1-x}Fe_xCr_2S_4$, M. R. Spender and A. H. Morrish, *Can. J. Phys.* **49**, 2659-70 (1971).

2363. Etude par Diffraction Neutronique des Perovskites $ErAlO_3$, $HoAlO_3$, $TbRhO_3$, etc., Jean Sivardiere, Simone Quezel-Ambrunaz, *Compt. Rend. B* **273**, 619 (1971).

2364. The Heat Capacity of $PrCo_2$ and $NdCo_2$ Between 8 and 300°K, C. Deenadas, R. S. Craig, N. Marzouk, and W. E. Wallace, *J. Solid State Chem.* **4**, 1-3 (1972).

2365. Magnetic Properties of $(Zr_{1-x}Ti_x) Zn_2$ and $(Zr_{1-x}Hf_x) Zn_2$ Alloys, T. Kato and M. Shimizu, *J. Phys. Soc. Japan* **31**(3), 723-29 (1971).

2366. Magnetische Eigenschaften des Quasibinären Systems $Y_2(Fe_{1-x}Co_x)_{17}$, W. Steiner, H. R. Kirchmayr, and W. Springs, *Z. Angew. Phys.* **32**, 146-49 (1971).

2367. Magnetic Structure and Exchange Interactions in $Bi_2Fe_4O_9$, V. A. Bokov, S. I. Yushchuk, G. V. Popov, N. N. Parfenova, and A. G. Tutov, *Soviet Phys.-Solid State* **13**(6), 1333 (1971).

2368. Magnetic Properties of Cobalt-Containing Iron Garnets, K. P. Belov, B. V. Mill', V. I. Sokolov, and T. D. Hien, *Soviet Phys.-Solid State* **13**(6), 1380 (1971).

2369. Magnetic Properties of Face-Centered Cubic $Fe_{65} (Pt_xIr_{1-x})_{35}$, M. Shiga, M. Kosaka, and Y. Nakamura, *Phys. Stat. Sol. B* **50**, 351 (1972).

2370. Magnetism and Phase Transition in Two-Dimensional Lattices; $M(HCOO)_2 \cdot 2H_2O$ (M; Mn, Fe, Ni, Co), Kazuyoshi Takeda and Kazuko Kawasaki, *J. Phys. Soc. Japan* **31**(4), 1026-36 (1971).

2371. Thermal and Magnetic Properties of Linear Ferro- and Antiferro-Magnetic Substances, $MCl_2 \cdot 2NC_5H_5$ (M=Co, Cu,), Kazuyoshi Takeda, Shun-ichi Matsukawa, and Tiichiro Haseda, *J. Phys. Soc. Japan* **30**(5), 1330-36 (1971).

2372. Specific Heat of $Ni (NH_3)_2 Ni(CN)_4 \cdot 2C_6H_6$ at Low Temperature, Shiochi Nagata, Tsuyoshi Maruyamaushi, and Takashi Watanabe, *J. Phys. Soc. Japan* **30**(4), 1054-58 (1971).

2373. Band Antiferromagnetism in fcc $(CoMn)_{1-x}Fe_x$, Kengo Adachi, Kiyoo Sato, Masaaki Matsui, and Yasuo Fujio, *J. Phys. Soc. Japan* **30**, 1201-1202 (1971).

2374. Temperature Dependence of the Paramagnetic Resonance Line-Widths in $RbNiF_3$ and $RbMnF_3$, Eiichi Toyota and Kinshiro Hirakawa, *J. Phys. Soc. Japan* **30**(3), 692-6 (1971).

2375. Thermal Expansion of Au_4V, Naoko Kasai and Shinji Ogawa, *J. Phys. Soc. Japan* **30**(3), 736-41 (1971).

2376. Magnetic Susceptibility of Antiferromagnetic $Co(HCOO)_2 \cdot 2H_2O$, Kazuyoshi Takeda and Shun-ichi Matsukawa, *J. Phys. Soc. Japan* **30**, 887 (1971).

2377. Exchange Scattering of Low-Energy Electrons from Antiferromagnetic NiO, Tadasu Suzuki, Noboru Hirota, Hiroshi Tanaka, and Hitoshi Watanabe, *J. Phys. Soc.* **30**, 888 (1971).

2378. Antiferromagnetism of γ-Phase Manganese Alloys Containing Ni, Zn, Ga, and Ge, Hidema Uchishiba, *J. Phys. Soc. Japan* **31**(2), 452-60 (1971).

2379. Neutron and Magnetic Studies of a Single Crystal of Fe_2TiO_4, Yoshikazu Ishikawa, Shoichi Sato, and Yasuhiko Syono, *J. Phys. Soc. Japan* **3**(2), 452-60 (1971).

2380. Magnetic Moment in Laves Phase Compound. II. $Zr(Fe_{1-x}Mn_x)_2$ and $Zr(Fe_{1-x}Co_x)_2$, Kazuo Kanematsu, *J. Phys. Soc. Japan* **31**(5), 1355-60 (1971).

2381. Thermal Expansion Coefficient and Spontaneous Volume Magnetostriction of Fe-Ni (fcc) Alloys, Yasunori Tanji, *J. Phys. Soc. Japan* **31**(5), 1366-73 (1971).

2382. NMR Study of the Magnetic Phase of EuSe, Takeshi Komaru, Tadamiki Hihara, and Yoshitaka Koi, *J. Phys. Soc. Japan* **31**(5), 1391-98 (1971).

2383. Phase Relation of Nonstoichiometric Chromium Sulfide CrS_x in the Range of $1.200 \leq x \leq 1.480$, Kenzo Igaki, Nobumitsu Ohashi, and Masao Mikami, *J. Phys. Soc. Japan* **31**(5), 1424-30 (1971).

2384. Magnetic Properties of Rare Earth Gallium Compounds RGa, Hironobu Fuji, Nobuaki Shohata, Tetsuhiko Okamoto, and Eiji Tatsumoto, *J. Phys. Soc. Japan* **31**, 1952 (1971).

2385. The Magnetic Properties of Pd_3-Rare Earth Phases, W. E. Gardner, J. Penfold, and I. R. Harris, *J. Phys.* (France), Colloq. **32**(2-3), Part 2, 1139-40 (1971).

2386. Spectroscopic Properties of Dysprosium Vanadate, C. J. Ellis, K. A. Gehring, M. J. M. Leask, and R. L. White, *J. Phys.* (France), Colloq. **32**(2-3), Part 2, 1024-5 (1971).

2387. Solid Helium Three at Very Low Temperatures, R. C. Richardson, *J. Phys.* (France), Colloq. **31**(3), 79-89 (1970).

2388. Magnetic Structure and Exchange Integral Measurement in TlMnF$_3$, L. Madhay Rao and N. S. Satya Murthy, *J. Phys.* (France), Colloq. 32(2-3), Part 2, 617-8 (1971).

2389. Experimental Study of Magnetic and Crystallographic Transitions in DyVO$_4$, F. Sayetat, J. X. Boucherle, M. Belakhovsky, A. Kallel, F. Tcheou, and H. Fuess, *Phys. Letters* 34(7), 361-2 (1971).

2390. Etude Magnetique et Diffraction Neutronique dans les Composés DyOCl et TbOCl, D. Elmaleh, D. Fruchart, and J. C. Joubert, *J. Phys.* (France), Colloq. 32(2-3), Part 2, 741-3 (1971).

2391. Neutron Diffraction Study of the Magnetic Spiral Structure of MnO$_2$, J. Z. Gonzalo and D. Cox, *R. Soc. Esp. Fis. Quim., An. Fis., Esp.* 66(11-12), 407-14 (1970).

2392. Magnetische Eigenschaften der Lanthanid-Nickel-Verbindungen LnNi$_5$, W. E. Wallace and M. Aoyagi, *Monatsch. Chemie* 102, 1455-1461 (1971).

2393. Raman Scattering and Phase Transitions in a KMnF$_3$ Crystal, Yu. A. Popkov, V. V. Eremenko, and V. I. Fomin, *Soviet Phys.-Solid State* 13(7), 1701-8 (1972).

2394. Weak Ferromagnetism of Molybdenum Triflouride, Ya. V. Vasil'ev, K. A. Khaldoyanidi, A. A. Opalovskii, and V. M. Yudin, *Soviet Phys.-Solid State* 13(6), 1543-5 (1971).

2395. Magnetic and X-Ray Studies on Bi$_{1-x}$TR$_x$O$_{1.5}$ Solid Solutions, A. G. Tutov, E. S. Sher, and V. M. Yudin, *Soviet Phys.-Solid State* 13(6), 1546-7 (1971).

2396. Renforcement du Terme de Chaleur Specifique Electronique à l'Apparition de l'Ordre Antiferromagnetique dans les Diborures des Metax de Transition, J. Castaing, R. Caudron, and P. Costa, *J. Phys.* (France), Colloq. 32(2-3) Suppl. C 1-638 (1971).

2397. Propriétés Magnetiques et Structures Magnetiques des Composés Tb$_3$Al$_2$ and Dy$_3$Al$_2$, B. Barbara, C. Becle, and R. Lemaire, *Z. Angew. Physik.* 32, 113-16 (1971).

2398. Magnetic Properties of BaCoF$_4$ Single Crystals, R. V. Zorin, B. I. Al'shin, D. N. Astrov, and N. A. Syunina, *Soviet Phys.-Solid State* 13(7), 1807 (1972).

2399. Präparation und Eigenschaften von Eu$_{1-x}$Gd$_x$S-Schichten, E. Bayer and W. Zinn, *Z. Angew. Physik.* 32, 83-8 (1971).

2400. The Magneto-Thermal Resistivity of α-MnCl$_2 \cdot$4H$_2$O at Low Temperatures, M. J. Metcalfe, *Phys. Letters* 32A(5), 373-75 (1971).

2401. Experimental Study of the Induced Magnetization of Cu(NH$_4$)$_2$Br$_4 \cdot$2H$_2$O Near the Curie Temperature, E. Velu, D. Cadoul, B. Lecuyer, and J. P. Renard, *Phys. Letters* 36A(5), 443-44 (1971).

2402. Electrical Resistivity of Cobalt Disulfide, S. Ogawa and T. Teranishi, *Phys. Letters* 36A(5), 407-08 (1971).

2403. Structure Magnetique et Rotation de Spin de Mn$_3$NiN, D. Fruchart, E. F. Dertaut, R. Madar, G. Lorthioir, and R. Fruchart, *Solid State Commun.* 9, 1793-1797 (1971).

2404. Transport Properties of Europium Oxide Thin Films, C. Paparoditis, R. Suryanarayanan, C. Llinares, E. Monteil, and G. Bordure, *Solid State Commun.* 9, 1871-1876 (1971).

2405. Critical Magnetic Neutron Scattering from Ferrous Flouride, M. T. Hutchings, M. P. Schulhof, and H. J. Guggenheim, *Phys. Rev. B* 5(1), 154-68 (1972).

2406. Sound Velocity and Magnetoelastic Coupling in MnTe, K. Walther and K. H. Sarges, *Phys. Letters* 36A(4), 309-10 (1971).

2407. Susceptibilite Magnetique des Nitrures Mixtes de Vanadium et de Chrome, F. Ducastelle and P. Costa, *Phys. Letters* 33A(7), 447-8 (1970).

2408. Shift of Néel Temperature and EPR Line width of KMnF$_3$ with Mg Doping, R. P. Gupta, M. S. Seehra, and W. E. Vehse, *Phys. Rev. B* 5(1), 92-95 (1972).

2409. Magnetic Properties of Cu-Mn and Cu-Fe Alloys, J. J. Hauser, *Phys. Rev. B* 5(1), 110-15 (1972).

2410. Magnetic Behavior of Metamagnetic DyAlO$_3$, L. M. Holmes, L. G. Van Uitert, R. R. Hecker, and G. W. Hull, *Phys. Rev. B* 5(1), 138-46 (1972).

2411. Magnetic Structure of HoFe$_2$, J-M. Moreau, C. Michel, M. Simmons, T. O'Keefe and W. J. James, in *Proceedings of the 8th Rare Earth Research Conference,* April 19-22, 1970, Reno, Nevada, 1, pp. 34-40, T. A. Henrie and R. E. Lindstrom, Eds., CONF-700403 (1970).

2412. Electrical Resistivity and Magnetic Susceptibility of Definite Compounds in the Tin-Samarium System, Annick Percheron, P. Lethuillier, J. L. Feron, and O. Gorochov, in *Proceedings of the 8th Rare Earth Research Conference,* April 19-22, 1970, Reno, Nevada, 1, 322-29, T. A. Henrie and R. E. Lindstrom, Eds., CONF-700403 (1970).

REFERENCES

2413. Saturation Magnetic Moment and Crystalline Anisotropy of Single Crystals of Light Rare Earth–Cobalt Compounds, RCo_5, E. Tatsumoto, T. Okamoto, H. Fuji, and C. Inoue, *J. Phys.* (France), 32 Suppl., 550-1 (1971). (Compte rendu Conference Internationale de Magnetisme, Grenoble, Sept., 14-19, 1970.)
2414. Magnetic Properties of $FeAsO_4 \cdot 2H_2O$ and $KFe_4(OH)_4(AsO_4)_3 \cdot 3H_2O$, M. Takano, T. Takada, T. Wada, and K. Okada, *J. Phys. Soc. Japan* **31**, 298 (1971).
2415. Study of the Mössbauer Effect, Magnetization, and Crystal Structures of the Pseudobinary Compounds $ThCo_{5-5x}Fe_{5x}$, A. M. van Diepen, K. H. J. Buschow, and J. S. van Wieringen, *J. Appl. Phys.* **43**(2), 645 (1972).
2416. Neutron Diffraction Study of Antiferromagnetism in Face-Centered Cubic Mn-Ir Alloys, Takashi Yamaoka, Mamoru Mekata, and Hideo Takaki, *J. Phys. Soc. Japan* **31**, 301 (1971).
2417. Antiferromagnetism of γ-Iron Mangane Alloys, Yasuo Endoh and Yoshikazu Ishikawa, *J. Phys. Soc. Japan* **30**(6), 1614-27 (1971).
2418. Magnetic Properties of the Diluted Ferromagnetic System: $xSrRuO_3 \cdot (1-x)LaRhO_3$, A. Kanbayasi, Tomonao Miyadai, H. Kondo, and S. Miyahara, *J. Phys. Soc. Japan* **30**, 1754 (1971).
2419. Anomalous Behavior of the Low-Temperature Magnetic Specific Heat of $NiCl_2 \cdot 2H_2O$, L. G. Polgar, A. Herweijer, and W. J. M. de Jonge, *Phys. Rev. B* **5**(5), 1957-60 (1972).
2420. Magnetic Susceptibility and Electrical Resistivity of Dilute Chromium-Titanium Alloys, C. H. Chiu, M. H. Jericho, and R. H. March, *Can. J. Phys.* **49**, 3010-3023 (1971).
2421. Destruction of Ferromagnetism in $ZrZn_2$ at High Pressure, T. F. Smith, J. A. Mydosh, and E. P. Wohlfarth, *Phys. Rev. Letters* **27**(5), 1732-35 (1971).
2422. Temperature Dependence of the Magnetic Phase Transitions of $CoCl_2 \cdot 2H_2O$, Marian A. Lowe, C. R. Abeledo, and A. A. Misetich, *Phys. Letters* **37A**(4), 274-76 (1971).
2423. Magnetic Susceptibility of Single Crystals Mn_3B_4, Mitsuo Kasaya, Tadamiki Hihara, and Yoshitaka Koi, *J. Phys. Soc. Japan* **30**, 286 (1971).
2424. Magnetic Properties of the Systems (Cr, Fe) Sb and (Cr, Co) Sb, Kazuyuki Yamaguchi, Hiroshi Watanabe, Hisao Yamamoto, and Yasuo Yamaguchi, *J. Phys. Soc. Japan* **31**(4), 1042-52 (1971).
2425. Magnetic Properties and Magnetic Structures of Tb_3Al_2 and Dy_3Al_2, D. Barbara, C. Decle, and R. Lemaire, (Centre National de La Recherche Scientifique, Grenoble, France), CONF-710337-1 (Oct. 22, 1971), 11 pp. (From Conference on Magnetism; Salzburg, Austria, Mar. 29, 1971.)
2526. Mössbauer Scattering in $^{141}_{59}Pr$, Joel Lee Groves, Ph.D. thesis, University of Illinois at Urbana, 1971. Available from University Microfilms, Ann Arbor, Michigan, Order No. 72-6941, 93 pp.
2427. A New Antiferromagnetic Phase In EuSe Below 1.8K, R. Griessen, M. Landolt, and H. R. Ott, *Solid State Commun.* **9**, 2219-2223 (1971).
2428. Paramagnetic Resonance and Magnetic Measurements on $GdNi_3$ Compound, E. Burzo and I. Ursu, *Solid State Commun.* **9**, 2289-2292 (1971).
2429. Spontaneous Magnetization and Microwave Susceptibilities of $GdCl_3$, S. Scheller, *Phys. Kondens. Materie* **13**, 268-272 (1971).
2430. Specific Heats of $LaIn_3$, $CeIn_3$, and $PrIn_3$ at Temperatures Between 1.5 and 4.2°K, S. Nasu, A. M. van Diepen, H. H. Neumann, and R. S. Craig, *J. Phys. Chem. Solids* **32**, 2773-2777 (1971).
2431. Atude de la Phase Fe_5Sn_3 et de Quelques Solutions Solides, C. Djege-Mariadassou, E. Both, and G. Trumpy, *Amm. Chim.* **5**, 505-516 (1970).
2432. Magnetism in Amorphous Fe-Pd-P Alloys, T. E. Sharon and C. C. Tsuei, *Phys. Rev. B* **5**(3), 1047-64 (1971).
2433. A Note On the Critical Behavior of $MnCO_3$ and NiF_2, I. Maartense, *Solid State Commun.* **9**, 2071-2076 (1971).
2434. The Paramagnetic Susceptibility of the Ferrite FeV_2O_4, P. de V. du Plessis, *J. Phys. C: Solid State Phys.* **4**, 2919-22 (1971).
2435. Magnetic Interactions in Antiferromagnetic $DyPO_4$, C. J. Ellis, M. J. M. Leask, D. M. Martin, and M. R. Wells, *J. Phys. C: Solid State Phys.* **4**, 2937-44 (1971).
2436. Magnetic Anisotropies in Antiferromagnetic Rare Earth Antimonide Single Crystals, G. Busch and O. Vogt, *J. Appl. Phys.* **39**(2), 1334-35 (1968).
2437. Magnetic and Structural Characteristics of Lanthanide–Nickel Compounds, R. E. Walline and W. E. Wallace, *J. Chem. Phys.* **41**(6), 1587-91 (1964).

2438. Cs^{133} and Cl^{35} NMR in Antiferromagnetic $CsNiCl_3$, R. H. Clark and W. G. Moulton, *Phys. Rev. B* **5**(3), 788-96 (1972).

2439. Electrical Resistivity and the Depression of the Néel Temperature in Cr-Mo and Cr-Fe, M. A. Mitchel and J. F. Goff, *Phys. Rev. B* **5**(3), 1163-70 (1972).

2440. Antiferromagnetism in the $V_xCr_{1-x}B_2$ System, R. J. Schoenberger and R. G. Barnes, pp. 252-55 in IS-2600 (Ames Lab., Iowa State Univ., Dec. 1971).

2441. Mössbauer Study of the Magnetic Properties of the Monoborides: $M_{1-x}Fe_xB$ (M = Ti, Cr, V, Mn, Co), D. DeYoung and R. G. Barnes, pp. 256-57 in IS-2600 (Ames Lab., Iowa State Univ., Dec. 1971).

2442. Low-Temperature Heat Capacity of $LaPt_2$ and $CePt_2$ and the Magnetic Susceptibility of $CePt_2$, R. R. Joseph, K. A. Gschneidner, Jr., and R. E. Hungsberg, *Phys. Rev. B* **5**(5), 1878-85 (1972).

2443. On the Formation of the Ferromagnetic Phase in $MnAs_{0.97}P_{0.03}$, K. Barner, *Phys. Stat. Sol. A* **8**, 119-24 (1971).

2444. Magnetoelectric Effect in Ferroelectric-Antiferromagnetic $Li(Fe_{0.5}Ta_{0.5})O_2F$, I. H. Ismailzade, R. G. Yakupov, and T. A. Melik-Shanazarova, *Phys. Stat. Sol. A* **8**, K63 (1971).

2445. Magnetic Phase Transitions in the UAS-USe System, J. Leciejewicz, R. Troc, A. Murasik, and T. Palewski, *Phys. Stat. Sol. B* **48**, 445 (1971).

2446. Magnetic Properties of Protactinium Tetrachloride, M. E. Hendricks, E. R. Jones, Jr., J. A. Stone, and D. G. Karraker, *J. Chem. Phys.* **55**(6), 2993-2997 (1971).

2447. Magnetic Susceptibility of α-Plutonium, Alain Blaise and Jean-Marc Fournier, *Solid State Commun.* **10**, 141-44 (1972).

2448. Structures Magnetiques des Composés MFe_2Se_4 (M = Ti, V, Cr, Fe, Co, Ni), B. Lambert-Andron, G. Berodias, and D. Babot, *J. Phys. Chem. Solids* **33**, 87-94, (1972).

2449. A Magnetization and Susceptibility Study of the Order-Disorder Transformation in VAu_4; Short Range Order, J. R. Asik, R. J. Gambino, and T. R. McGuire, *J. Phys. Chem. Solids* **33**, 391-399 (1972).

2450. Etude des Composés Antiferromagnetiques Yb_2O_2S et Yb_2O_2Se Par Effet Mössbauer Sur Yb^{170}, F. Gonzalez Jimenez and P. Imbert, *Solid State Commun.* **10**, 9-13 (1972).

2451. Ferro- and Metamagnetism of Rare Earth Compounds, G. Busch, P. Junod, and O. Vogt, *Phys. Letters* **6**(1), 79-83 (1963).

2452. Crystallographic, Optical, and Magnetic Properties of (Monoclinic) Eu_2SiO_4, R. Verreault, *Phys. Kondens. Materie* **14**, 37-54 (1971).

2453. Spin Fluctuation Effects in Early Antiferromagnetic Vanadium and Chromium Diborides, J. Castaing, P. Costa, M. Heritier, and P. Lederer, *J. Phys. Chem. Solids* **33**, 533-547 (1972).

2454. Magnetic Transitions in Manganese Sulfates $MnSO_4$ and $MnSO_4 \cdot H_2O$, M. Lecomte, J. de Gunzbourg, M. Teyrol, A. Miedan-Gros, and Y. Allain, *Solid State Commun.* **10**, 235-237 (1972).

2455. New Antiferromagnetic Intermetallic Compound in the Mn-Pd System: $MnPd_2$, G. Kadar, E. Kren, and M. Marton, *J. Phys. Chem. Solids* **33**, 212-215 (1972).

2456. The Specific Heat of Dilute Palladium-Manganese Alloys; Critical Behavior and Magnetic-Field Dependence, B. M. Boerstoel, J. J. Zwart, and J. Hansen, *Physica* **57**, 397-420 (1972).

2457. Magnetic Properties of UAsY Compounds (Y = S, Se, Te), A. Zygmunt and M. Duczmal, *Phys. Stat. Sol. A* **9**(2), 659-65 (1972).

2458. Magnetic Properties of Some Uranium Compounds of AlB_2 Type, A. Misiuk, J. Mulak, A. Czopnik, and W. Trzebiatowski, Reprint dated Oct. 27, 1971; to be published in *Bull. Acad. Polon. Sci., Ser. Sci. Chim.*

2459. Temperature Dependence of the Electric Resistance of Mn_3B_4, V. N. Novogrudskii, V. I. Timoshchuk, and I. G. Fakidov, *Soviet Phys.-Solid State* **13**(9), 2372-3 (1972).

2460. Anomaly in the Thermoelectric Power of Cobalt at the Curie Point, Ya. A. Kraftmakher and T. Yu. Pinegina, *Soviet Phys.-Solid State* **13**(9), 2345 (1972).

2461. Pressure-Temperature Phase Diagram and Crystal Structure of NiS, D. B. McWhan, M. Marezio, J. P. Remeika, and P. D. Dernier, *Phys. Rev. B* **5**(7), 2552-56 (1972).

2462. Magnetic Properties of $FeCl_2$ in Zero Field. I. Excitations, R. J. Birgeneau, W. B. Yelon, E. Cohen, and J. Makovsky; II. Long Range Order, W. B. Yelon and R. J. Birgeneau, *Phys. Rev. B* **5**(7) 2607-21 (1972).

2463. EuTe. I. Magnetic Behavior of Insulating and Conducting Single Crystals, N. F. Oliveira, Jr., S. Foner, and Y. Shapira, *Phys. Rev. B* **5**(7), 2634-46 (1972).

REFERENCES

2464. EuTe. II. Resistivity and Hall Effect, Y. Shapira, S. Foner, and N. F. Oliveira, Jr., *Phys. Rev. B* **5**(7), 2647-57 (1972).
2465. EuTe. III. Ultrasonic Behavior, Y. Shapira, and T. B. Reed, *Phys. Rev. B* **5**(7), 2657 (1972).
2466. Short-Range Magnetic Order in CoO, M. D. Rechtin and B. L. Averbach, *Phys. Rev. B* **5**(7), 2693-704 (1972).
2467. Magnetic Behavior and Microstructure of Vanadium Phosphate Glasses, E. J. Friebele, L. K. Wilson, and D. L. Kinser, *J. Am. Ceram. Soc. Vol.* **55**, 164-68 (March, 1972).
2468. Weak Ferromagnetism in $NaMnF_3$, I. Maartense, (Department of Physics, University of Manitoba Winnipeg, Canada) Reprint, to appear in *Intern. J. Magnetism*.
2469. Magnetisme—Temperature de Néel de l' Hydroxyde Cuivrique, Michel Diot, Puerre Turlier, Jean Claude Volta, and M. Marcel Prettre, *Compt. Rend. B* **274**, 225-8 (1972).
2470. Magnetism In Rhombohedral Iron-Titanium Oxides, M. F. Westcott-Lewis and L. G. Parry, *Australian J. Phys.* **24**, 719-34 (1971).
2471. Etude des Propriétés Magnetiques du Nitrure de Chrome CrN, J. Chaussy, R. Georges, P. Mollard, and J. Voiron, *J. Phys.* (France), Colloq. 32(2-3) Suppl., C1-977-C1-979 (1971).
2472. Nuclear Magnetic Resonance Study of Transition Metal Monoborides. I. ^{11}B Nuclear Electric Quadrupole Interaction Parameters, R. B. Creel and R. G. Barnes, *J. Chem. Phys.* **56**(4), 1549-62 (1972).
2473. Magnetic and Electrical Properties of Ba_2MReO_6 Ordered Perovskites, A. W. Sleight and J. F. Weiher, *J. Phys. Chem. Solids* **33**, 679-687 (1972).
2474. Electronic and Magnetic States in Metallic Compounds I. Kondo Effect and Ferromagnetism in CoAl, *J. Phys. Chem. Solids* **33**, 561-576 (1972).
2475. Magnetism, Dimensional Changes, and Magnetic Transitions in Hydrated Cesium Manganese Chloride, John A. Wollam and Paul R. Aron, Lewis Research Center, NASA, Cleveland, Ohio, NASA TN D-6652, (February, 1972), 22 pp.
2476. Study of Magnetic Ordering in Uranium Monochalcogenides By Neutron Diffraction, F. A. Wedgwood and Moshe Kuznietz, in *Rare Earths and Actinides,* **168** (short papers from the conference held at the University of Durham, from 5-7 July 1971), (The Institute of Physics, London and Bristol).
2477. Verlauf der Optischen Doppelbrechung bei Antiferromagnetischen Phasenubergangen, Irmin-Rudolf Jahn, Ph.D. thesis Eberhard-Karls-Universitat zu Tubingen, (1971).
2478. An Anomaly in the Specific Heat of $CoF_2 5HF,6H_2O$ and $CoSiF_6,6H_2O$ in the Region 240-275°K, S. K. Dutta Roy, B. Ghosh, S. Kar, and S. N. Kaul, *J. Phys. Chem. Solids* **33**, 755-757 (1971).

Recent Reviews, Conferences, and Compilations on Magnetic Materials

1972

Nuclear Magnetic Resonance Study of Transition Metal Monoborides. I. ^{11}B Nuclear Electric Quadrupole Interaction Parameters
    R. B. Creel and R. G. Barnes
    J. Chem. Phys. 56:1549 (1972)

Magnetism and Magnetic Materials 1971
    C. D. Graham, Jr. and J. J. Rhyne, eds.
    Proceedings of the 17th Annual Conference on Magnetism and Magnetic Materials, held at Chicago, Nov. 16-19, 1971
    AIP Conference Proceedings No. 5
    American Institute of Physics, New York (1972), 1616 pp.

Physics of Electronic Ceramics, Part B
    L. L. Hench and D. B. Dove, eds.
    Marcel Dekker, Inc., New York (1972)
    Electronic Phenomena in Ceramics Conference, University of Florida, 1969; in two parts: Theory of antiferromagnetism and ferrimagnetism, J. B. Goodenough, pp. 777-832; Microstructure and processing of ferrites, F. J. Schnettler, pp. 833-856; Microwave garnet compounds, G. R. Harrison and L. R. Hodges, Jr., pp. 857-914

Magnetic Resonance Reviews
    Charles P. Poole, Jr., ed.
    New quarterly review journal will survey magnetic resonance literature on a yearly basis, with the main field subdivided into a number of categories which, according to the volume of literature put out in each, will be reviewed annually or every two or three years. Among the articles scheduled for the first issues is "Guide to the Magnetic Resonance Literature," (H. A. Farach and C. P. Poole, Jr.).

Magnetism of Rare-Earth Elements, Alloys, and Compounds
    J. J. Rhyne and T. R. McGuire
    IEEE Trans. Magnetics, MAG-8:105-130 (1972)

1971

IVth International Colloquium on Magnetic Thin Films, Prague, Czechoslovakia, Sept. 21-23, 1970
    CONF-700963
    Czech. J. Phys. Vol. 21B, Nos. 4-5 (1971), entire issue

Symposium on Magneto Optical Effects, Dec. 11-12, 1969
    Symposia Faraday Soc., No. 3 (1969, published Aug. 1971), entire issue

Applications of Micromagnetics
    Amikam Aharoni
    CRC Critical Reviews in Solid State Sciences 2(2):121-180 (1971)
    266 refs.

Domain Walls and Micromagnetics
    A. Aharoni
    J. Phys. Colloq. 32(2-3)(Pt. 2):966-971 (1971).

Magnetic Properties of the Platinum Metals and Their Alloys
    H. J. Albert and L. R. Rubin
    Platinum Group Metals and Compounds, Adv. Chem. Series 98, American Chemical Soc., Washington, D. C. (1971), pp. 1-16
    Symposium sponsored by the Division of Inorganic Chemistry at the 158th Meeting of the American Chemical Society, New York N. Y., Sept. 8-9, 1969

Unusual Magnetic Field Effects in Organic Solids
    P. Avakian and A. Suna
    Mat. Res. Bull. 6:891-898 (1971)

Magnetoelectricity in Theory and Experiment
    E. F. Bertaut and M. Mercier
    Mat. Res. Bull. 6:907-922 (1971)
    Theoretical and experimental aspects of magnetoelectricity in magnetically ordered systems

Magnetomechanical Effects in the Rayleigh Region
    R. R. Birss
    IEEE Trans. Magnetics, MAG-7:113-133 (1971)

Tables of the Frequency Dependence of Permeability for the Polivanov Domain Model
    J. E. L. Bishop
    J. Phys. D: Appl. Phys. 4:1235-1238 (1971)
    Tabulated as functions of frequency and domain size

A Review of Methods for the Examination of Magnetic Domain Structure
    M. J. Bowman and A. D. Booth
    Metallography 4(2):103-131 (1971)

Materials for Magnetic Functions
    F. N. Bradley
    Hayden Series in Materials for Electrical and Electronic Design, A. E. Javitz, ed., Hayden Book Co., Inc., New York (1971)

Magnetic Properties of Actinides
    M. B. Brodsky
    Rare Earths and Actinides (Short papers from the Conference held at the University of Durham, July 5-7, 1971) The Institute of Physics, London and Bristol, pp. 76-83

Review of New Materials for the Generation of Very Low Temperatures by Hyperfine-Enhanced Nuclear Adiabatic Cooling
    E. Bucher, K. Andre, J. P. Maita, A. S. Cooper, and L. D. Longinotti
    J. Phys. (Paris), Colloq. Vol. 32, No. 2-3, Cl. 115-116 (1971)
    From International Conference on Magnetism, Grenoble, France, Sept. 14, 1970, CONF-700940

Rare-Earth-Cobalt Intermetallic Compounds
    K. H. J. Buschow
    Philips Res. Repts. 26:49-64 (1971)
    Compounds formed and dependence of stability on RE metallic radius

Intermetallic Compounds of Rare Earth Elements and Ni, Co, or Fe
    K. H. J. Buschow
    Phys. Stat. Sol. 7A:199-210 (1971)
    Survey of the composition and crystal structure

Structure Electronique et Propriétés Magnetiques des Diborures des Metaux de Transition
    Jacques Castaing
    Publication No. 140 (1971), Office National d'Etudes et de Recherches Aerospatiales, 29, Avenue de la Division Leclerc, 92 Chatillon, France, 61 pp.
    116 refs.

Review of Magnetic Properties of Fe-Ni Alloys
    G. Y. Chin
    IEEE Trans. Magnetics, MAG-7:102-113 (1971)
    90 refs.

Fundamentals of Magnetostriction
    B. D. Cullity
    J. Metals 23:35-41 (1971)

Etude Magnetique et Magneto-Optique d'Antiferromagnetiques dans des Champs trés Intenses (500 kOe)
    J. de Gunzbourg
    Ph. D. thesis, University of Paris, CNRS No. 5701 (1971), 146 pp.

Diffraction Method of Magnon Investigations at the Pulsed Neutron Source
    L. Dobrzynski and H. Kepa
    Joint Institute for Nuclear Research, Dubna, USSR
    JINR-P3-5944 (1971), 16 pp.
    In Russian

Magnetism in the Rare Earth Metals - A Review
    R. J. Elliott
    Oxford University, England
    CONF-711001-(Vol. 1), pp. 200-207 (1971)
    From Ninth Rare Earth Research Conference, Blacksburg, Virginia, Oct. 10, 1971

Magnetic Materials Analyses by Nuclear Spectrometry: A Joint Approach to Mössbauer Effect and Nuclear Magnetic Resonance
    Dieter Feldmann, H. R. Kirchmayr, Albert Schmolz, and Mircea Velicescu
    IEEE Trans. Magnetics, MAG-7:61-91 (1971)
    Elements and compounds tabulated; measurement methods surveyed;
    276 refs.

Review on Domains in Thin Magnetic Samples
    E. Feldtkeller
    J. Phys., Colloq. 32(2-3) (Part 1):452-456 (1971)

Proceedings of the 9th Rare Earth Research Conference
    P. E. Field, ed.
    CONF-711001-(Vol. 1), 1971, 442 pp.
    Held in Blacksburgh, Virginia, October 10-14, 1971

Electron-Magnon Interaction in the Itinerant Electron Ferromagnet
    Isoris S. Gergis
    Thesis Case Western Reserve University, Cleveland, Ohio, Sept. 1971
    Tech. Rept. 71-72

Anomalous Properties of the Vanadium Oxides
    J. B. Goodenough
    Annual Rev. Mater. Sci. 1:101-138 (1971)
    Annual Reviews, Inc., Palo Alto, Calif.

Cobalt Alloy Permanent Magnets
    J. E. Gould
    Permanent Magnet Association Central Research Lab., Sheffield, England
    Cobalt Monograph Series (Centre d'Information du Cobalt, Brussels, 1971)
    A review of the current technology of the alnicos and other cobalt alloy permanent magnets; 82 pp., author and subject index

Proceedings of the Sixteenth Annual Conference on Magnetism and Magnetic Materials
    C. D. Graham, Jr., R. J. Joenk, Jr., and J. J. Rhyne, eds.
    J. Appl. Phys. Vol. 42, No. 4 (1971)
    Held Nov. 17-20, 1970, Miami Beach, Florida

Mössbauer Spectroscopy
    N. N. Greenwood
    Chapman & Hall, London (1971), 672 pp.

Mössbauer Effect Methodology, Vol. 6
    Irwin J. Gruverman, ed.
    Plenum Press, New York, 1971, 230 pp.
    Proceedings of the Sixth Symposium on Mössbauer Effect Methodology, held in Chicago, Illinois, Jan. 25, 1970

Contribution à l'Etude et à l'Interpretation des Processus d'Aimantation des Ferromagnetiques
    J. Guilmart

Ph. D. Thesis, Grenoble, CNRS No. 5861 (May 19, 1971), 130 pp.
Review of experimental methods

Etude des Structures Cristallines et Magnetiques par Diffraction de Neutrons
J. Guyader
Rev. Chim. Miner. 8:185-213 (1971)
417 refs.

Magnetostrictive Phenomena in Metallic Materials and Some of Their Device Applications
R. M. Hornreich, Harvey Rubinstein, and R. J. Spain
IEEE Trans. Magnetics, MAG-7:29-48 (1971)
104 refs.

A New Technique for Neutron Diffraction
C. R. Hubbard, C. O. Quicksall and R. A. Jacobson
Nucl. Instr. Methods 94:185-186 (1971)

Verlauf der optischen Doppelbrechung bei antiferromagnetischen Phasenübergängen
Irmin-Rudolf Jahn
Ph. D. thesis, Eberhard-Karls-Universität zu Tübingen, 1971
Binary metal fluorides

The Presently Known Electronic and Magnetic Properties of Uranium Monopnictides and Monochalcogenides
M. Kuznietz
Rare Earths and Actinides, (Short papers from the conference held at the University of Durham, July 5-7, 1971) The Institute of Physics, London and Bristol (1971), pp. 162-167

Metal-Deficient Sulphospinels
F. K. Lotgering and G.H.A.M. van der Steen
J. Solid State Chem. 3:574-581 (1971)

Magnetic Film Materials
F. E. Luborsky and R. O. McCary
Proc. IEEE 59(10):1474-1480 (1971)

Theory of Thermal Neutron Scattering — The Use of Neutrons for the Investigation of Condensed Matter
W. Marshall and S. W. Lovesey
International Series of Monographs on Physics
Oxford University Press, New York (1971)

Neutron Radiography
R. S. Matfield
Atom, No. 174 pp. 84-99 (1971)

Conduction Electron Scattering and the Resistance of the Magnetic Elements
G. T. Meaden
Contemp. Phys. 12:313-337 (1971)
Rare-earth elements and some transition metals

Fourier Techniques for Analysis of Magnetic Structure Factors: Application to 3D Metals
R. M. Moon
Intern. J. Magnetism 1:219-231 (1971)

Electron-Phonon Interaction and Paramagnetic Susceptibilities for Transition Metals
Nobuo Mori
J. Phys. Soc. Japan 31:359-367 (1971)

Magnetism and Local Molecular Field
Louis Neel
Science 174:985-992 (1971)

Crystal Field Effects in Rare Earth Hexaborides
John C. Nickerson III
Thesis, Stanford University, Stanford, California, 1971, 235 pp.
Available from University Microfilms, Ann Arbor, Michigan, Order No. 71-19,736

Tables of Magnetic Structures Determined by Neutron Diffraction
Part I — Cubic System (Report No. 1/PS, 1970)
Part II — Tetragonal System (Report No. 4/PS, 1970)
A. Oles, A. Bombik, M. Kucab, W. Sikora, and F. Kajzar
Part III — Rhombohedral and Hexagonal Systems. Elaboration and Redaction (Report INT-7; NP-18845, 1971)
Part V — Monoclinic system (Report INT-11; NP-19010, 1971)
A. Oles A. Bombik, F. Kajzar, and M. Kucab
Institute of Nuclear Techniques, Cracow, Poland
Tables contain: chemical symbols, space groups, cation distribution, crystal distortion temperatures, magnetic transition temperatures, descriptions of magnetic structures including figures, magnetic moments, magnetic space groups for some structures

Classifications of Magnetic Structures
W. Opechowski and Tommy Dreyfus
Acta Cryst. A27:483 (1971)

Nuclear Magnetic Resonance in Ferromagnets and Antiferromagnets
M. P. Petrov and E. A. Turov
Appl. Spectros. Revs. 5(2):265-331 (1971)

A Review of Rare Earth — Cobalt Alloy Systems
A. E. Ray
Determination of Certain Properties and Characteristics of Rare Earth Metals, Laser Crystals, Semiconductors, and Dielectric Materials, A. E. Ray and G. I. Hoffer (University of Dayton Research Institute), AFML-TR-71-36 (May 1971), pp. 11-28
Phase diagrams; melting points; alloying behavior; crystal structures

Phase Diagrams for the Ce — Co, Pr — Co, and Nd — Co Alloy Systems
A. E. Ray and G. I. Hoffer
Determination of Certain Properties and Characteristics of Rate Earth Metals, Laser Crystals, Semiconductors, and Dielectric Materials, A. E. Ray and G. I. Hoffer (University of Dayton Research Institute), AFML-TR-71-36 (May 1971), pp. 3-9

Magnetic and Inelastic Scattering of Neutrons by Metals
T. J. Rowland and Paul A. Beck, eds.
Gordon and Breach, London (1971), 256 pp.

Review of the Structures and Properties of the Rare Earth-Iron Compounds of Formula R_2Fe_{17}
J. E. Schweizer
Determination of Certain Properties and Characteristics of Rare Earth Metals, Laser Crystals, Semiconductors, and Dielectric Materials, A. E. Ray and G. I. Hoffer (University of Dayton Research Institute), AFML-TR-71-36 (May 1971), pp. 30-58
Methods of preparation, phase diagrams, crystal structures, magnetic properties
55 refs.

Observation of Antiferromagnetic Phase Transitions by Ultrasonic Techniques
  Y. Shapira
  J. Appl. Phys. 42:1588-1594 (1941)

Mössbauer Effect Data Index Covering the 1969 Literature
  John G. Stevens and Virginia E. Stevens, eds.
  Plenum Publishing Corp., New York (1971), 292 pp.
  Cumulative index to the experimental data and literature of Mössbauer effect research from 1969 is an up-dating of Mössbauer Effect Data Index, 1958-1965, published in 1967

The Development of Permanent Magnets Based on Cobalt and Rare Earths
  K. J. Strnat and A. E. Ray
  Determination of Certain Properties and Characteristics of Rare Earth Metals, Laser Crystals, Semiconductors, and Dielectric Materials, A. E. Ray and G. I. Hoffer (University of Dayton Research Institute), AFML-TR-71-36 (May 1971), pp. 61-95

Investigations of Ferromagnetic Domain Walls by Lorentz Microscopy
  T. Suzuki
  Z. Angew. Phys. 32:75-80 (1971)

Quantitative Analysis of Inelastic Scattering in Two-Crystal and Three-Crystal Neutron Spectrometry; Critical Scattering from $RbMnF_3$
  A. Tucciarone, H. Y. Lau, L. M. Corliss, A. Delapalme, and J. M. Hastings
  Phys. Rev. 4B:3206-3245 (1971)
  The use of the neutron spectrometer in the quantitative determination of neutron scattering cross sections is discussed in detail

Some Recent Achievements in Magnetic Thin Films
  L. Valenta
  J. Phys., Colloq. 32(2-3) (Part 1):15-18 (1971)

Magnetism of Intermetallic Compounds
  W. E. Wallace
  Vol. 6 in Progress in Solid State Chemistry, Pergamon Press (1971)

Mössbauer Study of Phosphides Containing Iron
  R. Wappling, L. Hoggstrom, S. Rundqvist, and E. Karlsson
  J. Solid State Chem. 3:276-292 (1971)

Magnetic Resonance in Metals
  J. Winter
  International Series of Monographs on Physics
  Oxford University Press, Oxford (1971), 304 pp.

Recent Mössbauer Effect Studies in Intermetallic and Rare-Earth Compounds
  W. Zinn
  J. Phys. Colloq. 32(2-3)(Part 2):724-730 (1971)

1970

Third All-Polish Seminar on Nuclear Magnetic Resonance and Its Use, Cracow, Poland, December 1-2, 1970
  Institute of Nuclear Physics, Cracow, INP-747; CONF-701216 (Dec. 1970), 172 pp.
  In Polish

Les Elements des Terres Rares, Paris-Grenoble, Mai 5-10, 1969
  Colloques Internationaux du Centre National de la Recherche Scientifique, No. 180, Vol. II
  Editions du Centre National de la Recherche Scientifique, Paris, (1970)
  Magnetisme des Elements et Composes, pp. 17-98; Ordre Magnetique dans les Alliages par Diffraction Neutronique, pp. 211-274; Ordre Magnetique dans les Oxydes et Autres Metalloides, pp. 291-374

1970 Digests of the International Magnetics Conference
  IEEE, New York (1970)
  Conference held at Washington, D. C., April 1970; sponsors: IEEE Magnetics Group, 165 pp.
  Emphasis is on applications of magnetic materials

Nuclear Magnetic Resonance. Vol. 1: A DCC Bibliography, January 1966-November 1969
  Defense Documentation Center, Alexandria, Va., AD-703400; DDC-TAS-70-18-1-Vol-1 (March 1970), 345 pp.

Izv. Akad. Nauk SSSR, Ser. Fiz. Vol. 34, No. 5 (1970), entire issue
  Ferrites: articles in Russian

L'Antiferromagnetisme Nucléaire. Un Nouveau Phenomène
  A. Abragam and M. Goldman
  Fiz.-Mat. Spis., Balg. 13(4):274-287 (1970)

Possible Species of Ferromagnetic, Ferroelectric, and Ferroelastic Crystals
  Keitsiro Aizu
  Phys. Rev. B2:754-772 (1970)
  A determination is made of all possible species of full ferromagnetics, partial ferromagnetics, full ferroelectrics, partial ferroelectrics, full ferroelastics, and partial ferroelastics, and it is found out in which of these species two or all of ferromagnetism, ferroelectricity, and ferroelasticity should couple completely or incompletely with each other

Magnetic Semiconductors
  I. G. Austin and D. Elwell
  Contemp. Phys. 11:455-476 (1970)

Rare-Earth-Compound Permanent Magnets
  J. J. Becker
  J. Appl. Phys. 41:1055-1064 (1970)
  78 refs.

La Physique des Transitions
  N. Boccara
  Presses Univ. France, Paris (1970), 127 pp.

Magnetization and Electrical Conductivity Studies of Some Transition Metal Oxide Glasses
  Robert M. Catchings III
  Thesis, Wayne State University, Detroit, Michigan, 1970, 143 pp.
  Available from University Microfilms, Ann Arbor, Michigan, Order No. 71-390

Magnetic Resonance
  C. K. Coogan, N. S. Ham, S. N. Stuart, J. R. Pilbrow, and G. V. H. Wilson, eds.
  Plenum Press, New York (1970), 386 pp.
  Proceedings of an International Symposium, Melbourne, August 1969

RECENT REVIEWS, CONFERENCES, AND COMPILATIONS

Film Ferromagnetiques
  P. Favard, ed.
  Proceedings of the 7th International Congress on Electron Microscopy, Grenoble, France, 1970, Vol. II, pp. 605-628
  Soc. Fr. Micr. Electron., Paris (1970)
  12 articles on studies of magnetic domains

Magnetostriction of Paramagnetic Transition Metals
  E. Fawcett
  Phys. Rev. B2:1604-1613 (1970)

A Table of Magnetic Hyperfine-Structure Anomalies
  G. H. Fuller and V. W. Cohen
  ORNL-4591 (September 1970), 26 pp.
  References to the original papers are given for the ratios listed, along with a code letter which indicates the experimental methods used; the cutoff date for the literature search is Dec. 1969

Proceedings of the Fifteenth Annual Conference on Magnetism and Magnetic Materials
  C. D. Graham, Jr. and R. J. Joenk, Jr., eds.
  J. Appl. Phys. Vol. 41, No. 3 (1970)
  Held Nov. 18-21, 1969, Philadelphia, Penn.

Landolt-Bornstein Numerical Data and Functional Relationships in Science and Technology, Group III: Crystal and Solid State Physics, Volume 4, Magnetic and Other Properties of Oxides and Related Compounds, Parts A and B
  K.-H. Hellwege and A. M. Hellwege, eds.
  Springer-Verlag, Berlin, Heidelberg, New York (1970)
  Crystallographic and magnetic data are given for all substances included; the optical, electrical, and technological data are also given for most of them, when of special importance and not yet contained in other volumes of the New Series; materials index.)

Molecular Engineering in the Design of Microwave Ferrimagnetic Garnets
  A. S. Hudson
  Rev. Phys. Tech. 1:9-26 (1970)

Theory of Magnetically Ordered Crystals with Impurities
  Yu. A. Izyumov and M. V. Medvedev
  Izdat. Nauka, Moscow (1970), 271 pp.

Magnetic Neutron Diffraction
  Yu. A. Izyumov and R. P. Ozerov
  Plenum Press, New York (1970), 598 pp.

Ferromagnetic Domains in Uniaxial Materials
  Jan Kaczer
  IEEE Trans. Magnetics, MAG-6:442-445 (1970)
  Brief survey of the essential features of the basically stable domain structures in uniaxial materials at remanence

Dielectric, Elastic and Magnetic Property-Symmetry and Phase Transition Representation
  E. T. Keve and S. C. Abrahams
  Ferroelectrics 1:243-246 (1970)

Table of Hyperfine Fields
  T. A. Koster and D. A. Shirley
  University of California, Berkeley
  UCRL-20411; CONF-700933-5 (Nov. 1970), 17 pp.
  From Conference on Hyperfine Interactions Detected by Nuclear Radiation, Rehovoth and Jerusalem, Israel

Sound Propagation near Magnetic Phase Transitions
  B. Luthi, T. J. Moran, and R. J. Pollina
  J. Phys. Chem. Solids 31:1741-1758 (1970)

Survey of the Field of Magnetic Semiconductors
  S. Methfessel
  IBM J. Res. Develop. 14:207-213 (1970)

Chalcogenide Spinel Magnetic Semiconductors
  Kazuo Miyatani
  Kotai Butsuri 5:1-20 (1970)
  More than one hundred compounds are summarized along with their magnetic, electrical and optical properties; in Japanese, tabulation in English

The Electrodynamics of Magneto-Electric Media
  T. H. O'Dell
  North-Holland, Amsterdam (1970)

Magnetic Symmetry and Transport Properties of Crystals
  P. V. Pantulu and E. Sudarshan
  J. Phys. C3:604-609 (1970)
  For the 122 classes of crystals; form of fourth order tensors

Magnetic Form Factor and Magnetization Densities by Means of Polarized Neutrons
  A. Paoletti
  Riv. Nuovo Com. 2:451-497 (1970)
  Review of method; examples of metals, alloys, and compounds

Ordre Magnetique et Champs Forts
  R. Pauthenet
  Ann. Phys. 5:61-62 (1970)
  73 refs.

Magnetostriction of Ferrites
  G. A. Petrakovskii
  Izv. Akad. Nauk SSSR Ser. Fiz. 34:1052-1063 (1970)
  In Russian
  English translation in Bull. Acad. Sci. USSR Phys. Ser., U.S.A.

Critical Attenuation of Sound at Magnetic Phase Transitions
  Richard James Pollina
  Ph. D. Thesis, Rutgers University, The State University of New Jersey, 1970

Magnetic Materials, 1970
  Marshall Sittig
  Noyes Data Corporation, Park Ridge, New Jersey (1970)
  Techniques for producing magnetic materials based on the U. S. Patent literature since 1960

Magnetically Ordered Ferroelectric Materials
  Selby M. Skinner
  IEEE Trans. on Parts, Materials and Packaging PMP-6:68-90 (1970)
  Theory; possible applications; microwaves

Propriétés et Applications des Materiaux Semi-Conducteurs Magnetiques
  Jacques P. Suchet
  Ann. Phys. 5:67-76 (1970)
  Crystallographic and magnetic transitions, magnetic dispersion, magnetoresistance, Faraday effect, Astrov effect, Hall effect, etc.

Kondo Effect in Some Intermetallic Compounds of Ce
   H. J. van Daal and K. H. J. Buschow
     Phys. Stat. Sol. A3:853 (1970)
     Review; 53 refs.

Anomalous Properties of Dilute Magnetic Alloys, A Critical Review of Some Experiments
   J. E. van Dam and G. J. van den Berg
     Phys. Stat. Sol. A3:11-26 (1970)
     Review; 85 refs.

Comments on Electronic Transport in Magnetic Semiconductors
   S. von Molnar and T. Kasura
     Proceedings of the Tenth International Conference on the Physics of Semiconductors, Cambridge, Massachusetts, August 17-21, 1970, pp. 233-242, CONF-700801

Thermal Neutron Diffraction
   B. T. M. Willis, ed.
     Harwell Postgraduate Series
     Oxford University Press (1970), 220 pp.
     Proceedings of International Summer School, Harwell, July 1-5, 1968, on the Accurate Determination of Neutron Intensities and Structure Factors

1969

Magnetism of Pyrite-Type Compounds
   Kengo Adachi
     Nippon Butsuri Gakkaishi 24:518-528 (1969)
     Review of the magnetic and other properties of transition-metal chalcogenides

Magnetic Resonance and Radiofrequency Spectroscopy
   P. Averbuch, ed.
     North-Holland, Amsterdam (1969)
     Proceedings of the XVth Colloque A.M.P.E.R.E., Grenoble, September 1968.

Neutron Scattering Techniques
   G. E. Bacon
     Techniques for the Direct Observation of Structure and Imperfections, Vol. II, Part 2 of Techniques of Metals Research, R. F. Bunshah, ed. (Interscience, New York, 1969), pp. 642-671

Magnetism and Metallurgy, Vol. 2
   A. E. Berkowitz and E. Kneller, eds.
     Academic Press, New York and London (1969)

Nuclear Magnetic Resonance in Magnetic Materials
   Monisha Bose
     Progr. Nucl. Magnet. Reson. Spectrosc. 4:335-444 (1969)
     279 refs.

Basic Research and Applications in Magnetism
   R. M. Bozorth
     IEEE Trans. Magnet. 5:692-700 (1969)
     123 refs.

Table of Antiferromagnetic Materials Studied by Neutron Diffraction
   D. E. Cox
     Brookhaven National Laboratory, Upton, N. Y.
     BNL-13822 (June 1969), 91 pp.

Nuclear Magnetic Resonance in Transition Metal Borides
   Roger B. Creel
     Ph. D. Thesis, Iowa State University, IST 323 (Nov. 1969)

Magnetic Interactions in Rare Earth Borides
   Zachary Fisk
     Thesis, University of California, San Diego, California, 1969, 96 pp.

Magnetic First Order Phase Transitions
   N. P. Grazhdankina
     Soviet Phys. Usp. 11:727-745 (1969)
     125 refs.

Spin Diffusion and Nuclear Magnetic Relaxation in a Crystal Containing a Magnetic Impurity
   G. R. Khutshshvili
     Soviet Phys. Usp. 11(6):802-815 (1969)
     Review; 50 refs.

Transparent Ferromagnets
   G. S. Krinchik and M. V. Chetkin
     Soviet Phys. Usp. 12:307-319 (1969)
     Review; 106 refs.

Theory of X-Ray and Thermal Neutron Scattering by Real Crystals
   Mikhail Krivoglaz
     Plenum Press, New York (1969)

Physics of Ferrites and Related Oxides
   Svatopluk Krupicka
     Academia, Prague (1969), 594 pp.
     In Czech

Magnetic Properties of Binary Rare Earth Alloys
   Arthur H. Millhouse
     Thesis, Virginia Polytechnic Institute, Blacksburg, Va., TID-25248 (Sept. 1969), 135 pp.

Etude de Structures Cristallines et Magnétiques par Diffraction de Neutrons
   A. Oles
     Akad. gorn.-hutn. Krakowie, Zesz. nauk 208:145-153 (1969)

Propriétés Cristallographiques et Magnétiques des Composés Intermétalliques du Nickel avec les Metaux de Terres Rares
   D. Paccard
     Thesis, Grenoble, 1969; Arch. orig. Centre Document. C.N.R.S., No. 1478, 28 April 1967
     73 refs.

First-Order Magnetic Phase Transitions
   L. Pal
     Hungarian Academy of Sciences, Central Research Institute for Physics, Budapest
     KFKI 4/1969

Magnetic Interactions in Some Ferri- and Antiferromagnetic Normal Spinels Studied by Neutron Diffraction
   Rene Plumier
     Ph. D. Thesis, University of Paris, 1968; also CEA-R-3756 (March 1969), 148 pp.

Methode Simple d'Enumeration des Classes Cristallines, Magnetiques et à Plusieurs Couleurs
   J. Sivardiere
     Bull. Soc. Fr. Mineral. Cristallogr. 92:3-8 (1969)

RECENT REVIEWS, CONFERENCES, AND COMPILATIONS

Magnetic Materials
R. S. Tebble and D. J. Craik
Wiley-Interscience, New York (1969)

Phase Transitions, Soft Modes, and Critical Fluctuations
H. Thomas
IEEE Trans. Magnet. 5:874-886 (1969)

Recent Developments in Conventional Soft Magnetic Materials
J. E. Thompson
Brit. J. Appl. Phys.: J. Phys. D, Ser. 2, 2:933-952 (1969)
Grain-oriented silicon-iron, high-permeability materials and thin ferromagnetic films

Nuclear Magnetic Resonance in Ferro- and Antiferro-Magnetics (Yadernyi Magnitnyi Rezonans v Ferro- i Antiferro-Magnetikakh)
E. A. Turov and M. P. Petrov
Nauka, Moscow (1969), 260 pp.
A table of NMR experimental data for individual nuclei— ^{1}H through ^{169}Tm— in ferro- and antiferromagnetics is included; 339 refs.

Heusler Alloys
Peter J. Webster
Contemp. Phys. 10:559-577 (1969)
A review is given of the principla theories that have been employed to explain their magnetic properties, and some of the more recent results are discussed in terms of the generalized molecular field theory

Review of Recent Work on the Magnetic and Spectroscopic Properties of the Rare-Earth Orthoferrites
R. L. White
J. Appl. Phys. 40:1061-1069 (1969)
Includes "bubble-domain" review; 64 refs.

Semiconducting Ferromagnetic Spinels
Peter J. Wojtowicz
IEEE Trans. Magnet. 5:840-848 (1969)
Review with 68 refs.; Cr chalcogenide spinels

1968

Magnetism in Metals and Alloys
Advanced Course, Kjeller, Norway, 1968
Institutt for Atomenergi, Kjeller, Norway, 1969
Kjeller Report No. 132

Neutron Inelastic Scattering, Vol. 2
International Atomic Energy Agency, Vienna, Austria, STI/PUB/187-Vol. 2 (1968), 471 pp.
Fourth IAEA Symposium, Copenhagen, May 20-25, 1968; magnetic systems, molecular dynamics, and experimental methods

Magnetism and Magnetic Materials 1968 Digest
Hsu Chang and Thomas R. McGuire, eds.
Academic Press, New York and London (1968), 315 pp.
Survey of the technical literature of the preceding year

Magnetic Properties of Rare Earth Metals
Bernard R. Cooper
Solid State Physics, Advances in Research and Applications, F. Seitz, D. Turnbull, and H. Ehrenreich, eds. Vol. 21, Academic Press, New York and London (1968), pp. 393-490

Localized Magnetic Impurity States in Metals: Some Experimental Relationships
M. D. Daybell and W. A. Stevert
Rev. Mod. Phys. 40(2):380-389 (1968)

Progress in Nuclear Magnetic Resonance Spectroscopy, Vol. 4
J. W. Emsley, J. Feeney, and L. H. Sutcliffe, eds.
Pergamon Publ. Company (1968)

A Bibliography on Nuclear Magnetic Resonance of Inorganic Fluorides
Jean Furnish, comp.
Los Alamos Scientific Laboratory, Los Alamos, N. M., LA-3871-MS (Feb. 1968), 17 pp.
1957-1966

Investigation of Magnetostrictive Ferrites as a Material for Ultrasonic Radiators
I. P. Golyamina
Dissertation, Akust. Inst. AN SSSR, Moscow, 1968

Low-Energy Neutron Physics
I. I. Gurevich and L. V. Tarasov
North-Holland Publishing Co., Amsterdam (1968), 607 pp.
Studies of magnetic crystals emphasized

Nuclear Magnetic Resonance Spectometry
J. P. Heeschen
Analyt. Chem. 40:560R-607R (1968)
Review; 2227 refs

Measurement Problems in Neutron Diffraction
E. A. Jumpertz
Messtechnik 76:1-9 (1968)
In German

Magnetostrictive Materials and Application on Magnetics
Y. Kikuchi
IEEE Trans. Magnetics 4:107-117 (1968)

Magnetic Semiconductors
S. Methfessel and D. C. Mattis
Handbuch der Physik (Encyclopedia of Physics), Vol. XVIII/1 (1968), pp. 389-560
All aspects of these materials, including band structure; theory of transport properties; indirect exchange, both theory and experimental evidence; optical properties; and electrical properties

Magnetic Susceptibility — Aspects of Instrumentation and Application to Chemistry and Solid State Science
L. N. Mulay
Anal. Chem. 40(5):440R-455R (1968)

New Magnetostrictive Materials and Transducers (I and II)
E. A. Neppiras
J. Sound Vibration 8(3):408-457 (1968)

Magnetism of the Rare Earths (Magnetismul Paminturilor Rare)
I. Pop
Editura Academiei Republicii Socialiste Romania, Bucharest (1968)
Comprehensive review of the rare-earth elements. The first chapter is concerned with basic definitions and behaviors, the magnetic properties of ionic compounds of the rare earths, magnetic resonance phenomena, energy levels, and crystal field effects. The next three chapters deal exclusively with the magnetic

behaviors of the metals, alloys and intermetallic compounds, and ferrites. The fifth and last chapter is on applications. 416 pages with almost 700 refs.

Catalog of Nuclear Quadrupole Interactions and Resonance Frequencies in Solids. Part I. Elements and Inorganic Compounds
  S. L. Segel and R. G. Barnes
  Ames Laboratory, Iowa
  IS-520 (Rev.), Spril 1968, 126 pp.

Neutron Diffraction — A General Tool in Physics
  C. G. Shull
    Proc. Symp. on Current Problems in Neutron Scattering, Centro Studi Nucleari Della Casaccia, Rome, Sept. 1968

Magnetically Ordered Ferroelectrics
  G. A. Smolenskii, V. A. Bokov, V. A. Isupov, N. M. Krainik, and G. H. Nedlin
    Helv. Phys. Acta 41:1187-1198 (1968)
    Review of thermodynamic theory and experimental data

Thermodynamic and Magnetic Properties of the Rare Earth Chalcogenides, Pnictides, Halides, and Binary Semi-Metallic Compounds
  Edgar F. Westrum, Jr.
    Progress in the Science and Technology of the Rare Earths, Vol. 3, Leroy Eyring, ed., Pergamon Press, New York (1968), pp. 459-574

Encyclopedia of Physics (Handbuch der Physik), Vol. 18, Part 1: Magnetism/Magnetismus
  H. P. J. Wijn, ed.
    Springer-Verlag, Berlin, Heidelberg, New York (1968)

1967

Spin Waves
  A. L. Akhiezer
    Ukr. Fiz. Zh. 12(10):1589-1605 (1967)
    In English

Magnetic Properties, II
  W. R. Angus, J. Favede, J. Hoarau, and A. Pacault
    Vol. 10 in Eigenschaften der Materie in Ihren Aggregatzustanden, Karl-Heinz Hellwege und Anne Marie Hellwege, eds., Springer-Verlag, Berlin, Heidelberg, New York (1967)

Space-Group Selection Rules for Magnetic Crystals
  A. P. Cracknell
    Progr. Theor. Phys., Japan 38:1252-1269 (1967)

Magnetism and Magnetic Materials 1967 Digest
  W. D. Doyle and A. B. Harris, eds.
    Academic Press, New York and London (1967), 280 pp.
    A survey of the technical literature of the preceding year

Magnetic Structure Studies by Mössbauer Spectroscopy
  U. Gonser
    ORNL-IIC-10, pp. 36-58
    Proceedings of Second Symposium on Low Energy X- and Gamma Sources and Applications held at University of Texas, Austin, Texas, March 27-29, 1967

Study of Magnetic Properties of Materials Using the Mössbauer Effect
  R. W. Grant
    Adv. in Chem. Series, No. 68, pp. 34-51 (1967)

Magnetic Scattering of Neutrons and Microscopic Magnetic Properties of Solids
  W. C. Koehler
    Trans. Amer. Crystallogr. Ass., 3:53-73 (1967)

Mössbauer Effect Data Index 1958-1965
  A. H. Muir, Jr., K. J. Ando, and Helen M. Coogan
    Interscience Publishers, New York, London, Sydney (1967)

Experimental Magnetochemistry, Nonmetallic Magnetic Materials, Vol. VIII, Selected Topics in Solid State Physics
  Michael M. Schieber
    Wiley-Interscience, New York (1967)
    Contains materials index and tabulations

1966

Magnetic Domains and Techniques for their Observation
  R. Carey and E. D. Isaac
    Academic Press, New York (1966)

Encyclopedia of Physics (Handbuch der Physik), Vol. 18, Part 2: Ferromagnetism/Ferromagnetismus
  H. P. J. Wijn, ed.
    Springer Verlag, Berlin, Heidelberg, New York (1966), 560 pp.
    437 pages in English

1965

The Mössbauer Effect, Bibliographical Series No. 16
    Philippine Atomic Energy Commission, Manila, STI/PUB/21/16 (1965), 140 pp.

The Physical Principles of Magnetism, Wiley Series on the Science and Technology of Materials
  A. H. Morrish
    Wiley-Interscience, New York (1965), 680 pp.

1964

Group Theory and Quantum Mechanics
  Michael Tinkham
    McGraw-Hill Book Co., New York (1964)
    Symmetries of magnetic structures, pp. 303ff

1963

Magnetism and the Chemical Bond
  J. B. Goodenough
    Wiley, New York (1963)

Magnetism, Vol. 1. Magnetic Ions in Insulators, Their Interactions, Resonances, and Optical Properties
  G. T. Rado and H. Suhl, eds.
    Academic Press, New York and London (1963)

1962

Neutron Magnetic Form Factors and X-Ray Atomic Scattering Factors for Rare-Earth Ions
  M. Blume, A. J. Freeman, and R. E. Watson
    J. Chem. Phys. 37:1245-1253 (1962)

RECENT REVIEWS, CONFERENCES, AND COMPILATIONS

Microwave Ferrites and Ferrimagnetics
: Benjamin Lax and Kenneth J. Button
  McGraw-Hill Book Co., New York, Toronto, London (1962)

Ferromagnetismus
: E. Kneller
  Springer-Verlag, Berlin (1962)

1961

The Magnetostrictive Equations and Their Coefficients
: Z. Kaczkowski
  Rozprawy Elektrotech. 7(2):235-275 (1961)

Information Centers and Other Sources

Alloy Data Center, Alloy Physics Section, Metallurgy Division, National Bureau of Standards
: Washington, D. C. 20234, telephone 301-921-2917
  Critically evaluated data on physical properties of metals and alloys; magnetic properties; Knight shift

CODATA International Compendium of Numerical Data Projects
: Springer Verlag, New York Heidelberg, Berlin (1969)
  Identifies, with address, international centers, data projects, and publications concerning (a) optical, x-ray, mass, and magnetic resonance spectra; (b) crystallographic, electrical, and magnetic properties; (c) thermophysical properties; (d) nuclear and isotopic data. Published form and availability of data are noted in each case.

NMR, NQR, EPR Current Literature Service
: Issued by the publishers of DMS-Documentation of Molecular Spectroscopy in association with the Institut für Spektroskopie, Dortmund, and DMS Scientific Advisory Board, London
  Butterworth and Company (Publishers) Ltd., London; Verlag Chemie, GmbH, Weinheim/Bergstr.
  Resonance literature from 1963; reference list and card index; 600 journals covered

Nuclear Magnetic Resonance Spectrometry Abstracts
: Science and Technology Agency, 3 Dyers Building, London, E.C. 1, England
  First issue January 1971

Monochromator and Analyzing Crystals

1971

Monochromatic X Rays
  Pierre Ballereau
  CEA-BIB-194, 1971, 101 pp.
  Commissariat à l'Energie Atomique, France
  Bibliography, to July 1968

Electron Beam Microanalysis — Part I. The Fundamental and Applications
  D. R. Beaman and J. A. Isasi
  Mater. Res. Std., pp. 8-78 (Nov. 1971)
  Contains section, and table, on crystals used in wavelength dispersive spectrometers, 2d spacing, reflecting planes, useful wavelength coverage

Solution Growth of Organic Analyzer Crystals for Ultrasoft X-Ray Spectroscopy
  G. Bohm and K. Ulmer
  J. Crystal Growth 10:175-78 (1971)
  Dioctadecyladipate (OAO, 2d = 93.8 Å), potassium acid phthalate (KAP, 2d = 26.6 Å) and thallium acid phthalate (TlAP, 2d = 25.9 Å)

Optimization of Multilayer Soap Crystals for Ultrasoft X-Ray Diffraction
  M. W. Charles
  J. Appl. Phys. 42(9):3329 (1971)
  2nd interplanar spacings listed for five Pb soap crystals; 61 refs.

Measurements of Absolute Reflectivities of Mosaic Crystals and Their Wavelength Dependence
  B. Dorner
  J. Appl. Cryst. 4(3):185-190 (1971)

Experimental and Theoretical Investigations of Dynamical Neutron Diffraction by Using Crystals with a Low Dislocation Density
  F. Eichhorn, M. Kosmowski, H.-G. Schopf, and G. E. R. Schulze
  Phys. Stat. Sol. A4:445-457 (1971)
  Double-crystal diffractometer; Ge

Neutron Diffraction by a Piezoelectric Resonator
  R. W. Englehart and A. M. Jacobs
  Nature 229 (5280):111-112 (1971)

Device for Bending Crystal Slabs for Use in Focusing X-Ray Spectrometers
  Erich Gilberg
  Rev. Sci. Instr. 42:1189-1190 (1971)
  Slabs can be bent into a curved cylindrical form having the curvature of a circle or of a logarithmic spiral, the latter giving ideal focusing for one wavelength

Neutron Reflection Properties of Bent Single Crystals
  W. Karas, H. Rauch, and E. Seidl
  Phys. Stat. Sol. A5:397-404 (1971)
  Effect of deformation on mosaic spread and rocking curves; Pb

Diffraction Polarization of Neutrons in a Polarized Crystal
  L. N. Korennaya
  Soviet Phys.-Solid State 13:727-730 (1971)
  The propagation of a coherent neutron wave through a magnetized specimen with polarized nuclei is examined

Misorientation Effects in a Monochromator Crystal
  T. L. Lowe
  J. Phys. E (Sci. Instr.) 4:1068-1071 (1971)
  Misorientation produced during bending of a monochromator crystal can lead to an x-ray beam highly nonuniform in angular distribution; oscillating the monochromator is employed to obtain a uniform beam

X-Ray Crystal Collimators Using Successive Asymmetric Diffractions and Their Applications to Measurements of Diffraction Curves. III. Type II Collimator
  Tadashi Matsushita, Seishi Kikuta, and Kazutake Kohra
  J. Phys. Soc. Japan 30(4):1136-1144 (1971)
  2 Si crystals

Neutron Diffraction Effects due to the Lattice Displacement of a Vibrating Quartz Single Crystal
  R. Michalec, L. Sedlakova, B. Chalupa, D. Galociova, and V. Petrzilka
  Acta Cryst. A27:410-414 (1971)

Neutron Reflectivity of Thermally Treated Zinc Single Crystals
    A. Modrzejewski and J. Kobla
    Institute of Nuclear Research, Warsaw, Poland
    INP-Report-129/II/PS (1971), 8 pp.

Pyrolytic Graphite as a Selective Neutron Filter
    H. Pinto and H. Shaked
    Nucl. Instr. Methods 97:71-74 (1971)

Tables Permettant le Depouillement des Diagrammes de Rayons X et Abaques de Reglage des Monochromateurs à Lame Courbe
    A. J. Rose
    Editions due Centre National de la Recherche Scientifique, 15, quai Anatole-France, Paris VII, France (1971), 144 pp.
    Includes charts for focal lengths of bent and ground quartz single crystal monochromators

Neutron Scattering Study of the Lattice Dynamical Phase Transition in V_3Si,
    G. Shirane, J. D. Axe, and R. J. Birgeneau
    Solid State Commun. 9:397-400 (1971)
    Oriented pyrolytic graphite crystals; bent graphite (002) as a monochromator and flat graphite as an analyzer

1970

Preparation of Very Large Bismuth Single Crystals for Neutron Filters
    S. Bednarski
    J. Cryst. Growth 6:193-194 (1970)

The Growth of Potassium Hydrogen Phthalate Crystals and Their Optical, Piezoelectric, and Elastic Properties
    L. M. Belyaev, G. S. Belikova, A. B. Gil'varg, and I. M. Sil'vestrova
    Soviet Phys.-Cryst. 14(4):544-549 (1970)
    Translated from Kristallografiya 14:645-651 (1969)

Collimators and Crystals in a Non-focusing X-Ray Spectrometer
    H. Bent
    Spectrochim. Acta B25(11):613-627 (1970)
    Comparison of different analyzing crystals

Growth and Test of Spectrometer Crystals for the Soft X-Ray Region (10-100 Å)
    G. Bohm and K. Ulmer
    Z. Angew. Phys. 29:287 (1970)

Single Crystal with a Gradient in the Lattice Spacing
    J. Burgeat, J. Primot, and F. Rustichelii
    Z. Phys. 238(2):140-147 (1970)
    Solid solution of $CoO-NiO$

A Contribution to the Theory of X-Ray Monochromators with Approximate Focusing of Johann's Type
    J. Cermak
    J. Phys. E (Sci. Instr.) 3:615-620 (1970)

X-Ray Spectrometric Characteristics of Potassium Acid Phthalate Crystals
    D. Chopra
    Rev. Sci. Instr. 41:1004-1006 (1970)

Observation of Structures in the Chromium $K\alpha_1$ Line with Spherically Bent Crystal Spectrometer
    K. Das Gupta and P. F. Gott
    Phys. Letters 33A:276-277 (1970)
    Mica crystal; to replace two-crystal spectrometer for higher orders of reflection

Neutron Diffraction on Piezoelectric Vibrating Bars
    D. Galociova, J. Tichy, J. Zelenka, R. Michalec, and B. Chalupa
    Phys. Stat. Sol. 2A:211-216 (1970)
    Natural and Synthetic quartz

Increasing the Resolution of the Graphite Monochromator Using Collimation
    R. W. Gould and P. Mehl
    Norelco Reptr. 17(2):29-30 (1970)

Back-Reflection of Neutrons from Mosaic Crystals
    P. Hiismaki
    Acta Polytech. Scand., Series No. 70, Helsinki 1970, 36 pp.
    Solved numerically with a digital computer for the (111)-reflection of Al

Method for Determining the Degrees of Polarization of Infrared Polarizers and Monochromators
    K. Kudo, T. Arai, and T. Ogawa
    J. Opt. Soc. Am. 60(8):1046-1050 (1970)

Studies of Crystal Perfection: Tantalum Silicide and Beryllium
    Technical Report, 1 June 1969-31 March 1970
    C. J. London and J. D. Meakin
    Franklin Institute, Philadelphia, Pa.
    NYO-3716-4; P-C1870-3-1 (March 31, 1970), 26 pp.
    To develop perfect single crystals large enough to be monochromators for neutron diffraction

Polarization of Three Vacuum-Ultraviolet Monochromators Measured with a Biotite Polarizer
    A. Matsui and W. C. Walker
    J. Opt. Soc. Am. 60:64-65 (1970)
    $K_2(Mg, Fe, Al)_6(Si, Al)_8O_{20}(OH)_4$

The Polarization of X Rays by Real Crystals and the Determination of their Mosaic Structure
    N. M. Olekhnovich
    Soviet Phys.-Cryst. 14(5):727-727 (1970)
    Transl from Kristallografiya 14:840-844 (1969)
    Ge and LiF

Neutron Diffraction on Lattice Planes Deformed by Flexural Vibrations of Quartz Single Crystals
    V. Petrzilka, J. Vrzal, R. Michalec, B. Chalupa, P. Mikula, and J. Zelenka
    Phys. Stat. Sol. 42:895-902 (1970)

Singly Bent Graphite Monochromators for Neutrons
    T. Riste
    Nucl. Instr. Methods 86:1 (1970)

A Double-Crystal Soft X-Ray Spectrometer
    E. Suoninen, M. Karras, and J. Levoska
    Acta Polytech. Scand., Physics Including Nucleonics Series No. 71, Ph 71 (1970), 20 pp.
    $CaCO_3$ analyzer crystals; 5 to 10 Å region

X-Ray Spectrometry with Bent Crystals Used in Analyzing Radiation Occurring Beyond Focal Range
    V. A. Trapeznikov and V. P. Sapozhnikov
    Pribory Tekh. Eksper., No. 3, p. 227 (1970)

X-Ray Emission Line Wavelength and Two-Theta Tables
    E. W. White, G. V. Gibbs, G. G. Johnson, Jr., and G. R. Zechman, Jr.
    ASTM Data Series DS 37, American Society for Testing and Materials, 1965
    Includes 2θ values for the 23 most commonly used analyzer crystal. ASTM DS 37A (1970) is second edition

1969

Perfect Crystal with Temperature Gradient as a Neutron Monochromator
    B. Alefeld
    Z. Phys. 228:454-464 (1969)
    In German
    CaF_2 gradient perpendicular to reflecting planes

Use of X-Ray Spectrometry of the (200) Reflection of Ammonium Dihydrogen Phosphate
    W. L. Baun
    Anal. Chem. 41(6):830-831 (1969)

Clinochlore: A Versatile New Analyzing Crystal for the X-Ray Region 5-27 Å
    W. L. Baun and E. W. White
    Anal. Chem. 41(6):821-822 (1969)
    $(Mg,Fe,Al)_3(Si, Al)_2O_5(OH)_4$

Angular Positions of X-Ray Emission Lines of the Elements for Common Analyzing Crystals
    Martin Berman and Sabri Ergun
    Bureau of Mines IC-8400 (1969), 309 pp.
    First, second, and third-order reflections; based on a set of wavelengths for 2346 observed x-ray transitions of the elements Li to Am

Neutron Diffraction on Quartz Crystals as a Function of Vibrating Frequencies
    B. Chalupa, R. Michalec, and D. Galociova
    Nucl. Instr. Methods 76:357-358 (1969)

X-Ray Crystal Monochromator With a Reflecting Surface That Conforms to Part of a Logarithmic Spiral
    W. K. de Jongh
    North American Philips Co., Inc.
    Patent U. S. 3439163 (Sept. 1965; publ. April 1969; prior Sept. 1964, Netherlands 6410514)

Using Germanium Single Crystals Produced by Stepanov Method in X-Ray Spectrometers
    Zh. E. Gorbulskaya, I. B. Staryi, and A. A. Khanonkin
    Izv. Akad. Nauk SSSR, Ser. Fiz. 33:2020-2021 (1969)
    In Russian

SiO_2 (Quartz), MgO, PbF_2, and Bi as Low-Pass Neutron Velocity Filters
    Sten Holmeryd and Donald Connor
    Rev. Sci. Instr. 40:49-56 (1969)

Coherent Neutron Scattering by Cobalt with Nuclear Polarization
    Y. Ito and C. G. Shull
    Phys. Rev. 185:961-967 (1969)

High-Resolution Monochromatization of Neutrons by Multiple Bragg Reflection in Hexagonal Close-Packed Crystals
    D. A. Kottwitz
    Acta Cryst. 25A:459-464 (1969)
    Be, Mg, Zn

Experimental Study of the Propagation of X-Rays in a Perfect Crystal Deformed Under the Action of a Thermal Gradient
    C. Malgrange
    Acta Cryst. 25A:356-363 (1969)
    Si crystals

Integrated Reflection and Reflecting Power of a Crystal Monochromator
    A. Z. Men'shikov, L. I. Novak, Yu. A. Dorofeed, V. G. Arkhipov, and S. K. Sidorov
    Inst. Exper. Tech., No. 3, p. 564 (1969)

Neutron Diffraction Properties of Machined Large Metallic Single Crystals
    A. Modrzejewski and J. Kobla
    Kristall u. Tech. 4(1):135-148 (1969)
    Preparation of crystal monochromators; effects of deformation or quenching in liquid N_2

A Simple Method of Increasing the Mosaic Spread of Germanium Neutron Monochromators
    O. P. Nikotin
    Nucl. Instr. Methods 72(1):77-78 (1969)

Efficiency of Neutron Crystal Monochromators
    M. Popovici, Z. Gheorghiu, and D. Gelberg
    Nucl. Instr. Methods 69(1):125-130 (1969)
    Pb crystals

An X-Ray Monochromator for Diffuse Scatter Measurement
    J. S. Reid and T. Smith
    J. Sci. Instr. 2(7):601-604 (1969)
    Doubly curved LiF

Measurements of Gamma Rays and X-Rays with a Bent-Crystal Spectrometer
    James Dutton Reierson
    Iowa State University, Ames, Iowa
    An abstract of a dissertation submitted to the Graduate Faculty in partial fulfillment of the requirements for the degree of Doctor of Philosophy (1969)
    Ge and quartz

Oriented Graphite as a Neutron Monochromator
    T. Riste and K. Otnes
    Nucl. Instr. Methods 75:197-202 (1969)

Analysis of Composite Neutron Monochromator Systems Consisting of Curved Crystalline Lamellas
    F. Rustichelli
    Nucl. Instr. Methods 74:219-223 (1969)

Studies on Neutron Monochromator Crystals Having a Lattice Spacing Gradient. Part I
    F. Rustichelli
    European Atomic Energy Community, Ispra, Italy
    EUR-4414 (Sept. 18, 1969), 30 pp.
    In Italian
    Includes methods of crystal production. Translated from the Italian for Oak Ridge National Laboratory, ORNL-TR-2401

Observation of Fine Structures of Chromium $K\alpha_{1,2}$ Lines with a High Resolution Three Crystal Spectrometer
    M. Shah and K. Das Gupta
    Phys. Letters 29A:570-571 (1969)
    Sets of quartz or Si crystals

New Crystal Spectrometer for Neutrons
    R. Stedman, L. Almqvist, G. Raunio, and G. Nilsson
    Rev. Sci. Instr. 40:249-255 (1969)
    MgO, Fe, Pb, Zn, Ge, Al, and Cu monochromators

Neutron Monochromator Studies at A.E.R.E.
    K. C. Turberfield
    United Kingdom Atomic Energy Authority, Harwell, England
    AERE-R-5647 (Feb. 1968), 35 pp.
    Mosaic widths of up to 30' to 40' have been achieved in copper crystals by doping with beryllium and up to 12' in germanium by distortion at high temperatures

Improving the Diffracting Properties of LiF: Comparison with Graphite
    J. Vierling, J. V. Gilfrich, and L. S. Birks
    Appl. Spectry. 23:342-345 (1969)
    It appears that one can approximate theoretical mosaic crystals by either distorting a nearly perfect crystal such as LiF or by orienting powder crystallites as in the case of graphite)

Focusing Monochromators
    J. Witz
    Acta Cryst. 25A:30-42 (1969)

1968

Tailoring Analysing Crystals to Specific Applications in X-Ray Spectroscopy
    W. L. Baun and D. W. Fischer
    XIII Coll. Spectrosc. International, Ottawa, 1967, p. 420
      (A. Hilger Ltd., London, 1968)
    (200) reflection of ADP

Diffraction of Neutrons on a Vibrating Quartz Crystal
    B. Chalupa, R. Michalec, V. Petrzilka, J. Tichy, and J. Zelenka
    Phys. Stat. Sol. 29:K51 (1968)

Coherent Crystal Radiation Affects the Measurement of the X-Ray Linewidths
    K. Das Gupta and Herbert Welch
    Phys. Rev. Letters 21:657-660 (1968)
    Three-crystal spectrometer; quartz crystals

X-Ray Monochromators and Resonators from Single Crystals
    R. D. Deslattes
    Appl. Phys. Letters 12:133-135 (1968)
    Si and Ge

A Small Set of Reference Crystals for Double-Crystal Topography
    R. D. Deslattes and B. Paretzkin
    J. Appl. Cryst. 1:176-178 (1968)
    Six reference crystals of Si

Focussing Monochromators and Transmission Techniques
    P. M. de Wolff
    Norelco Reptr. 15:44-49 (1968)

Study of Beryllium Monocrystal
    L. Manas Diaz
    An. Fis. 64(3/4):81-92 (1968)
    In Spanish
    The [010], [110], [210] and [310] zones and their crystallographic planes have been determined. Their interplanar distances and structure factors have been calculated

Untersuchungen zur Fokussierung von Röntgenstrahlen mit gebogenen Kristallen
    K. Feser and A. Faessler
    Z. Physik 209:1-12 (1968)
    Bent quartz

2θ Values for X-Ray Analysis with Calcium Sulphate, (020), 2d = 15185 Å, as the Analyzing Crystal
    C. Freiburg and W. Reichert
    Kernforschungsanlage, Zentrallabor für Chemische Analyse, Julich, Germany
    JUL-530-CA (June 1968), 87 pp.
    In German

2θ Values for X-Ray Analysis with Quartz, (10$\bar{1}$1), 2d = 6,6863 Å, as the Analyzing Crystal
    D. Freiburg and W. Reichert
    Kernforschungsanlage, Zentrallabor für Chemische Analyse, Julich, Germany
    JUL-533-CA (June 1968), 77 pp.
    In German

2θ Value for the X-Ray Spectral Analysis with Lithium Fluoride, (200), 2d = 4.0276 Å as the Analyzer Crystal
    C. Freiburg and W. Reichert
    Kernforschungsanlage, Julich, Germany
    JUL-534-CA (June 1968), 58 pp.
    In German

2θ Value for the X-Ray Spectral Analysis with Lithium Fluoride, (200), 2d = 2.848 Å as the Analyzer Crystal
    C. Freiburg and W. Reichert
    Kernforschungsanlage, Julich, Germany
    JUL-535-CA (1968)
    In German

2θ Value for X-Ray Spectral Analysis with Topaz (303), 2d = 2.712 Å, as the Analyzer Crystal
    C. Freiburg and W. Reichert
    Kernforschungsanlage, Julich, Germany
    JUL-536-CA (June 1968), 45 pp.
    In German

2θ Value for the X-Ray Spectral Analysis with Ammonium Dihydrogen Phosphate, (011), 2d = 10.648 Å, as the Analyzer Crystal
    C. Freiburg and W. Reichert
    Kernforschungsanlage, Julich, Germany
    JUL-531-CA (1968)
    In German

2θ Value for the X-Ray Spectral Analysis with Pentaerythrit, (332), 2d = 8.742 Å as the Analyzer Crystal
    C. Freiburg and W. Reichert
    Kernforschungsanlage, Julich, Germany
    JUL-532-CA (1968)
    In German

Application of the Graphite Monochromator to Light Element X-Ray Spectroscopy
R. W. Gould, S. R. Bates, and C. J. Sparks
Appl. Spectry. 22:549-551 (1968)

"Perfect Crystals." A Study of Their Structural Defects
Michael Hart
Sci. Prog., Oxf. 56:429-447 (1968)
X-ray moire interferometry and double crystal diffraction topography in almost perfect crystals of silicon, germanium, and quartz

The $K\alpha_2$ Radiation Monochromated by a New Filter and Its Topographical and Other Applications
Sukeaki Hosoya
Japan. J. Appl. Phys. 7:1-6 (1968)
Ru; not single crystal

The Effect of Temperature and Pressure on the Mosaic Spread in a Ge Monocrystal
J. Konstantinovic, M. Zivanovich, and M. Davidovic
Nucl. Instr. Methods 65:233-234 (1968)

X-Ray Monochromator Configurations and a Monochromator Design for X-Ray Single Crystal Diffractometers
A. McL. Mathieson
Rev. Sci. Instr. 39:1834-1837 (1968)
Application as a polarization analyzer to assess the extinction characteristics of the specimen and monochromator crystals

The Activity of the Crystal Growth Laboratory of N.R.C. "Democritus"
K. Papathanossopoulos and A. Kypriotis
Greek Atomic Energy Commission
DEMO 68/5 (June 1968)
Zn, Sn, Pb, and Al for neutron monochromators

Experimental Studies of Polarization Effects in Spin-Wave Scattering of Neutrons
T. Riste, R. M. Moon, and W. C. Koehler
Phys. Rev. Letters 20:997-999 (1968)
$Li_{0.5}Fe_{2.5}O_4$

Studies of the Subgrain Structure of Curved Aluminum Crystal Analyzers of X-Ray Spectrographs
I. B. Staryi and A. A. Khanonkin
Soviet Phys.-Cryst. 12:633 (1968)
Translated from Kristallografiya 12:724-726 (1967)

The Achromatization of Crystal-Optical Components for Rotating the Vibration Direction of Plane Polarized Light
A. Weyrauch and K. Rantsch
Optik 28(3):235-241 (1968)
In German

Production of a Monoenergetic Polarized Neutron Beam
Yu. G. Abov, O. P. Aleshko-Ozhevskii, O. N. Ermakov, and I. I. Yamzin
Soviet Phys.-Cryst. 11:597 (1967)
Single crystal of Co + 8% Fe

Optimization of Lead Stearate Crystals for the Diffraction of Ultra Soft X Rays
M. W. Charles and B. A. Cooke
J. Sci. Instr. 44:976 (1967)

Methods of Obtaining Monochromatic X-Rays and Neutrons, Bibliography 3
F. H. Herbstein, ed.
International Union of Crystallography, Commission on Crystallographic Apparatus
NP-17314 (1967), 84 pp.
Monochromation of x-rays by balanced filters, monochromation of x-rays by crystals, and monochromation of neutrons

Two-Crystal Neutron Diffractometer with Movable Monochromator
B. I. Khrushchev and L. S. Sharipova
Pribory Tekh. Eksper. 12(6):48-50 (1967)
In Russian

Some Suggestions for the Application of Constricted Monochromator Crystals for Neutron Diffraction and Scattering Measurements
H. Maier-Leibnitz
Ann. Acad. Sci. Fennicae A VI 267:3-17 (1967)
In German
Construction of monochromator crystals from lamellae

Crystal Mosaic Spread Measurement by Means of a Neutron Two-Axis Spectrometer
H. Rauch
Atomkernenergie 12(11/12);:395-398 (1967)
Large Pb single crystal

Dependence of the Resolution of an X-Ray Spectrometer on the Radius of Curvature of the Crystal and the Slit Width
V. A. Trapeznikov and V. A. Trofimova
Instr. Exper. Tech. No. 6, (1967)
Translated from Pribory Tekh. Eksper. No. 6, p. 191 (1967)

1966

Single Crystal Diffractometry
U. W. Arndt and B. T. M. Willis
University Press, Cambridge (1966)
Neutron and x-ray

Method of Making an Optical Tool by Chalmers Growth of Aluminum Single Crystals, Then Electroetching and Replicating the 110 Plane
P. E. Doherty, H. H. Blau, and R. S. Davis
Arthur D. Little, Inc.
U. S. Patent 3,235,630 (1966), 7 pp.

Modulation of Diffracted Neutrons with a Piezoelectric Crystal
T. F. Parkinson and M. W. Moyer
Nature 211:400-401 (1966)

Reflectivities of Some Metallic Crystals Used in Neutron Spectrometers
K. R. Rao
Dept. of Atomic Energy, Bombay, India
AEET-259 (1966), 29 pp.
Al, Cu, Pb; tables of absolute and effective reflectivities

Analysis of Experiments in Single Crystal X-Ray Diffractometry
Robert E. Zumwait
Ph. D. thesis, Delaware University, Neward, 1966, 126 pp.
Available from University Microfilms, Inc., Ann Arbor,

Mich., Order No. 67-11753
Compendium of the current technology in crystal monochromatization of the diffracted beam is included

1965

A Polycrystalline X-Ray Analyzing Crystal
M. Cannon
Advances in X-Ray Analysis, Vol. 8, William M. Mueller, Gavin Mallett, and Marie Fay, eds., Plenum Press, New York (1965), pp. 285-300 (Proceedings of the Thirteenth Annual Conference on Applications of X-Ray Analysis held in Denver, 1964
Graphite

Relative Intensities of X-Ray Reflections of Various Orders for Ten Crystal Monochromators
Pol Duwez
W. M. Keck Laboratory of Engineering Materials, California Institute of Technology
Contract AT(04-3)-221, Tech. Report No. 32, CALT-221-227 (Jan. 1965)
$CaCO_3$, ADP, KAP, KDP, EDDT, Si, SiO_2, Mica, Topaz

Removal of Second Order Neutrons by Oriented Single Crystal Filters
P. K. Iyengar, J. N. Soni, et al.
Nucl. Instr. Methods 37:121-124 (1965)

Efforts to Grow Very Large Beryllium Single Crystals for Neutron Spectrometers
J. F. Kirn
Virginia Institute for Scientific Research, Richmond, Va.
Contract AT(40-1)-2860, TID-21852 (Feb. 1965), 20 pp.

Producing Single-Crystal Plates of Cobalt—Iron Alloy for the Polarization of Neutrons
V. P. Ovcharov and O. P. Aleshko-Ozhevskii
Soviet Phys.-Cryst. 10:76 (1965)
Translated from Kristallografiya 10:96-98 (1965)

Secondary Extinction and Optical Properties of Neutron Crystal Spectrometers
M. Popovici and D. Gelberg
Rev. Rum. Phys. 10:1035-1050 (1965)
Nucl. Instr. Methods 40:77-83 (1965)
In French
The dependence of the luminosity on the mosaic spread of the crystal is calculated

A Crystal Neutron Monochromator
J. M. A. Reichelt
Atomic Weapons Research Establishment, Aldermaston, England
AWRE-O-98/65 (Dec. 1965), 20 pp.
Improvement in Al crystal growth technique, resulting in a single crystal with small mosaic spread

Analyzer Crystals for X-Ray Spectroscopy in the Region 25-100 Å
I. W. Ruderman, K. J. Ness, and J. C. Lindsay
Appl. Phys. Letters 7:17 (1965)

Single Crystal Filters for Attenuating Epithermal Neutrons and γ-Rays in Reactor Beams
B. M. Rustad, J. Als-Nielsen, A. Bahnsen, C. J. Christensen, and A. Nielsen
Rev. Sci. Instr. 36:48-54 (1965)
Bi-filter

Propagation of Bragg-Reflected Neutrons in Large Mosaic Crystals and Efficiency of Monochromators
S. A. Werner and A. Arrott
Phys. Rev. 140:675A-686A (1965)
Effects of absorption and of crystal-cutting angle on the efficiency of monochromating crystals are discussed in some detail

1964

X-Ray Fluorescence Tables: Lithium Fluoride Crystal
W. P. Amsbury, W. W. Lee, J. H. Rowan, and G. E. Walden
Union Carbide Corp., Y-12 Plant, Oak Ridge, Tenn.
Contract W-7405-eng-26, Y-1470-C (1964)
Tables arranged by element, order, and line, and also angular tabulations for a specific analyzing crystal, were derived from a critical amalgamation of three standard reference tables of x-ray fluorescence lines

Study and Production of Polarized Monochromatic Thermal Neutron Beams
H. Beil
Submitted as a thesis to University of Paris, 1964, CEA-R-2329 (1964), 70 pp.
92Co—8Fe crystals as polarizers and analyzers

Higher Order Contamination in Crystal Monochromatization of Neutrons
O. W. Dietrich and J. Als-Nielsen
Research Establishement RISO, Denmark
RISO-73 (1964), 11 pp.

The Use of Germanium Dendrites for the Monochromatization of X-Rays
A. Fingerland and L. Cervinka
Acta Cryst. 17:508-511 (1964)

A Curved Crystal Monochromator for the X-Ray Diffractometer
D. K. Koffman and S. H. Moll
Norelco Reptr. 11:95 (1964)

Changes in the Mosaic Structure in Connection with the Distribution of Impurities during the Growth of Single and Double (Aluminum) Crystals from a Melt
A. A. Kralina and V. O. Esin
Mekhanism i Kinetika Kristallizatsii (Mechanism and Kinetics of Crystallization), Izd. Nauka i Tekhnika, Minsk, (1964), pp. 438-452

Reflection of Collimated Neutrons by a Mosaic Single Crystal
N. Kunitomi, M. Sakamoto, Y. Hamaguchi, and H. Betwuyaka
J. Phys. Soc. Japan 19:2280-2285 (1964)
Pb and LiF

Polarization Factor for X-Ray Monochromator Crystals
S. Miyake, S. Togawa, and S. Hosaya
Acta Cryst. 17:1083-1084 (1964)

X-Ray and Neutron Studies of the Lineage Structure in Large Al Single Crystals
A. Modrzejewski and S. Szarras
Phys. Stat. Sol. 5:K23-26 (1964)

Preliminary Studies for a Curved-Crystal Neutron Spectrometer
   R. R. Spencer, L. G. Miller, and M. S. Moore
   IDO-17022 (1964), 24 pp.
   Studies of both plane and curved crystals of various materials

Reflecting Power of Elastically-Curved Crystals to X-Rays
   D. Taupin Compt. Rend. 258:1212-1214 (1964)
   Compt. Rend. 258:1212-1214 (1964)
   In French
   Si

A Vertically Rotating Double-Crystal X-Ray Spectrometer
   M. C. Wittels, F. A. Sherrill, and A. C. Kimbrough
   Advances in X-Ray Analysis, Vol. 7, William M. Mueller, Gavin Mallett, and Marie Fay, eds., Plenum Press, New York (1964), pp. 265-280 (Proceedings of the Twelfth Annual Conference on Applications of X-Ray Analysis held August 7-9, 1963)

Germanium as a Neutron Monochromator
   C. S. Barrett, M. H. Mueller, and L. Heaton
   Rev. Sci. Instr. 34:847 (1963)

High-Precision Cutting of Co−Fe Single Crystal For Use in a Neutron Spectrometer
   H. Beil, P. Carlos, and R. Genin
   J. Phys. 24 (Suppl. 3):15A-18A (1963)
   In French

Crystal Spectrometer for Polarized Neutrons
   H. Beil, P. Carlos, R. Genin, C. Signarbieux, R. Joly, and J. Matuszek
   J. Phys. 24:89-93 (1963)
   In French
   Fe_3O_4 crystals

Production and Study of Monochromatic Polarized Beams of Slow Neutrons
   H. Beil, P. Carlos and J. Matuszek
   J. Phys. 24:359-366 (1963)
   In French
   Series of Co−Fe single crystals as polarizers and analyzers

Measurement of the Polarization of Thermal-Neutron Beams by Magnetized Mirrors
   M. T. Burgy, G. R. Ringo, and others
   ANL-6797 (1963), pp. 35-44
   93Co−7Fe

Structure and Magnetic Properties of Electrolytic Layers of Cobalt Used for Neutron Polarization
   A. Calusaru and S. Rypjanu, and L. Benes
   Acad. Rep. Populare Romine, Studii Cercetari Fiz. 14, 239-247 (1963)
   In Rumanian

Combination of Two Bent Crystal Monochromators for Precision Work on Scattering and Diffraction
   K. Das Gupta
   Technical Report No. 20, U. S. Atomic Energy Comission, Contract No. AT(04-3-)-221 (quartz) (1963)

Some Experience in the Work with the Neutron Crystal Spectrometer
   Z. Dimitrijevic, S. Krasnicki, R. Maglic, V. Markovich, and J. Todorovic
   PAN-270 (1963), 29 pp.
   Al and Ge single crystals as monochromators and analyzers

Mosaic Structure of Single Crystal Plate of Iron
   T. Fujiwara, S. Dohi, and T. Takeda
   Mem. Defense Acad. Japan III, No. 2, 117-124 (1963)

Performance of Curved Crystals in the Range 3 to 12 Å
   J. L. Jones, K. W. Paschen, and J. B. Nicholson
   Appl. Opt. 2, 955-961 (1963)
   X-ray fluorescence spectroscopy

Rentgenovskaya Difraktometria
   D. M. Kheiker and L. S. Zevin
   Fizmatgiz, Moscow (1963), pp. 134-162
   Crystal monochromators; balanced filters, differential discrimination

The Transmission of Thermal Neutrons Through a Large Single Crystal of Bismuth at Liquid Nitrogen Temperature
   S. Menardi, R. Haas, and W. Kley
   Inelastic Scattering of Neutrons in Solids and Liquids, Vol. 1, International Atomic Energy Agency, Vienna, (1963), pp. 139-146

The Reflection Ability of Mosaic Crystals with Regard to Their Use as Monochromators in Neutron Diffraction Spectrometers
   H. Stehr
   Z. Krist. 118:263-272 (1963)
   In German

Long-Spacing Metal Organic Crystals for X-Ray Spectrometer
   L. S. Birks and J. M. Siomkajlo
   Spectrochim. Acta 18:363-366 (1962)

Resolution and Luminosity of Crystal Spectrometers for Neutron Diffraction
   G. Caglioti and F. P. Ricci
   Nucl. Instr. Methods 15:155-163 (1962)
   SiO_2, Al, Pb monochromators

Natural Silicates of Large Parameter for Weak X-Ray Spectrographs and Monochromators
   Y. Cauchois, C. Bonnelle, and J. Orcel
   Bull. Soc. Franc. Mineral. Crist. 85:188-189 (1962)
   In French

The Use of Flat and Curved Monochromator Crystals in Neutron Spectrometers for Investigating Single Crystals
   H. Dachs and H. Stehr
   Z. Krist. 117:135-145 (1962); also NP-tr-923 (1962)
   In German

Removal of Higher Orders in the Thermal Region from a Neutron Crystal Spectrometer
   V. P. Duggal and C. O. Thaper
   Rev. Sci. Instr. 33:49-50 (1962)
   Be single crystal

Production of a Monochromatic Beam of Polarized Neutrons by Bragg Reflection from Co−Fe Alloy
   G. A. Ferguson, Jr.
   PB-181078 (1962), 50 pp.

Methods of Polarization of Cold, Thermal and Resonance Neutrons
   R. Michalec

Cesk. Casopis Fys. 12A:272-291 (1962)
In Czech
62 refs.

Preparation of Bent Crystals for Monochromatizing X-Rays
F. B. Riggs, Jr.
Rev. Sci. Instr. 33:875-876 (1962)

International Tables for X-Ray Crystallography
B. W. Roberts and W. Parrish, eds. (General editor, K. Lonsdale, Kynoch Press, Birmingham, 1962)
Volume III, pp. 73-88
A general review is given, including discussion of the following points: spectral distribution at the target and as observed; filter techniques, including single and balanced filters; crystal monochromators and their uses. Many useful tables and a bibliography are given

Polarized Thermal Neutron Beams Obtained from Magnetized Mirrors and Single Crystals of Magnetite and Cobalt – Iron Alloy
O. Steinsvoll, K. Abrahams, and T. Riste
KR-37, 121 pp. (1962)

Composition of Neutron Beams from Crystal Monochromators
R. R. Spencer, J. R. Smith, and R. M. Brugger
Pile Neutron Research in Physics, International Atomic Energy Agency, Vienna, (1962), pp. 433-444

1961

Production of Polarized Neutrons by Reflection from Cobalt Mirrors
Yu. G. Abov, V. A. Beketov, A. D. Gulko, O. N. Ermakov, P. A. Krupchitskii, Yu. V. Taran, and N. S. Shatlovskaya
Pribory Tekh. Eksper. 4:51-55 (1960)
Instr. Exper. Tech. 4:571-576 (1961)

Preparation of Single Crystals of High Melting Temperature Materials
S. Bednarski and A. Modrzejewskii
Polish Academy of Sciences Inst. of Nucl. Res., Report No. 245/1-B (Warsaw, Poland, June 1961)
Co-8%Fe

A Polarized Neutron Beam Produced by Bragg Reflection from Co – Fe Alloy
M. A. Clark and J. M. Robson
Can. J. Phys. 39:1-12 (1961)

X-Ray Polarizer
H. Cole, F. W. Chambers, and C. G. Wood
J. Appl. Phys. 32:1942-1945 (1961)
Ge

Die Forming Doubly Focusing X-Ray Monochromators
H. J. Garrett and H. A. Lipsitt
Rev. Sci. Instr. 32:942-945 (1961)

The Intrepretation of X-Ray Diffraction Photographs
N. F. M. Henry and H. Lipson and W. A. Wooster, eds.
Macmillan, 2nd edition, London (1961)
P. 33: crystal-reflection radiation; p. 240: harmonics from crystal monochromators

Neutron Spectrometer for Producing Pure Monochromatic Beams in the Thermal Region (Quartz)
H. Bjerrum Muller and F. J. Shore and V. L. Sailor
Rev. Sci. Instr. 32:654-658 (1961)

The Use of Cobalt Mirrors for the Polarization of Thermal Neutrons
Yu. A. Murav'ev, B. A. Obinyakov, and Yu. A. Prokofev
Proc. Working Conf. on Slow Neutron Physics. Dec. 7-12, 1961
Obedinennyi Inst. Yaderaykh Issledovanii, Dubna; AEC-tr-5734, p. 94

Use of a Germanium Single Crystal for Monochromating X-Radiation
Y. P. Simanova and V. K. Trunov
Zavod. Lab. 27:180-181 (1961)
In Russian

Crystal Structure Analysis
M. J. Buerger
John Wiley and Sons, Inc., New York (1960), pp. 139-188
Balanced filters and crystals monochromators

Construction of a Lead Crystal Monochromator for Neutron Diffraction Studies
D. F. Litvin
Soviet Phys.-Cryst. 4:623-624 (1960)
60-80 mm dia., 220 mm long, Czochralski method

X-Ray Diffraction by Polycrystalline Materials
H. S. Peiser, H. P. Rooksby, and A. J. C. Wilson, eds.
Institute of Physics, London (1960)
Chapter 4: G. W. Brindley, pp. 122-144: This chapter deals mainly with the curved-crystal monochromators used as focusing devices in x-ray powder analysis; there are short sections on plane monochromators and plane concentrating monochromators

α-Fe_2O_3 as a Crystal Monochromator
A. Wanic and T. Riste
Rev. Sci. Instr. 31:214-215 (1960)

Ref
Z
7144
S58
S65
v.5